CU00708742

EXPLORING THE ORIGIN, EXTENT, AND FUTURE OF LIFE

Philosophical, Ethical, and Theological Perspectives

Where did we come from? Are we alone? Where are we going? These are the questions that define the field of astrobiology. New discoveries about life on Earth, the increasing numbers of extrasolar planets being identified, and the technologies being developed to locate and characterize Earth-like planets around other stars, are continually challenging our views of nature and our connection to the rest of the universe.

In this book, philosophers, historians, ethicists, and theologians provide the perspectives of their fields on the research and discoveries of astrobiology. A valuable resource for graduate students and researchers, the book provides an introduction to astrobiology, and explores subjects such as the implications of current origin-of-life research, the possible discovery of extraterrestrial microbial life, and the possibility of altering the environment of Mars.

CONSTANCE M. BERTKA is an Adjunct Professor at the Wesley Theological Seminary, Washington DC. She was Director of the Program of Dialogue on Science, Ethics, and Religion of the American Association for the Advancement of Science (AAAS) from 2002 to 2008.

Cambridge Astrobiology

Series Editors

Bruce Jakosky, Alan Boss, Frances Westall, Daniel Prieur, and Charles Cockell

Books in the series

EXPLORING THE ORIGIN, EXTENT, AND FUTURE OF LIFE

Philosophical, Ethical, and Theological Perspectives

Edited by

CONSTANCE M. BERTKA

American Association for the Advancement of Science

CAMBRIDGE UNIVERSITY PRESS
Cambridge, New York, Melbourne, Madrid, Cape Town, Singapore, São Paulo, Delhi

Cambridge University Press
The Edinburgh Building, Cambridge CB2 8RU, UK

Published in the United States of America by Cambridge University Press, New York

www.cambridge.org
Information on this title: www.cambridge.org/9780521863636

© Cambridge University Press 2009

This publication is in copyright. Subject to statutory exception
and to the provisions of relevant collective licensing agreements,
no reproduction of any part may take place without
the written permission of Cambridge University Press.

First published 2009

Printed in the United Kingdom at the University Press, Cambridge

A catalog record for this publication is available from the British Library

Library of Congress Cataloguing in Publication data
Exploring the origin, extent, and future of life : philosophical, ethical, and theological
perspectives / [edited by] Constance M. Bertka.
p.   cm. – (Cambridge astrobiology; 4)
Includes bibliographical references and index.
ISBN 978-0-521-86363-6 (Hardback)
1. Exobiology.   2. Life–Origin–Philosophy.   I. Bertka, Constance M.
QH326.E97 2009
576.8′3–dc22
2009027281

ISBN 978-0-521-86363-6 Hardback

Cambridge University Press has no responsibility for the persistence or
accuracy of URLs for external or third-party internet websites referred to
in this publication, and does not guarantee that any content on such
websites is, or will remain, accurate or appropriate.

To Fei

Contents

Contributors

Constance M. Bertka

Program of Dialogue on Science, Ethics, and Religion, American Association for the Advancement of Science; Wesley Theological Seminary, 4500 Massachusetts Avenue, NW, Washington, DC 20016, USA

Francisca Cho

Department of Theology, Georgetown University, 108 New North, 37 and O streets NW, Washington, DC 20057, USA

Cynthia S. W. Crysdale

School of Theology, Sewanee: The University of the South, 735 University Ave., Sewanee, TN 37383, USA

Celia Deane-Drummond

Centre for Religion and the Biosciences, Theology and Religious Studies Department, University of Chester, Parkgate Road, Chester, CH1 4BJ, UK

Steven J. Dick

History Division, National Aeronautics and Space Administration Headquarters, 300 E Street SW, Washington, DC 20546, USA

Iris Fry

The Cohn Institute for the History and Philosophy of Science and Ideas, Tel Aviv University, P.O. Box 39040, Ramat Aviv, Tel-Aviv, 69978, Israel

Robert M. Hazen

Carnegie Institution of Washington, Geophysical Laboratory, George Mason University, 5251 Broad Branch Rd NW, Washington, DC 20015, USA

Jack J. Lissauer
Space Science and Astrobiology Institute, Ames Research Center, National Aeronautics and Space Administration

Mark Lupisella
Goddard Space Flight Center, National Aeronautics and Space Administration, 8800 Greenbelt Rd, Greenbelt, MD 20771, USA

Christopher P. McKay
Ames Research Center, National Aeronautics and Space Administration, Mail Stop 245-3, Moffett Field, CA 94035, USA

Ernan McMullin
Program of History and Philosophy of Science, University of Notre Dame, 100 Malloy Hall, Notre Dame, IN 46556, USA

Carl B. Pilcher
Astrobiology Institute, Ames Research Center, National Aeronautics and Space Administration, Moffett Field, CA 94035, USA

Margaret S. Race
SETI Institute, 515 N. Whisman Road, Mountain View, CA 94043, USA

Richard O. Randolph
Bioethics Department, Kansas City University of Medicine and Biosciences, 1750 Independence Avenue, Kansas City, Missouri 641006-1453, USA

Lynn J. Rothschild
Ames Research Center, National Aeronautics and Space Administration, Mail Stop 245-3, Moffett Field, CA 94035, USA

Kelly C. Smith
Department of Philosophy and Religion, Clemson University, 126D Hardin Hall, Clemson, SC 29634, USA

James E. Strick
Program in Science, Technology, and Society, Franklin and Marshall College, P.O. Box 3003, Lancaster, PA 17604, USA

Acknowledgements

The editor is grateful to all the authors for both their contributions to this volume and for their participation in the workshop series from which this volume grew. The editor is also grateful to the American Association for the Advancement of Science (AAAS) for its support of the AAAS Program of Dialogue on Science, Ethics, and Religion, the program through which this work was undertaken with the encouragement of Al Teich, Audrey Chapman, and James Miller. The editor is also indebted to Jessica Huls for her assistance in proofing and formatting the volume. This work was completed with support from the National Aeronautics and Space Administration, NAG5–12966, and a grant from the John Templeton Foundation, #1860.

The interpretations and conclusions contained in this book are those of the authors and participants, and reflect their views and not necessarily those of the AAAS, its Council, Board of Directors, Officers, or the views of the institutions with which the authors and participants are affiliated.

1

Astrobiology in a societal context

Constance M. Bertka

1.1 Introduction

In 2003 the American Association for the Advancement of Science, Program of Dialogue on Science, Ethics, and Religion, invited over twenty scholars from diverse fields, scientists active in astrobiology, as well as philosophers, historians, ethicists, and theologians, to explore together the philosophical, ethical, and theological implications of research and discoveries in astrobiology. A major motivation for this effort was the recognition that the very questions that define astrobiology as a discipline – Where do we come from? Are we alone? Where are we going? – are multidisciplinary in nature and have broad appeal to the public-at-large.

It is unavoidable that the science of astrobiology will intersect with, and inevitably challenge, many deeply held beliefs. Exploration possibilities, particularly those that may include the discovery of extraterrestrial life, will continue to challenge us to reconsider our views of nature and our connection to the rest of the universe. Much work has already been done in this area.[1] What is unique about our present circumstance is that past theoretical musings may soon benefit from a renewed urgency that is awakened both by new discoveries and by technological advances. Many of the astrobiologists assembled for this workshop have in common another interest, working proactively to provide more opportunities for non-scientists to both share in the excitement of this field, and to be informed participants in a public dialogue that considers the opportunities and challenges associated with astrobiology in the near future. With that goal in mind participants were asked that their work together have a pragmatic focus on the implications of, for example, current origin-of-life research, the discovery of extraterrestrial microbial life in the solar

---

[1] See, for example, S.J. Dick. *Plurality of Worlds: The Origins of the Extraterrestrial Life Debate from Democritus to Kant* (Cambridge: Cambridge University Press, 1982).

Exploring the Origin, Extent, and Future of Life: Philosophical, Ethical, and Theological Perspectives, ed. Constance M. Bertka. Published by Cambridge University Press. © Cambridge University Press 2009.

system (in contrast to the implications of contact with extraterrestrial intelligent life), and the possibility of terraforming Mars.

The astrobiologists who participated in the workshops were charged with laying the groundwork for each topical discussion by presenting a review of the work of their field in a manner that would be accessible to scholars outside of their field. These scientific presentations were followed by presentations that reflected on the topic from a philosophical, ethical, theological or historical perspective. The perspectives included here are not intended to be exhaustive and the volume is skewed toward Christian theological responses. Given the religious demographics in this country (see below) that focus is understandable. Hopefully future work will serve to broaden the dialogue.

Following the general astrobiology themes, the three workshop topics explored were the "Origin of Life," the "Extent of Life," and the "Future of Life," all with the pragmatic focus noted above. The workshops were held over a two-year period and the majority of participants attended all three. This level of commitment and interest in the series enabled a progressively rich and ongoing level of dialogue to take place among the participants. This volume has certainly benefited from that extended dialogue. The chapters collected here were prepared in light of those discussions and each was reviewed by at least one workshop participant outside of the author's field of expertise. A general summary of the key points of each of the three parts of the volume is presented below. I also offer my own thoughts on the unique contributions that astrobiology may bring to the scientific community as it works at fostering a positive relationship between science and society, particularly here in the United States.

1.2 Astrobiology and public engagement

As a scientific discipline, astrobiology works from the assumption that the origin and evolution of life can be accounted for by natural processes, that life could emerge naturally from the physical materials that make up the terrestrial planets. Whether or not life will be a rare or common occurrence on other terrestrial planets is yet to be determined. However, with the growing number of extrasolar planets being discovered, and the development of technologies and missions to specifically search for Earth-like planets, we are progressing ever closer to determining how common or rare life is in the universe [1]. For the time being the greatest possibility of discovering past or present life on a planet other than Earth probably lies with our exploration of Mars. Indeed the focus of Mars exploration has been to understand the history of water on Mars, as the assumption is that this history will be intimately linked to the history of life on Mars, if life has ever been there or is present there now.

The focus on water reflects the decision to search for life "as we know it." The only life we currently know of is the life found on Earth, and for the scientific community the shared common ancestry of all Earth life, and its astounding diversity, is explained by the theory of evolution. The work of astrobiology, at its very core, is fueled by the theory of evolution. However, the dubious position that science holds with at least half of the American public is perhaps nowhere better illustrated than by the controversy over the teaching of evolution in this country.

A survey by the Pew Forum on Religion and Public Life (2005) [2] revealed that 42% of those surveyed believe that "life has existed in its present form since the beginning of time" (pp. 7–11). This answer persists nearly 150 years after the publication of Charles Darwin's *On the Origin of Species*, the landmark work in which Darwin proposed that living things share common ancestors and have "descended with modification" from these ancestors through a process of natural selection [3]. Perhaps even more distressing is the fact that these numbers have not changed in decades, despite the astounding advancements in science that have resulted over this same time period, i.e. from space exploration, both manned and robotic, to sequencing the human genome and imaging the brain in action. Knowledge about the physical world gained through science has certainly increased, but at least half the public is failing when it comes to science literacy, or at least evolution literacy.

It may appear that the other half is doing a bit better, after all 48% reported that "life has evolved over time," but only 26% of those were willing to credit natural selection as the process responsible [2] (p. 7). At least 18% of this group, who preferred "evolution guided by a supreme being" might welcome the latest challenge to evolution, the so-called Intelligent Design theory or ID. The basic premises behind ID are that some structures or processes associated with life are irreducibly complex, or are evidence of complex specified information, both of which imply that an intelligent designer is responsible for their production, and that evolution through natural selection simply could not have done the job.[2] An astounding majority of those surveyed, 64%, regardless of their own acceptance or rejection of evolution, feel that both evolution and creation science or ID should be taught in the public science classroom because that is the "fair" solution [2] (p. 10). How will these facts bear on the usefulness of astrobiology as a tool for encouraging a US public to share in the excitement of scientific discovery and be informed participants in a public dialogue concerning next steps? The answer, I believe, depends in large part on whether or not astrobiology as a discipline has something unique to offer society when it comes to working on the relationship between science and religion. We

---

[2] For a critique of Intelligent Design as well as other creationist positions see E.C. Scott. *Evolution vs. Creationism* (Westport, CT: Greenwood Press, 2004).

know precisely where the evolution illiteracy problem is rooted, in the 2005 Pew survey quoted above when people were asked "to identify the biggest influence on your thinking about how life developed," the response chosen most frequently, 42%, was "my religious beliefs" [2] (p. 10). Before considering how astrobiology might contribute to public engagement issues, we must first turn to a brief review of the religious landscape in the United States.

A few years ago I accepted an invitation to speak at a local meeting of the Secular Humanist Association. The group was interested in the general topic of science and religion, and more specifically, why the American Association for the Advancement of Science sponsored a program of Dialogue on Science, Ethics, and Religion. The secular humanists were puzzled by the fact that the scientific community would even put energy into engaging religious communities. As our discussion continued, it became clear to me that this group viewed religion as a relic on its way out and contributed its demise in large part to the success of science in revealing "the truth." Clearly one of the ideological roots of this group was the secularization theory. Simply put, this theory holds that as society becomes increasingly modernized and secular, religious belief, and the influence of religious institutions, will decline. The theory was a popular one with sociologists of religion in the 1960s but has now largely been abandoned because it is not supported by the empirical data [4, 5]. If secularization is not on the rise, how has the religious landscape, particularly in the USA, but also globally, changed?

A 2005 survey conducted by the Baylor Institute for Studies of Religion and the Department of Sociology at Baylor University directly addresses the question in terms of American piety [6]. The goal of the Baylor Religion Survey is to understand religion in America. It is a work in progress with additional surveys to be conducted over the next several years. Initial results speak directly to the question of secularization. The survey's authors note that previous national studies seemed to indicate that over the past quarter century there was an increase in the percentage of the population subscribing to no religion, rising from 8% to over 14%. However, they note that these previous studies asked people to identify their religious affiliation based on a list of possible dominations. Given the growth in non-denominational congregations, as well as congregations that do not emphasize their denominational identity, the authors suspected that this type of survey may have incorrectly equated declining denominational ties with declining religion. To test this assumption, in addition to requesting information on denomination, their survey also asked for the name and address of the respondent's place of worship. In this way the researchers could use the information to locate the place of worship within a denominational structure, even if the respondent did not recognize this connection.

Using this approach, only 10.8% of the population is determined to be religiously unaffiliated, versus the 14% of previous surveys, and the previously unaccounted for 3% belong to Evangelical denominations. The denominations with the greatest affiliation include Catholic (21.2%), Evangelical Protestant (33.6%), and Mainline Protestant (22.1%). Furthermore, of the 10.8% that report being unaffiliated with organized religion, 62.9% report believing in God or some higher power [6] (pp. 7–8, 12). Religion is not in decline in the US and is experiencing growth similar to that found in other areas of the world, namely an increase in Evangelical Protestantism, Charismatic Christianity and "spirituality" [7].

Within the USA the growth of Charismatic Christianity can be seen within both Catholicism as well as within Protestantism. In its earliest phase the Charismatic movement was referred to as "Pentecostalism" and within the US it originated in the early twentieth century within a marginalized urban community in Los Angeles, the Asuza Street Apostalic Faith Mission. "Charismatic" is the more general term used to describe the movement in the later twentieth century. Charismatic Christianity is characterized by an emphasis on a direct personal experience of the Spirit of God, often evidenced in worship services by "speaking in tongues" and "healings." In addition to the rise of Charismatic Christianity within the US, another notable, and related trend, is the increase in "spirituality." Recall that even of the 10.8% of individuals in the Baylor Survey who report being unaffiliated with organized religion, 62.9% report believing in God or some "higher power." This finding is revealed even more dramatically in the results of a recent Gallup Poll, *The Spiritual State of the Union: The Role of Spiritual Commitment in the United States* [8]. This survey, consisting of 1004 interviews with US adults conducted in February and March 2006, reports that 40% of the respondents described themselves as "spiritual but not religious" and this percentage has increased by 10% since a similar survey conducted in 1999. The majority of respondents (49%) still define themselves as "religious"; however, another 7% describe themselves as "both" and only 3% as "neither." When asked more specifically about their beliefs, 82% reported believing in God whereas 13% chose instead belief in a "universal Spirit." Note then that of the 40% who chose to describe themselves as "spiritual but not religious" 26% still identify with theism.

Spirituality is also a factor in the UK, where Steve Bruce has shown that over a 40-year time span belief in a personal God may have fallen by 8%, but at the same time belief in "some sort of spirit or vital force which controls life" has risen by 3% [9]. Paula Heelas uses this work as well as data on religious belief in Sweden and the United States to argue that the middle ground, between those who regularly attend church and those who are declared atheists and agnostics, is increasing, and this increase is due at least in part to the growth of spirituality [10]. What exactly do we mean by "spirituality"? Heelas defines the term by contrasting it with religion, which she describes as being centered on a transcendent God whose will is mediated through tradition. Spirituality, in contrast, is not concerned with institutionalized authority, but on a direct experience of the divine immanent in life:

"Spirituality" has to do with the personal; that which is interior or immanent; that which is one's experienced relationship with the scared; and that wisdom or knowledge which derives from such experiences. At heart, spirituality has come to mean "life" [10] (p. 358).

It appears that those who define themselves solely as spiritual actually share a common ground with the Charismatic branches of Evangelical Christians, Mainline Protestants, and Catholics, and it is this common ground that is growing within the USA. Heelas refers to this common ground as the "HS" factor, which for non-traditionalists, or non-theist spiritualists, implies "the 'higher self' within" that is pursued. For more traditional or theistic spiritualities, "HS" refers to the quest for the indwelling of the Holy Spirit; for both the focus is on empowering this life in the here and now. Can astrobiology actually provide a bridge to engaging a religious US public that increasingly, both within and outside of theistic traditions, describes itself as spiritual? Maybe, if we are willing to build more than one type of bridge.

How would the goals of astrobiology be viewed by those subscribing to theistic spirituality, specifically the increasing numbers of Charismatic Christians in the United States? As this group places more emphasis on interpreting the Bible as an authoritative historical document than liberal Christian communities, we can predict that the greatest tension might be between what astrobiology learns about the origin and evolution of life and the Biblical creation story. The same would be true for conservative Evangelical Protestants. However, it would be incorrect to assume that either all Charismatic Christians or all Evangelical Protestants will insist on a literal reading of Genesis; not all are fundamentalists and in fact Charismatic Christianity in the United States is characterized by the importance it places on a direct experience of the divine, over a strict adherence to doctrine. For more liberal Charismatic Christians and Evangelical Protestants the challenging theological questions will be the same as those for liberal mainline Christians: what was God's role in an origin of life that arose naturally, or, put another way, how does God act in nature?

Emphasizing the nature of science and the differences between science and religion (see below) is one effort the scientific community can make to help alleviate this tension, but serious theological work remains to be done as well. Whereas some of this work is beginning to take place in liberal mainline Christian circles,[3]

---

[3] See, for example, C. Baker. *The Evolution Dialogues: Science, Christianity, and the Quest for Understanding*, ed. J. B. Miller (Washington DC: American Association for the Advancement of Science, 2006).

the very nature of Charismatic Christianity, with its focus on individual experience at the expense of institutionalized doctrine, will be counterproductive in terms of building a theological framework that incorporates what science has learned about the world. Much work remains for the Evangelical Protestant community as well.

As for the other questions astrobiology poses, concerning the extent of life and the future of life, I would argue that these are less threatening, assuming that our greatest possibility for discovering life in the near future is microbial life on Mars, and that they can provide unique opportunities for outreach. The discovery of a second genesis of life on Mars most likely would ignite origin and evolution issues; microbial life related to Earth life would undoubtedly be an exciting discovery, but not one that is particularly challenging to religious beliefs. Future of life questions, in the broadest interpretation, might appear as an irrelevant question for Charismatic Christians and Evangelical Protestants in general, as the Biblical account of history promises a future salvation. On the other hand, interpreting this question in terms of sustainability issues on Earth can resonate with both liberal and more conservative Christians, as recently evidenced by the agreement initiated by prominent members of the Evangelical community to work collaboratively with the scientific community on global climate change [11].

I suggest above that we should not expect Charismatic Christianity to be encouraging of developing theologies that incorporate scientific findings, not because of a negative attitude toward science, but because the focus of spirituality is on individual experience, not the development of doctrine. However, I also suggest that particular assumptions and possible discoveries of astrobiology will most likely be met with resistance by members of this group, as well as Evangelical Protestants in general, and even many liberal Christians, because of the challenges they raise concerning God's role in nature and creation. That said, I am still intrigued by the identification of "spirituality" with "life," quoted above, and what this might imply for a research and exploration program focused on life. Does astrobiology present a unique opportunity for the scientific community to engage with religious communities, whose members are increasingly defining themselves as spiritual, through a shared interest in "life"?

There is evidence that the answer for non-theist spiritualists, admittedly a smaller percentage than the theist spiritualists (13% vs. 26%), is yes. Consider, for example, the work of Ursula Goodenough [12]. Her main premise, eloquently expressed in her work, *The Sacred Depths of Nature*, is that a modern understanding of nature can give rise to religious emotions, which she defines as "shared cosmology and shared morality." Through the story that science tells about the cosmos, with an emphasis on the origin and interconnectedness of life on this planet, a planetary ethic may emerge that will provide a guideline for addressing global concerns, including sustainability issues. For the non-theist spiritualist the story about life that astrobiology is both telling, and striving to discover, might serve to encourage an interior experience of the divine, that is, "The Sacred Depths of Nature." Can the same be true for a theist spiritualist, particularly if their worldview stems from conservative Christianity? Perhaps the answer depends on how the astrobiology community engages these individuals.

The scientific community at large, despite unfortunately the opinion of some of its most vocal members, is taking great effort to remind the public that science and religion do not need to be in conflict. One recent example is the new publication by the National Academy of Science, *Science, Evolution, and Creationism* [13]. This publication emphasizes the "non-overlapping magesteria" or contrast view of science and religion which was popularized within the scientific community especially by Stephen J. Gould and his book *Rock of Ages: Science and Religion in the Fullness of Life* [14]. Simply put, this model assumes that there can be no conflict between science and religion since they respond to different questions; science tells us how, religion tells us why. The Clergy Letter Project is another recent effort that promotes the contrast view of science and religion [15]. The letter was initiated on the part of the scientific community (in response to the controversy over the teaching of evolution in public science classrooms) but seeks the help of clergy to spread the message. To date over 11,000 clergy members have signed the letter, which includes the plea, "We ask that science remain science and that religion remain religion, two very different, but complementary forms of truth." This is a message the scientific community is comfortable with and one that fits nicely within a broader program that encourages the scientific community at large to pay careful attention to how they "frame" their public statements [16].

Certainly, in regards to engaging a religious public around what science is learning about the world, this model is preferable to one more commonly assumed, a conflict model, which depicts science and religion as two endeavors inherently at odds with one another (a model which historians of science and religion have shown to be oversimplified).[4] When we speak as science educators we use the contrast model when we desperately try to help people understand what science is and can do, and what it isn't and can't do. This is an important message and one that scientists need to keep conveying. That said, I would argue that another approach has its time and place as well and that the astrobiology community would do well to embrace that approach – namely one that looks for dialogue and interaction.[5] In order to participate in that approach, we need to be able to listen, and to recognize

---

[4] For a review of the interaction between science and Christianity through history see D.C. Lindberg and R. Numbers (eds.). *God and Nature: Historical Essays on the Encounter between Christianity and Science* (Berkeley, CA: University of California Press, 1986).

[5] For a review of models relating Science and Religion see, for example, J.F. Haught. *Science and Religion: From Conflict to Conversation* (New York: Paulist Press, 1995).

that different religious constituencies will require different engagement models – we must be building multiple bridges.

Let me illustrate that point with a quote from the *Washington Post Magazine* in January 2006. The cover from that edition read, "Darwin vs. God, what the war between evolution and Intelligent Design is really about." I believe the author, who interviewed students, teachers, and scientists for this story, expresses quite nicely why we need more than a contrast approach in dealing with science and religion, "If intelligent design advocates have generally been blind to the overwhelming evidence for evolution, scientists have generally been deaf to concerns about evolution's implication" [17].

1.2.1 Who is it that feels we are not listening and what can we do to reach them?

The 2005 Pew Survey responses, when broken down according to the broad Christian religious affiliation categories of "Evangelical," "Mainline Protestant," and "Roman Catholic," illustrate that all affiliations, even secular, have a percentage of individuals who believe that "life has existed in its present form since the beginning of time." This view is, however, predominant among those who classified themselves as Evangelicals (70% of Evangelicals). Mainline Protestants and Catholics are more accepting of life evolving over time (60%) with nearly equal amounts attributing the process to natural selection or guidance by a supreme being [2]. The contrast view of science and religion is particularly appealing to Mainline Protestants as is demonstrated, for example, by the positive response these congregations have given to another outreach effort associated with the Clergy Letter Project, "Evolution Sunday."

Evolution Sunday, held near Darwin's birthday, encourages clergy to devote one Sunday to emphasizing that science and religion complement one another [15]. Assigning denominations to the same broad religious affiliations used in the Baylor Survey [6], we find that of the 558 congregations that participated in Evolution Sunday in 2007 the majority (78%) represented Mainline Protestant churches; the next largest category (18.1%) consisted predominantly of Unitarian Universalist congregations (a liberal religion with Judeo-Christian roots). Evangelical Protestant congregations accounted for less than 2% of the participating congregations.

The success story of the Clergy Letter Project, Evolution Sunday, and the contrast approach in general, is that mainline Christian denominations are being vocal about supporting the teaching of evolution. This is good news. A remaining and serious challenge is to engage evangelical communities on this issue and a contrast approach to science and religion is unlikely to accomplish this end. We need to honestly ask ourselves, is our goal limited to keeping creationism out of the science classroom, perhaps by majority rule, or are we interested in working long term towards a widespread acceptance of evolution? A widespread acceptance of evolution in the United States will require that the scientific community go beyond a simple contrast approach and be willing to encourage, and participate in, a program of in-depth and long-term engagement with theologians and religious community leaders.

We are correct in helping people understand that science can't by itself answer questions about meaning. But in doing only this, without a broader conversation, are we also suggesting that what science learns about the world, what the world is, has absolutely no relevance for our thoughts about what the world means? Are we condoning a focus on life as an interior experience, disconnected from the larger context? If astrobiology has done anything for the scientific community it has at minimum encouraged us to expand our vision beyond that of our immediate discipline, to see the larger context and to struggle to find a way to participate in a multidisciplinary scientific dialogue. This is the very contribution that I suggest makes astrobiology the ideal discipline for expanding the dialogue into an even broader context with a specific invitation to theologians and religious leaders to participate. We may even be able to build on the interest of theistic spiritualists in "life" so that through astrobiology in particular we may engage their interest in what science has learned about the world.

Securing a religious public's support for space science research and exploration, so that it merits a high enough priority to claim resources, though vital, is only the beginning. It is in the sharing of what we learn from the exploration that the process comes full circle, and then, only when the discoveries are owned by both the religious and the spiritual, both the liberal and conservative. Hopefully this volume provides one example of the broadening of context that can occur for all involved when members of the astrobiology community invite theologians, ethicists, historians and philosophers to learn about astrobiology, to reflect on its work from the perspective of their own discipline, and to share their findings with the scientific community and the public at large. Summarized below are the key points of the contributors to this volume, organized according to astrobiology themes.

1.3 Origin of life

Where did we come from? Astrobiology includes the study of the origin and evolution of life on Earth and draws on the results of these studies to explore the possibility of life arising and thriving on other planetary bodies. The authors of the first part of this volume, an origins researcher, a historian, a philosopher, and two theologians, specifically address the origins question from the perspective of their discipline. They provide an introduction to the science of life's origin; describe attitudes towards the definition of life as a function of historical context; use the study of life's origin as an example to illustrate the nature of science; offer an early Christian perspective on life's origin which can reflect favorably on current scientific understanding; and argue for the unique perspective Theology can bring to astrobiology's pursuits around the questions of origin and future of life.

The focus of origins research is elucidating how the first living cell arose from non-living matter, a formidable task. Both Hazen and Fry emphasize that although the specific, historical pathway that gave rise to life on Earth may never be known, the scientific community harbors great optimism for discovering the general chemical pathways that led to this outcome. Hazen suggests that the study of emergence, "the natural process by which complexity arises" can provide a unifying theme for origins research and that the emergent systems approach is particularly amenable to experimentation.

Fry prefers an "arch and scaffolding model" as a unifying theme for origins research. She argues that this model excels at providing an answer to the question of apparent design, a question posed by the fact that both metabolism and replication of genetic material are required for cell functioning. If these two distinct systems are dependent on one another, how then can evolution account for their appearance? Fry suggests that the various solutions offered to answer this dilemma all have in common the analogy of a supporting scaffold used to build a stone arch; the scaffold is present during construction but not obvious in the final product.

Both Hazen and Fry note that origin-of-life research is a relatively recent endeavor propelled forward by the Miller–Urey experiments 50 years ago. Fry also argues that this is the time during which the theory of evolution "came into maturity" opening the door to the assumption that life formed in the past through a natural process from inorganic matter. Origin-of-life research, she argues, should be granted the same scientific status as evolution research and not viewed as a weak point in the scientific explanation of the history of life.

Strick's review of changes in the definition of life from a historical perspective illustrates how the origin-of-life question was handled gingerly both by Darwin and his supporters. Darwin was well aware that the theory of evolution would be challenging for a religious public and in the first edition of *Origins* took great care to leave the possibility of divine intervention open in regards to the appearance of the first life form. Strick's review of the reactions of Darwin's supporters, both religious and non-religious to the definition of life and its first origins, and changing perspectives on the definition of life by scientists up until the present, demonstrates the important role cultural experience plays in the scientific endeavor.

Cultural experience also plays an important role in Theology. McMullin's contribution to this volume traces the influence of the prevailing philosophies of Augustine's time on his views on creation. Augustine, a prominent theologian in the early Christian church whose work is foundational for Christian theology, offers a view of creation that allows for a gradualist account of life's origin and diversity. McMullin argues that Augustine's view of creation is also supportive of a natural transition from non-life to life on early Earth. He finds in Augustine's work an alternative to the divine intervention that Intelligent Design advocates call for. Unfortunately, as McMullin notes, Augustine's views on creation were not supported by Church authorities and also fell victim to the tendency for literal interpretation of the Bible that flourished with the Protestant reformation.

The origin-of-life discussion in this volume concludes with the thoughts of a contemporary theologian, Deane-Drummond, whose contribution reviews the theological themes associated with origin-of-life and ethical issues raised by both origin and future of life research. She concludes by arguing that Theology has a distinctive contribution to make when considering future discoveries and research. Her work outlines why naturalistic explanations are not threatening to theists and how theological ethics can particularly encourage a reflection on our responsibility to future generations.

Perhaps the time is right for Augustine's views on creation to be revisited and for the views of theologians that embrace and welcome naturalistic explanations of life's origin and diversity to be heard both within and outside of the church. If this were to occur, then origin-of-life research might be more broadly described as a challenging but fruitful frontier, rather than a weak and controversial point in a scientific narrative. And as we approach a clearer understanding of the origin of the first cell, perhaps duplicating a possible scenario in the laboratory, scientists will, I believe, benefit from the encouragement and insight of theologians equipped with an understanding and acceptance of the scientific worldview, yet ready to offer the insights of a different tradition.

1.4 Extent of life

Are we alone? Dialogue participants were asked to consider this question with a focus on searching for and discovering extraterrestrial (ET) non-intelligent life, specifically microbial life, either on a planetary body within our solar system or one beyond our solar system. The scientific contributions to this part describe how life on Earth can provide a model for searching for life elsewhere, and the status of our search for habitable extrasolar planets. A historical perspective that traces how the philosophical, ethical, and theological issues associated with ET life have been raised and addressed through time is included, as is a contribution that categorizes the main philosophical issues associated with the search and discovery of ET life. Unique to this part of the volume, there is also a specific discussion of policy issues that need to be addressed in regards to searching for and discovering

ET life. The final contribution to this part explores if, and how, views of God will change if ET life is discovered.

Where else in the solar system might we look for life? Rothschild reminds us how the possibilities have been expanded by both our growing understanding of life's tenacity on Earth, gleaned from studies of life in extreme environments, and by the exploration of our solar system. The search for life is guided, by necessity, by what we know about terrestrial life. Terrestrial life is carbon and water based, diverse, dominated by microbes and can survive in extreme environmental conditions. Extrapolating from the terrestrial life model Mars, for example, is judged one likely candidate for harboring existing or extant ET life because of the evidence for liquid water on its surface in the past, and likely underground in the present. Pilcher and Lissauer's contribution outlines another major development in the search for ET life, the hundreds of planets now known to be orbiting stars other than our own Sun, and the technical advances currently taking place that will expand the ability of astronomers to detect Earth-sized planets in the habitable zones of their parent stars (distances at which liquid water at the surface of the planets is plausible). Pilcher and Lissauer suggest that in the next couple of decades we will know whether planets like these are common or rare. We may even be able to search for biosignatures, evidence of life, in their atmospheres.

The possibilities of exploration and research currently underway are certainly exciting, and I imagine they will naturally result in a renewed interest in pondering humanity's place in the universe, but in this volume Dick reminds us that throughout history the philosophical, ethical, and theological issues associated with ET life have been considered. He notes that the possibility of other worlds, though not necessarily raising challenges for theism in general, does pose interesting questions for specific Christian doctrines. He argues that one result of Darwin's work was that these discussions were broadened to include the idea of cosmic evolution, and a universe characterized by its distinct orientation towards biology. He suggests the latter idea, dubbed the "biological universe," has attained the status of a worldview.

In categorizing the philosophical issues raised by the search for ET life, Lupisella also notes the impact on worldviews the search may produce. Will the search confirm a "biological," or in Lupisella's terminology "deliberate," universe or will we conclude that life is a rare and random event? Under these different scenarios how will humanity's views on value and meaning fare? Other philosophical issues he highlights include all of the questions raised by the fact that the exploration occurs in the light of only one known data point – life on Earth. For example, if we discover ET life in the solar system how will we discern whether or not it represents an independent origin of life? Another category of philosophical issues noted by Lupisella, and covered in greater detail in the final part of this volume, is the ethical considerations due to ET life, specifically microbial life. Where is our "prime directive"?

The "prime directive" or policy issues, are specifically noted in Race's contribution to this volume. Race outlines how the different types of potential discoveries will in fact require different types of policies. For example, the discovery of intelligent life rather than microbial life, or the discovery of a planet that could potentially be habitable by life as we know it, would most certainly call for different types of policies to be put into practice. Unfortunately, she notes that the science and technology are far ahead of current policies, which are focused largely on issues of cross-contamination and protecting future scientific exploration.

The final contribution to this part of the volume turns again to the question of philosophical issues, with a specific focus on theology that is framed by the question – "How will our views of God change if we discover ET life?" To answer this question Crysdale reviews the theological issues raised by an evolving universe, noting that the challenges posed by astrobiology are related to issues surrounding evolution. Ideas of divine determinism and intervention, as well as a view of humanity as the apex of meaning, have to be reconsidered. She offers an example of a revised classical theism that accepts both continuity and novelty in God. Crysdale concludes her contribution with a reminder that humanity lives within an ethic of risk, not control, and this reality should inform our exploration.

The dominant motif of this discussion is that the questions, particularly the philosophical and theological ones, raised by the search for and potential discovery of ET microbial life are not necessarily new, but there is a renewed urgency to consider the possible answers. Scientific and technological advancements are such that we are cycling closer than we have ever been before to discovering whether or not ET life exists. At minimum the question has shifted from one of speculation to one that has hope of being tested empirically. The proximity of an answer also heightens the need for sustaining an ethical and policy discussion that extends beyond a consideration of protecting future scientific discoveries. The final part of this volume provides insight into the types of topics such a discussion will encompass.

1.5 Future of life

Where are we going? Beyond understanding the history of life and the expanse of life, astrobiology is also concerned with the next chapter in life's story, both here on Earth and possibly other planetary bodies. Questions that could fall under the future of life topic are numerous, but the works collected in this volume specifically address one: is planetary ecosynthesis on Mars, commonly referred to as terraforming, technologically feasible, and if so what are the ethical issues raised by this possibility? This part of the volume includes a description of the suitability of Mars for supporting either transported terrestrial life or indigenous Martian life, and the technological steps that would be necessary to create a sustained biosphere on Mars. This introduction is followed by the presentation of three ethical frameworks for considering whether or not what might be done should be done. One framework explicitly draws on perspectives from the Christian tradition to consider these options and the other two are secular in nature. This final part of the volume also includes a contribution that considers how the role that stories play in Asian religions can offer new insight on all three astrobiology topics: the origin, extent, and future of life.

McKay begins this part of the volume by making a compelling argument that, given current technology and the natural resources available on the planet, Mars could be warmed, in about 100 years, to support a biosphere. He draws reference to a fact that we have recently become all too painfully familiar with – humans know how to increase the temperature of a planet and have unintentionally done just that to our own. The biosphere McKay is referring to, however, is not one that could support human life, but one that is similar to the biosphere that existed on Earth prior to the rise of O_2. Planetary ecosynthesis of this nature on Mars is not limited to the realm of science fiction. McKay then goes on to argue that utilitarian and intrinsic worth ethical arguments would both support a decision to move forward with ecosynthesis on Mars. Intrinsic worth arguments that include valuing the diversity of life support an argument that life should be encouraged to thrive on Mars, but indigenous life gets first priority. As McKay points out, this conclusion will place very specific demands on guidelines for conducting the current exploration of Mars.

Smith takes exception to McKay's use of intrinsic value arguments to encourage planetary ecosynthesis on Mars. He recognizes that it is common practice in environmental ethics to extend intrinsic value beyond humans, but questions what criteria are used to distinguish between an entity that possesses intrinsic value and one that does not. He argues that defining these criteria is essential and offers one solution, a "ratio-centric" approach. In this system the ability to reason is the defining criterion by which intrinsic value is assigned. For the sake of clarity, note that this assignment can be based on membership of a larger group and is not dependent on individual merit, i.e. mentally impaired individuals are not excluded as they are members of humanity. Smith suggests that having such a specific criterion in place will result in a "clear mechanism" for decision making. This does not mean that Smith is opposed on ethical grounds to encouraging life, either terrestrial or indigenous, to thrive on Mars, but that he is opposed to using a process to reach a decision on this matter that depends on intrinsic value being assigned to non-rational entities in the absence of a defining criterion.

Randolph, drawing on the Christian tradition, offers an alternative to using the ability to reason as the criterion for assigning intrinsic value. He begins his chapter by asking why a specifically Christian perspective on the issue is pertinent at all, and answers in part from a pragmatic standpoint that I find persuasive. Christians represent a significant proportion of the US population and the astrobiology community would do well to understand their perspectives. Randolph is not suggesting, however, that this is the only perspective that deserves consideration and in fact argues for the value of inviting a diverse collection of religious views to the discussion. The main point of his essay though is to present a specifically Christian ethical framework. He accomplishes this by emphasizing two themes in scripture that he suggests should be used as guides for contemporary thought. The first is that God has a preferential option for life and the second is that although all of life has intrinsic value, humanity alone has been entrusted with a responsibility to serve all of nature. Using these themes as the foundation, his contribution then explores the protections due to Earth life, and ET life (with a focus on microbial life), and the ethical questions raised by the possibility of planetary ecosynthesis on Mars.

These three contributions together make the point more forcibly than any single one alone could, how important it will be to ensure that an inclusive dialogue precedes our actions as a space-faring culture. The ethical frameworks presented by McKay, Smith, and Randolph could all be used to argue for planetary ecosynthesis on Mars. The possibility of the same outcome is intriguing, but in my opinion less important than the reality that each framework has the potential to draw our attention to the potential consequences of our actions, or inactions, that we might have previously ignored if not for the insights of traditions beyond our own. Technological advances will continue to guarantee that we face new choices. I believe we stand better equipped to respond to these choices if we are also challenged to move beyond the comfort of our own frameworks. The final contribution to this volume, a chapter that describes Asian religious perspectives on storytelling, provides a concrete example of how common Western motifs for the relationship between science and religion are called into question by Asian perspectives.

Cho argues that science and religion are often described as telling different stories about the world and that a Western perspective calls for the proper relationship between these two stories to be defined. Beyond adopting a view of inherent conflict between the two stories, she argues that the most common Western responses are to describe them as either complementary stories, where each is needed to complete the whole, or to seek to integrate the two stories into one. From an Asian perspective the reality of many stories, without the constraints of either wholeness or integration, is welcomed as it provides an opportunity to draw from a wealth of stories to choose the most useful for a given context. She encourages us to think about how the relationship between science and religion, and the consequences that result from assumptions about that relationship, might be altered if an Asian perspective was adopted.

1.6 Concluding thoughts

Cho's contribution to this volume is a vital reminder that the work presented here is only a beginning, focused from a pragmatic perspective on Christian theological responses, but by no means meant to be a conclusive work. The papers collected here cannot claim to be conclusive as to Christian theological views, let alone other philosophical, or ethical views, and it was never the intent of this work to approach that goal. Instead, I hope the contributions collected here will encourage a continued interest in exploring the work of astrobiology from a multidisciplinary perspective. I believe that astrobiology as a discipline is particularly burdened, perhaps blessed, with the responsibility to speak to society at large. After all, humanity itself may be inherently defined by the ability we collectively possess to ask "Where did we come from?," "Are we alone?," and "Where are we going?" I have found that the answers proposed in community, especially a diverse one, are far more intriguing than those I derive on my own. I hope this volume provides a similar experience for all who are generous enough to consider its content.

References

[1] J. C. Tarter, *et al.* A reappraisal of the habitability of planets around M dwarf stars. *Astrobiology*, **7** (2007), 30–65.
[2] The Pew Forum on Religion and Public Life. Public Divided on Origins of Life (August 30, 2005), available online at: http://pewforum.org/docs/?DocID=115.
[3] C. Darwin. *On the Origin of Species* (London: John Murray, 1859).
[4] P. L. Berger. Reflections on the sociology of religion today. *Sociology of Religion*, **62** (2001), 443–454.
[5] R. Stark. Secularization, R.I.P. *Sociology of Religion*, **60** (1999), 249–273.
[6] C. D. Bader, K. Dougherty, P. Froese *et al.* American Piety in the 21st Century: New Insights to the Depth and Complexity of Religion in the US (September 2006), available online at: www.baylor.edu/content/services/document.php/33304.pdf.
[7] L. Woodhead. The desecularization of the world: a global overview. In *The Desecularization of the World: Resurgent Religion and World Politics,* ed. P. L. Berger (Michigan: Ethics and Public Policy Center and Wm. B. Eerdmans Publishing Co., 1999), pp. 1–18.
[8] The Gallup Organization. *The Spiritual State of the Union: The Role of Spiritual Commitment in the United States* (Washington, DC: Spiritual Enterprise Institute, 2006).
[9] S. Bruce. *Religion in Modern Britain* (Oxford: Oxford University Press, 1995).
[10] P. Heelas. The spiritual revolution: from "religion" to "spirituality." In *Religions in the Modern World*, eds. L. Woodhead, P. Fletcher, H. Kawanami, and D. Smith (New York: Routledge, 2002), pp. 357–377.

[11] J. Heilprin. Evangelicals, scientists join on warming. *The Washington Post* (January 17, 2007).
[12] U. Goodenough. *The Sacred Depths of Nature* (New York: Oxford University Press, 1998).
[13] F. J. Ayala, *et al. Science, Evolution and Creationism* (Washington, DC: The National Academies Press, 2008).
[14] S. J. Gould. *Rock of Ages: Science and Religion in the Fullness of Life* (New York: The Ballantine Publishing Group, 1999).
[15] M. Zimmerman. *The Clergy Letter Project* (2004), http://www.butler.edu/clergy-project/rel_Evol_sun.htm.
[16] M. C. Nisbet and C. Mooney. Framing science. *Science*, **316** (2007), 56.
[17] S. Vedantum. Darwin versus God: what the war between evolution and intelligent design is really about. *Washington Post Magazine* (February 5, 2006), 8–15, 21–26.

Note added in proof: In March 2009 the results of the 2008 American Religious Identification Survey (ARIS) were released by the Institute for the Study of Secularism in Society and Culture at Trinity College (available online at: http://www.americanreligionsurvey-aris.org/reports/ARIS_Report_2008.pdf). This study, which began with an open-ended question "What is your religion if any?" and did not present respondents with a list of answers to choose from, reports that 15% of respondents still fall into the "None" category in 2008. However, like the Baylor study [6] (see also R. Stark. *What Americans Really Believe* (Texas: Baylor University Press, 2008)), the ARIS study does not indicate a significant rise in atheism since the last survey in 2001 and also reports that non-denominational Christianity is on the rise. According to the ARIS report, "Another notable finding is the rise in the preference to self-identify as 'Born Again' or 'Evangelical' rather than with any Christian tradition, church or denomination" (p. 6).

Part I

Origin of life

2

Emergence and the experimental pursuit of the origin of life

Robert M. Hazen

> The origin of life involved many, many emergences.
>
> Harold Morowitz [1]

The experimental investigation of life's origin commenced in earnest more than a half-century ago with the pioneering work of Miller [2], who synthesized many of life's molecular building blocks under plausible prebiotic conditions. Despite an initial euphoric sense that the origin mystery would soon be solved, scientists quickly realized that the transition from a geochemical to a biochemical world would not easily be deduced by the scientific method.

The great challenge of origins research lies in replicating in a laboratory setting the extraordinary increase in complexity that is required to evolve from isolated molecules to a living cell. The principal objective of this review is to describe some of the efforts by origin-of-life researchers to induce such increases in complexity. A unifying theme of these studies, and hence a useful organizing framework for this review, is the principle of emergence – the natural process by which complexity arises.

2.1 Emergence as a unifying concept in origins research

The origin of life may be modeled as a sequence of so-called "emergent" events, each of which added new structure and chemical complexity to the prebiotic Earth. Observations of numerous everyday phenomena reveal that new patterns commonly emerge when energy flows through a collection of many interacting particles [1, 3, 4, 5, 6, 7].

Exploring the Origin, Extent, and Future of Life: Philosophical, Ethical, and Theological Perspectives, ed. Constance M. Bertka. Published by Cambridge University Press. © Cambridge University Press 2009.

In the words of John Holland, an influential leader in the study of emergent systems, "It is unlikely that a topic as complicated as emergence will submit meekly to a concise definition, and I have no such definition" [6]. Nevertheless, emergent systems display three distinctive characteristics: (1) they arise from the interactions of many particles or "agents"; (2) energy flows through those systems of particles; and (3) they display new patterns or behaviors that are not manifest by the individual agents. Sand grains, for example, interact under the influence of wind or waves to form dune and ripple structures [8, 9]. Similarly, ants interact to form colonies, neurons interact to produce consciousness, and people interact to form societies [10, 11].

The history of the universe may be viewed as a progression of emergent events, from hydrogen to stars, from stars to the periodic table of chemical elements, from chemical elements to planets and life, from life to consciousness [1, 7, 12]. The inexorable stepwise transition from simplicity to emergent complexity is an intrinsic characteristic of the cosmos. In such a pregnant universe, one need not resort to divine intervention nor to Intelligent Design for life's origin. Thus, in spite of the lack of a precise definition, the recognition and description of these varied emergent systems provides an important foundation for origins of life research, for life is the quintessential emergent phenomenon. From vast collections of interacting lifeless molecules emerged the first living cell. Understanding the underlying principles governing such emergent systems thus provides insights to our experimental and theoretical efforts to understand life's origins.

Framing the origin-of-life problem in terms of emergence is more than simply providing a new label to parts of the origin story we don't understand. Emergent systems share key characteristics that can inform origin models. For example, all emergent systems, including collections of sand or cells or stars, require a minimum number of interacting agents before new patterns arise [7, 10]. These systems also require a source of energy within specific limits – too little energy and no patterning occurs, but too much energy and patterns are destroyed [7, 12]. Experimental observations of emergent systems also reveal that cycling of energy commonly enhances emergent complexity [7, 13, 14]. An understanding of these and other general characteristics of emergent systems thus informs experimental design and the theoretical analysis of life's origins.

One key to understanding life's origin is to recognize the critical role of disequilibrium. Many familiar natural systems lie close to chemical and physical equilibrium and thus do not display emergent behavior. For example, water gradually cooled to below the freezing point equilibrates to become a clear chunk of ice, whereas water gradually heated above the boiling point similarly equilibrates by converting to steam. Dramatically different behavior occurs far from equilibrium. Water subjected to the strong temperature gradient of a boiling pot displays complex turbulent convection [3, 4]. Water flowing downhill in the gravitational gradient of a river valley interacts with sediments to produce the emergent landforms of braided streams, meandering rivers, sandbars and deltas. Emergent systems seem to share a common origin; they arise away from equilibrium when energy flows through a collection of many interacting agents. Such systems tend spontaneously to become more ordered and display new, often surprising behaviors. The whole is more than the simple sum of the parts.

The overarching problem with studying life's origins is that even the simplest known life form is vastly more complex than any non-living components that might have contributed to it. How does such astonishing, intricate complexity arise from lifeless raw materials? What now appears to us as a yawning divide between non-life and life reflects the fact that the chemical evolution of life must have occurred as a stepwise sequence of successively more complex stages of emergence [7]. When modern cells emerged, they quickly consumed virtually all traces of these earlier stages of chemical evolution.

The challenge, therefore, is to establish a progressive hierarchy of emergent steps that leads from a prebiotic ocean enriched in organic molecules, to functional clusters of molecules perhaps self-assembled or arrayed on a mineral surface, to self-replicating molecular systems that copied themselves from resources in their immediate environment, to encapsulation and eventually cellular life. The nature and sequence of these steps may have varied in different environments and we may never deduce the exact sequence that occurred on the early Earth. Yet many researchers suspect that the inexorable direction of the chemical path is similar on any habitable planet or moon [1, 7, 15].

Such a stepwise scenario informs attempts to define life, because the exact point at which such a system of gradually increasing complexity becomes "alive" is intrinsically arbitrary. The evolutionary path to cellular life must have featured a rich variety of intermediate, complex, self-replicating emergent chemical systems. Each of those steps represents a distinctive, fundamentally important stage in life's molecular synthesis and organization. Each step requires independent experimental study, and perhaps a distinctive name in a taxonomy richer than "living" versus "non-living" [7, 16].

Ultimately, the key to defining the progressive stages between non-life and life lies in experimental studies of relevant chemical systems under plausible geochemical environments. The concept of emergence simplifies this experimental endeavor by reducing an immensely complex historical process to a more comprehensible succession of measurable steps. Each emergent step provides a tempting focus for laboratory experimentation and theoretical modeling.

This view of life's origin, as a stepwise transition from geochemistry to biochemistry, is of special relevance to the search for life elsewhere in the universe.

Before an effective search can be undertaken, it is necessary to know what to look for. It is plausible, for example, that Mars, Europa, and other bodies progressed only part way along the chemical path to cellular life. If each emergent step in life's origin produces distinctive and measurable isotopic, molecular, and structural signatures in its environment, and if such markers can be identified, then these chemical features represent important observational targets for planned space missions. It is possible, for example, that characteristic isotopic, molecular, and structural "fossils" survive only if they have not been eaten by more advanced cells. We may find that those distinctive molecular structures can serve as extraterrestrial "abiomarkers" – clear evidence that molecular organization and evolution never progressed beyond a certain pre-cellular stage. As scientists search for life elsewhere in the universe, they may be able to characterize extraterrestrial environments according to their degree of emergence along this multi-step path [7, 17].

2.2 The emergence of biomolecules

The first vital step in life's emergence on Earth must have been the synthesis and accumulation of abundant carbon-based biomolecules. In the beginning, life's raw materials consisted of water, rock, and simple volcanic gases – predominantly carbon dioxide and nitrogen, but with local concentrations of hydrogen, methane, ammonia, and other species. Decades of experiments have revealed that diverse suites of organic molecules can emerge from geochemical environments.

The experimental pursuit of geochemical organic synthesis, arguably the best understood aspect of life's origin, began a half-century ago with the pioneering studies of University of Chicago graduate student Stanley Miller and his distinguished mentor Harold Urey [1, 18, 19]. Together they established the potential role of organic synthesis that occurred in Earth's primitive atmosphere and ocean as they were subjected to bolts of lightning and the Sun's intense radiation.

This experimental investigation of life's origin is a surprisingly recent pursuit. Two centuries ago most scientists accepted the intuitively reasonable idea that life is generated spontaneously all around us, all the time. The question of life's ancient origins was not asked, at least not in the modern experimental sense [20, 21, 22].

By the early twentieth century many scientists would have agreed that life's origins, wherever and however it occurs, depends on three key resources. First, all known life forms require liquid water. All living cells, even those that survive in the driest desert ecosystem, are formed largely of water. Surely the first cells arose in a watery environment.

Life also needs a ready source of energy. The radiant energy of the Sun provides the most obvious supply for life today, but bolts of lightning, impacts of asteroids,

Earth's inner heat and the chemical energy of minerals have also been invoked as life-triggering energy sources.

And, thirdly, life depends on a variety of chemical elements. All living organisms consume atoms of carbon, oxygen, hydrogen, and nitrogen, with a bit of sulfur and phosphorus and other elements as well. These elements combine in graceful geometries to form essential biomolecules.

In spite of the intrinsic importance of the topic, it was not until the 1920s that such scientific speculation took a more formal guise. Most notable among the modern school of origin theorists was the Russian chemist Alexander Oparin [23]. In 1924, while still in his twenties and under the scrutiny of the atheist, Marxist Soviet hierarchy, Oparin elaborated on the idea that life arose from a body of water that gradually became enriched in organic molecules – the so-called "Oparin Ocean" or the "primordial soup." Somehow, he posited, these molecules clustered together and became self-organized into a chemical system that could duplicate itself. In many other cultures, where religious doctrine colored thinking on origins, these revolutionary ideas would have been seen as heretical but Oparin's ideas resonated with the materialist worldview of the Leninist leadership.

Many of Oparin's postulates were echoed in 1929 by the independent ideas of British biochemist and geneticist J. B. S. Haldane, whose brief, perceptive article focused on the production of large carbon-based molecules under the influence of the Sun's ultraviolet radiation [24]. Given such a productive chemical environment, Haldane envisioned the first living objects as self-replicating, specialized molecules.

2.2.1 The Miller–Urey experiment

Oparin and Haldane offered original and intriguing ideas that were subject to experimental testing, but Oparin and his contemporaries didn't try to replicate experimentally the prebiotic formation of biomolecules. Not until the years after World War II were the landmark experiments of Miller and Urey devised [2, 19, 25]. They mimicked Earth's early surface by sealing water and simple gases into a tabletop glass apparatus and subjecting the contents to electric sparks. When the experiment began the water was pure and clear, but within days the solution turned yellowish and a black residue had begun to accumulate near the electrodes. Reactions of water and volcanic gases had produced organic molecules, most notably a suite of amino acids (the building blocks of proteins), in abundance. The Miller–Urey experiment transformed the science of life's chemical origins. For the first time an experimental protocol duplicated a plausible life-forming process.

Given such an exciting finding, other groups jumped at the chance to duplicate the amino acid feat. More than a dozen amino acids were synthesized from scratch, along with other key biomolecules: membrane-forming lipids, energy-rich sugars and other carbohydrates, and metabolic acids. Enthusiasm grew as other scientists discovered promising new chemical pathways. In 1960 chemist John Oró demonstrated that a hot, concentrated hydrogen cyanide solution produces adenine, a crucial biomolecule that plays a role in both genetic material and in metabolism [26, 27]. Other chemists conducted similar experiments starting with relatively concentrated solutions of formaldehyde, a molecule thought to be common in some prebiotic environments [28]. Their simple experiments produced a rich variety of sugar molecules, including the critical compound ribose. Gradually, researchers filled in gaps in the prebiotic inventory of life's molecules.

As exciting and important as these results may have been, seemingly intractable problems remain. Within a decade of Miller's triumph serious doubts began to arise about the true composition of Earth's earliest atmosphere. Miller exploited a highly reactive, chemically reducing atmosphere of methane, ammonia and hydrogen – an atmosphere distinctly lacking in oxygen. But by the 1960s, new geochemical calculations and data from ancient rocks pointed to a much less reducing early atmosphere composed primarily of nitrogen and carbon dioxide – unreactive gases that do almost nothing of interest in a Miller apparatus [29].

For decades Miller and his supporters have countered with a pointed argument: the molecules of life match those of the original Miller experiment with great fidelity. Miller's advisor, Harold Urey, is said to have often quipped, "If God did not do it this way, then he missed a good bet" [25] (p. 41). Most geochemists discount the possibility of more than a trace of atmospheric methane or ammonia at the time of life's emergence (except perhaps in local volcanic environments) [30, 31], though Tian *et al.* recently proposed that hydrogen may have comprised as much as 30% of the Hadean atmosphere [32, 33].

Added to this atmospheric concern is the fact that the molecular building blocks of life created by Miller represent only the first step on the long road to life. Living cells require that such small molecules be carefully selected and then linked together into vastly more complex structures – lipid cell membranes, protein enzymes, and other so-called "macromolecules." Even under the most optimistic estimates, the prebiotic ocean was an extremely dilute solution of countless thousands of different kinds of organic molecules, most of which play no known role in life. By what processes were just the right molecules selected and organized?

The Miller–Urey scenario suffers from another nagging problem. Macromolecules tend to fragment rather than form when subjected to lightning and the Sun's ultraviolet light [29]. These so-called "ionizing" forms of energy are useful for making highly reactive molecular fragments that combine into modest-sized molecules like amino acids. Combining many amino acids into an orderly chain-like enzyme, however, is best accomplished in a less destructive energy domain. Emergent complexity relies on a flow of energy, to be sure, but not too much energy. Could life have emerged in the harsh glare of daylight, or was there perhaps a different, more benign origin environment?

2.2.2 Deep origins

All living cells require a continuous source of energy. Until recently, scientists claimed that the metabolic pathways of virtually all life forms rely directly or indirectly on photosynthesis. This view of life changed in February 1977, with the discovery of an unexpected deep-ocean ecosystem [34]. These thriving communities, cut off from the Sun, subsist on geothermal energy supplied by Earth's inner heat. Microbes serve as primary energy producers in these deep zones; they play the same ecological role as plants at Earth's sunlit surface. These one-celled vent organisms exploit the fact that the cold oxygen-infused ocean water, the hot volcanic water, and the sulfur-rich mineral surfaces over which these mixing fluids flow are not in chemical equilibrium. The unexpected discovery of this exotic ecosystem quickly led to speculation that a hydrothermal vent might have been the site of life's origin [35].

New support for the idea gradually consolidated, as hydrothermal ecosystems were found to be abundant along ocean ridges of both the Atlantic and Pacific. Researchers realized that at a time when Earth's surface was blasted by a continuous meteorite bombardment, deep ocean ecosystems would have provided a much more benign location than the surface for life's origin and evolution [36, 37, 38]. New discoveries of abundant primitive microbial life in the deep continental crust further underscored the viability of deep, hot environments [39, 40, 41]. By the early 1990s, the deep-origin hypothesis had become widely accepted as a viable, if unsubstantiated, alternative to the Miller surface scenario.

Following the revolutionary hydrothermal origins proposal, numerous scientists began the search for life in deep, warm, wet environments. Everywhere they look, it seems – in deeply buried sediments, in oil wells, even in porous volcanic rocks more than a mile down – microbes abound. Microbes survive under miles of Antarctic ice and deep in dry desert sand [42, 43]. These organisms seem to thrive on mineral surfaces, where water-rock interactions provide the chemical energy for life.

Another incentive exists for looking closely at the possibility of deep origins. If life is constrained to form in a sun-drenched pond or ocean surface then Earth, and perhaps ancient Mars or Venus, are the only possible places where life could have begun in our solar system. If, however, cells can emerge from deeply buried wet zones, then life may be much more widespread than previously imagined. The possibility of deep origins raises the stakes in our exploration of other planets and moons.

The hydrothermal origins hypothesis received increased attention with the publication of a seminal paper in 1988 – "Before enzymes and templates: theory of surface metabolism" by German chemist Günter Wächtershäuser [44]. Wächtershäuser proposed a theory of organic evolution in which minerals, mostly iron and nickel sulfides that abound at deep-sea volcanic vents, provide catalytic surfaces for the synthesis and assembly of life. Wächtershäuser's model incorporates several assumptions about the nature of the first living entity [44, 45, 46, 47, 48]. First, he rejects the "primordial soup" concept, arguing instead that the prebiotic emergence of biomolecules by Miller–Urey processes was irrelevant because the primordial soup was much too dilute and contained too many useless molecular species. Second, the first life form made its own molecules: most other workers assume that the first living entity scavenged amino acids, lipids, and other useful molecules from its surroundings – a strategy called "heterotrophy" (from the Greek for "other nourishment"). Wächtershäuser denies that possibility, since the soup was so dilute and unreliable. He counters that the first life must have been "autotrophic" ("self nourishment") – manufacturing its own molecular building blocks from scratch.

A third assumption is that the first life form relied on the chemical energy of minerals, not the Sun. Sunlight and lightning are too violent and they are uncharacteristic of the energy sources that most cells use today. Photosynthesis, furthermore, is an immensely complex chemical cascade requiring numerous proteins and other specialized molecules. Surface reactions on minerals, by contrast, are simple and similar to many cellular processes. Wächtershäuser also assumes that metabolism – a simple cycle of chemical reactions that duplicates itself – comes first. Many other researchers claim that life began with a self-replicating genetic molecule like RNA, but even the simplest genetic molecule is vastly more complex than anything in the soup.

These original ideas are amplified by one additional assumption common to all origin theories – that of biological continuity. Today's biochemistry, no matter how intricate and dependent on specialized enzymes, has evolved in an unbroken path from primordial geochemistry. Thus, for example, Wächtershäuser's scheme builds on the observation that many of the molecules that enable living cells to process energy have, at their very core, a mineral-like cluster of metal and sulfur atoms [49]. Could this feature be a kind of biomolecular fossil, a remnant of primordial chemistry?

The central chemical postulate of Wächtershäuser's theory is that iron and nickel minerals served as templates, catalysts, and energy sources for biosynthesis, all in one. In Wächtershäuser's view, simple molecules like carbon monoxide and hydrogen react on mineral surfaces to produce larger molecules. These molecules tend to have negative charges, and so stick to the positively charged mineral surfaces, where more reactions build larger and larger molecules. This collection of surface-bound molecules begins to feed off each other, eventually forming a chemical cycle that copies itself.

Recent experiments bolster Wächtershäuser's hypothesis. The most fundamental biological reaction is the incorporation of carbon atoms (starting with the gas carbon dioxide) into organic molecules. Many common minerals, including most minerals of iron, nickel, or copper, promote carbon addition [50, 51, 52, 53, 54, 55, 56]. One conclusion seems certain. Mineral-rich hydrothermal systems contributed to the early Earth's varied inventory of bio-building blocks.

It now appears that anywhere energy and simple carbon-rich molecules are found together, a suite of interesting organic molecules is sure to emerge [29]. In spite of the polarizing advocacy of one favored environment or another [25, 57, 58], experiments point to the likelihood that there was no single dominant source. By four billion years ago Earth's oceans must have become a complex, albeit dilute, soup of life's building blocks. Though not alive, this chemical system was poised to undergo a sequence of increasingly complex stages of molecular organization and evolution.

2.3 The emergence of macromolecules

Prebiotic processes produced a bewildering diversity of molecules. Some of those organic molecules were poised to serve as the essential starting materials of life, but most of that molecular jumble played no role whatsoever in the dawn of life. The emergence of concentrated suites of just the right mix remains a central puzzle in origin-of-life research.

Life's simplest molecular building blocks – amino acids, sugars, lipids, and more – emerged inexorably through facile, inevitable chemical reactions in numerous prebiotic environments. A half-century of synthesis research has elaborated on Miller's breakthrough experiments. Potential biomolecules from many sources must have littered the ancient Earth.

What happened next? Individual biomolecules are not remotely lifelike. Life requires the assembly of just the right combination of small molecules into much larger collections – "macromolecules" with specific functions. Making macromolecules is complicated by the fact that for every potentially useful small molecule in the prebiotic soup, dozens of other molecular species had no obvious role in biology. Life is remarkably selective in its building blocks; the vast majority of carbon-based molecules synthesized in prebiotic processes have no biological use whatsoever.

2.3.1 Life's idiosyncrasies

Consider sugar molecules, for example [59]. All living cells rely on two kinds of 5-carbon sugar molecules, ribose and deoxyribose (the "R" and "D" in RNA and

DNA, respectively). Several plausible prebiotic synthesis pathways yield a small amount of these essential sugars, but for every ribose molecule produced many other 5-carbon sugar species also appear – xylose, arabinose, and lyxose, for example. Adding to this chemical complexity is a bewildering array of 3-, 4-, 6-, and 7-carbon sugars, in chain, branch, and ring structures. What's more, many sugar molecules, including ribose and deoxyribose, come in mirror-related pairs. These left- and right-handed varieties possess the same chemical formula and many of the same physical properties, but they differ in shape like left and right hands. All known prebiotic synthesis pathways yield equal amounts of left- and right-handed sugars, but cells employ only the right-handed sugar varieties. Consequently, many origins researchers have shifted their focus to the processes by which just the right molecules might have been selected, concentrated, and organized into the essential structures of life.

The oceans are of little help because they are so vast – a volume greater than 320 million cubic miles. No matter how much organic matter was made, the oceans formed a hopelessly dilute soup. In such a random, weak solution, it would have been difficult for just the right combination of molecules to bump into one another and make anything useful in the chemical path to life. By what process were the right molecules selected?

2.3.2 Molecular selection

Lipid molecules, the building blocks of cell membranes, accomplish the selection trick in a striking way. One end of these long, slender molecules is hydrophilic (water loving), whereas the rest of the molecule is hydrophobic (water hating). Consequently, when placed in water, life's lipids spontaneously self-organize into tiny cell-like spheres. This selection process is rapid and spontaneous [60, 61, 62].

Most molecules don't self-organize. Consequently, many scientists have focused on surfaces as an alternative solution. Chemical complexity can arise at surfaces, where different molecules can congregate and interact. The surface of the ocean where air meets water is one promising interface, where a primordial oil slick might have concentrated organic molecules [63]. Evaporating tidal pools where rock and water meet and cycles of evaporation concentrate stranded chemicals provide another appealing scenario for origin-of-life chemistry [64]. Deep within the crust and in hydrothermal volcanic zones mineral surfaces may have embraced a similar role, selecting, concentrating, and organizing molecules on their periodic crystalline surfaces [38].

Solid rocks provide especially attractive surfaces for concentration and assembly of molecules. Experiments reveal that amino acids concentrate and polymerize on clay particles to form small, protein-like molecules, while layered minerals also have the ability to adsorb and assemble the building blocks of RNA or soak up small organic molecules in the rather large spaces between layers [65, 66, 67, 68, 69, 70]. Once confined and concentrated, these small molecules tend to undergo reactions to form larger molecular species that aren't otherwise likely to emerge from the soup.

In the most extreme version of this approach, Scottish researcher Graham Cairns-Smith has speculated that fine-grained crystals of clay, themselves, might have been the first life on Earth [71, 72, 73, 74]. The crux of the argument rests on a simple analogy. Cairns-Smith likens the origin of life to the construction of a stone archway, with its carefully fitted blocks and crucial central keystone that locks the whole structure in place. But the arch cannot be built by just piling one stone atop another. He proposes that a simple support structure like a scaffolding facilitates the construction and then is removed. "I think this must have been the way our amazingly 'arched' biochemistry was built in the first place. The parts that now lean together surely used to lean on something else – something low tech" [73] (p. 92). That something, he argues, was a clay mineral.

2.3.3 Right and left

One of the most intriguing and confounding examples of prebiotic molecular selection was the incorporation of handedness. Many of the most important biomolecules, amino acids and sugars included, come in "chiral," or mirror image, pairs. These left- and right-handed molecules have virtually the same energies and physical properties, and all known prebiotic synthesis pathways produce chiral molecules in essentially 50:50 mixtures. Thus, no obvious inherent reason exists why left or right should be preferred, yet living cells display the most exquisite selectivity, choosing right-handed sugars over left, and left-handed amino acids over right [7, 75, 76, 77].

Some analyses of chiral amino acids in carbonaceous meteorites point to the possibility that Earth was seeded by amino acids that already possessed a left-handed bias [78, 79, 80]. According to one scenario, left-handed molecules could have been concentrated if circularly polarized synchrotron light from a rapidly rotating neutron star selectively photolyzed right-handed amino acids in the solar nebula [81, 82, 83]. However, it is also difficult to eliminate entirely the possibility of a left-handed overprint imposed in the laboratory during the difficult extraction and analysis of meteorite organics.

Alternatively, many origin-of-life researchers argue that the chirality of life occurred as a chance event – the result of an asymmetric local physical environment on Earth. Such local chiral environments abounded on the prebiotic Earth, both as chiral molecules, themselves, and in the form of asymmetric mineral surfaces

[84, 85]. Experiments show that left- and right-handed mineral surfaces provide one possible solution for separating a 50:50 mixture of L and D molecules [77, 86]. Minerals often display chiral crystal faces, which might have provided templates for the assembly of life's molecules.

In spite of these insights, many aspects of prebiotic molecular selection and organization remain uncertain and thus represent topics of intense ongoing research. Even so, the emergence of macromolecular structures is but one increment in the stepwise progression from geochemistry to biochemistry. Life requires that macromolecules be incorporated into a self-replicating system.

2.4 The emergence of self-replicating systems

Four billion years ago the seeds of life had been firmly planted. The Archean Earth boasted substantial repositories of serviceable organic molecules. These molecules must have become locally concentrated, where they assembled into vesicles and polymers of biological interest. Yet accumulations of organic molecules, no matter how highly selected and intricately organized, are not alive unless they also possess the ability to reproduce. Devising a laboratory experiment to study chemical self-replication has proven to be vastly more difficult than the prebiotic synthesis of biomolecules or the selective concentration and organization of those molecules into membranes and polymers. In a reproducing chemical system, one small group of molecules must multiply again and again at the expense of other molecules, which serve as food.

2.4.1 Metabolism versus genetic mechanisms

A fundamental debate on the origin of life relates to the timing of two essential biological processes, metabolism versus replication of genetic material [15, 25, 87, 88, 89, 90, 91, 92]. Metabolism is the ability to manufacture biomolecular structures from a source of energy (such as sunlight) and matter scavenged from the surroundings (usually in the form of small molecules). An organism can't survive and grow without an adequate supply of energy and matter. Genetic mechanisms, by contrast, involve the transfer of biological information from one generation to the next – a blueprint for life *via* the mechanisms of DNA and RNA. An organism can't reproduce without a reliable means to pass on this genetic information.

The problem for understanding origins is that metabolism and genetic mechanisms constitute two separate, chemically distinct systems in cells. Nevertheless, metabolism and genetic mechanisms are inextricably linked in modern life. DNA holds genetic instructions to make hundreds of molecules essential to metabolism, while metabolism provides both the energy and the basic building blocks to make

DNA and other genetic materials. Like the dilemma of the chicken and the egg, it's difficult to imagine back to a time when metabolism and genetic mechanisms were not intertwined. Consequently, origin-of-life researchers engage in an intense ongoing debate about whether these two aspects of life arose simultaneously or independently and, if the latter, which one came first.

Most experts seem to agree that the simultaneous emergence of metabolism and genetic mechanisms is unlikely. The chemical processes are just too different, and they rely on completely different sets of molecules. It's much easier to imagine life arising one small step at a time, but what is the sequence of emergent steps?

Those who favor genetics first argue on the basis of life's remarkable complexity; they point to the astonishing intricacy of even the simplest living cell. Without a genetic mechanism there would be no way to insure the faithful reproduction of all that complexity. Metabolism without a genetic mechanism may be viewed as nothing more than a collection of overactive chemicals.

Other scientists are persuaded by the principle that life emerged through stages of increasing complexity. Metabolic chemistry, at its core, is vastly simpler than genetic mechanisms because it requires relatively few small molecules that work in concert to duplicate themselves. In this view the core metabolic cycle – the citric acid cycle that lies at the heart of every modern cell's metabolic processes – survives as a biochemical fossil from life's beginning. This comparatively simple chemical cycle is an engine that can bootstrap all of life's biochemistry, including the key genetic molecules.

Origin-of-life scientists aren't shy about voicing their opinions on the metabolism- vs. genetics-first problem, which will probably remain a central controversy in the field for some time. Meanwhile, as this debate fuels animated discussions, several groups of creative researchers are attempting to shed light on the issue by devising self-replicating chemical systems – metabolism in a test tube.

2.4.2 Self-replicating molecular systems

The simplest imaginable self-replicating system consists of one type of molecule that makes copies of itself [93]. Under just the right chemical environment, such an isolated molecule will become two copies, then four, then eight molecules, and so on in a geometrical expansion. Such an "autocatalytic" molecule must act as a template that attracts and assembles smaller building blocks from an appropriate chemical broth. Single self-replicating molecules are intrinsically complex in structure, but organic chemists have managed to devise several varieties of these curious beasts, including small peptides (made of amino acids) and short strands of DNA [94, 95, 96, 97, 98].

Nevertheless, these self-replicating molecules don't meet anyone's minimum requirements for life on at least two counts. First, such a system requires a steady input of smaller highly specialized molecules – synthetic chemicals that must be supplied by the researchers. Under no plausible natural environment could these idiosyncratic "food" molecules arise independently in nature. Furthermore – and this is a key point in distinguishing life from non-life – these self-replicating molecules can't change and evolve, any more than a Xerox copy can evolve from an original.

More relevant to biological metabolism are systems of two or more molecules that form a self-replicating cycle or network. In the simplest system, two molecules (call them AA and BB) form from smaller feedstock molecules A and B. If AA catalyzes the formation of BB, and BB in turn catalyzes the formation of AA, then the system will sustain itself as long as researchers maintain a reliable supply of food molecules A and B. Theorists elaborate on such a model with networks of many molecules, each of which promotes the production of another species in the system [99, 100].

What separates living systems from simple self-replicating collections of molecules? In part it's complexity; living systems require numerous interacting molecules. In addition, a truly living metabolic cycle must incorporate a certain degree of sloppiness. Only through such variations can the system experiment with new, more efficient reaction pathways and thus evolve.

A dramatic gap exists between plausible theory and actual experiment. Metabolism is a special kind of cyclical chemical process with two essential inputs. Living cells undergo chemical reactions, not unlike burning in which two chemicals (oxygen and fuel) react and release energy. However, the trick in metabolism, unlike an open fire, is to capture part of that released energy to make new useful molecules that reinforce the cycle. So metabolism requires a sequence of chemical reactions that work in concert.

2.4.3 *The iron-sulfur world*

The theoretical and experimental pursuit of the metabolism-first viewpoint is exemplified by Günter Wächtershäuser's iron-sulfur world hypothesis [44, 45, 46, 47, 48]. This origin-of-life scenario incorporates a strikingly original autotrophic, metabolism-first model. According to Wächtershäuser, all of life's essential biomolecules are manufactured in place, as needed from the smallest of building blocks: CO, H_2O, NH_3, H_2S, and so forth. Chemical synthesis is accomplished stepwise, a few atoms at a time.

The contrast between a heterotrophic versus an autotrophic origin is profound, and represents a fundamental point of disagreement among origin-of-life researchers. On the one hand, supporters of a heterotrophic origin point out that it's much easier for an organism to use the diverse molecular products that are available in the prebiotic soup, rather than make them all from scratch. Why go to the trouble of synthesizing lots of amino acids if they're already available in the environment? In this sense, heterotrophic cells can be simpler than autotrophic cells because they don't require all of the biochemical machinery to manufacture amino acids, carbohydrates, lipids, and so forth. One can thus argue that simpler heterotrophy should come first.

Autotrophic advocates are equally insistent that true simplicity lies in building molecules a few atoms at a time with just a few simple chemical reactions. Such a mechanism, furthermore, possesses a significant philosophical advantage: auto-trophism is deterministic. Rather than depending on the idiosyncrasies of a local environment for biomolecular components, autotrophic organisms make them from scratch the same way every time, on any viable planet or moon, in a predictable chemical path. Such a philosophy leads to a startling conclusion. For advocates of autotrophy the prebiotic soup is irrelevant to the origin of life. With autotrophy, biochemistry is hard wired into the universe. The self-made cell emerges from geo-chemistry as inevitably as basalt or granite.

As with any metabolic strategy, the iron-sulfur world model requires a source of energy, a source of molecules, and a self-replicating cycle. Wächtershäuser's model relies on the abundant chemical energy of minerals that find themselves out of equilibrium with their environment. He begins by suggesting that iron monosulfide (FeS), a common mineral deposited in abundance at the mouths of many deep-sea hydrothermal vents, is unstable with respect to the surrounding seawater. As a con-sequence, iron sulfide combines with the volcanic gas hydrogen sulfide (H_2S) to produce the mineral pyrite (FeS_2) plus hydrogen gas (H_2) and energy:

$$FeS + H_2S \rightarrow FeS_2 + H_2 + energy$$

Given that energetic boost, hydrogen reacts immediately with carbon dioxide (CO_2) to synthesize organic molecules such as formic acid (HCOOH):

$$energy + H_2 + CO_2 \rightarrow HCOOH$$

Wächtershäuser envisions cascades of these reactions coupled to build up essential organic molecules from CO_2 and other simple gases.

Recent experiments lend support to the iron-sulfur world hypothesis, which makes the unambiguous prediction that minerals promote a variety of organic reactions. Experiments by Wächtershäuser and colleagues in Germany used iron, nickel, and cobalt sulfides to synthesize acetate, an essential metabolic compound with two carbon atoms that plays a central role in countless biochemical processes [52]. They expanded on this success by adding amino acids to their experiments and making peptides – yet another essential step to life [53, 101]. These and other experiments lead to a firm conclusion: common sulfide minerals have the ability to promote a variety of interesting organic synthesis reactions.

A centerpiece of the iron-sulfur world hypothesis is Günter Wächtershäuser's conviction that life began with a simple, self-replicating cycle of compounds similar to the one that lies at the heart of every cell's metabolism – the "reductive citric acid cycle." But how did the first metabolic cycle operate without enzymes? Some of the requisite steps, such as combining carbon dioxide and pyruvate to make oxaloacetate, are energetically unlikely; if one step fails then the whole cycle fails. One of the clever proposals in Günter Wächtershäuser's model is that sulfide minerals promote primordial metabolic reactions. In fact, many modern metabolic enzymes have at their core a small cluster of iron or nickel and sulfur atoms – clusters that look like tiny bits of sulfide minerals. Perhaps ancient minerals played the role of enzymes.

Wächtershäuser's iron-sulfur world was manifest as flat life, which predated the emergence of cells. The first self-replicating entity was a thin layer of chemical reactants on a sulfide mineral surface. The film grew laterally, spreading from mineral grain to mineral grain as an invisibly thin organic film. Bits of these layers could break off and reattach to other rocks to make clone-like colonies. Given time, different minerals might induce variations in the film, fostering new "species" of flat life.

2.4.4 The RNA world

In spite of Wächtershäuser's elaborate iron-sulfur world scenario and several competing variants of that model [102, 103], most origin experts dismiss a purely metabolic life form in favor of a genetics-first scenario. Even the simplest known cell must pass volumes of information from one generation to the next, and the only known way to store and copy that much information is with a genetic molecule like DNA or RNA.

California chemist Leslie Orgel stated that the central dilemma in understanding a genetic origin of life is the identification of a stable, self-replicating genetic molecule – a polymer that simultaneously carries the information to make copies of itself *and* catalyzes that replication [91]. Accordingly, he catalogs four broad approaches to the problem of jumpstarting such a genetic organism.

One possibility is a self-replicating peptide – a short sequence of amino acids that emerged first and then "invented" DNA. That's an appealing idea because amino acids are thought to have been widely available in the soup. The problem is that, while cells have learned how to form ordered chain-like polymers, amino acids on their own link together in irregular clusters of no obvious biological utility. Alternatively, the simultaneous evolution of proteins and DNA seems even less likely because it requires the emergence of two improbable molecules. Cairns-Smith's clay world scenario provides an intriguing third option, though one totally unsupported by experimental evidence.

Consequently, the favored genetics origin model is based on a nucleic acid molecule like RNA – a polymer that acted both as a carrier of information and as a catalyst that promoted self-replication [104, 105, 106]. This idea received a boost in the early 1980s with the discovery of ribozymes – genetic material that also acts as a catalyst [107, 108, 109]. Modern life relies on two complexly intertwined molecules, DNA to carry information and proteins to perform chemical functions. This interdependence leads to the chicken–egg dilemma: proteins make and maintain DNA, but DNA carries the instructions to make proteins. Which came first? RNA, it turns out, has the potential to do both jobs.

The "RNA world" theory that quickly emerged champions the central role of genetic material in the dual tasks of catalyst *and* information transfer [110, 111, 112]. Over the years "RNA world" has come to mean different things to different people, but three precepts are common to all versions: (1) once upon a time RNA carried genetic information, (2) RNA replication followed the same rules as modern DNA, by matching pairs of bases such as A-T and C-G, and (3) proteins were not involved in the process. The first life form in this scenario was simply a self-replicating strand of RNA, perhaps enclosed in a protective lipid membrane. According to most versions of this hypothesis, metabolism emerged later as a means to make the RNA replication process more efficient.

The RNA world model is not without its difficulties. Foremost among these problems is the exceptional difficulty in the prebiotic synthesis of RNA [25, 111, 112, 113]. Many of the presumed proto-metabolic molecules are easily synthesized in experiments that mimic prebiotic environments. RNA nucleotides, by contrast, have never been synthesized from scratch. Furthermore, even if a prebiotic synthetic pathway to nucleotides could be found, a plausible mechanism to link those nucleotides into an RNA strand has not been demonstrated. It is not obvious how useful catalytic RNA sequences would have formed spontaneously in any prebiotic environment. Perhaps, some scientists speculate, a simpler nucleic acid preceded RNA [7, 114, 115, 116].

Whatever the scenario – metabolism first or genetics first, on the ocean's surface or at a deep ocean vent – the origin of life required more than the replication of chemicals. For a chemical system to be alive, it must display evolution by the process of natural selection.

2.5 The emergence of natural selection

Molecular selection, the process by which a few key molecules earned key roles in life's origin, proceeded on many fronts. Some molecules were inherently unstable or highly reactive and so they quickly disappeared from the scene. Other molecules easily dissolved in the oceans and so were effectively removed from contention. Still other molecular species may have sequestered themselves by bonding strongly to surfaces of chemically unhelpful minerals or clumped together into gooey masses of little use to emerging biology.

In every geochemical environment, each kind of organic molecule had its reliable sources and its inevitable sinks. For a time, perhaps for hundreds of millions of years, a kind of molecular equilibrium was maintained as the new supply of each species was balanced by its loss. Such an equilibrium features non-stop competition among molecules, to be sure, but the system does not evolve.

The first self-replicating molecules changed that equilibrium. Even a relatively unstable collection of molecules can be present in significant concentration if it learns how to make copies of itself. The first successful metabolic cycle of molecules, for example, would have proven vastly superior to its individual chemical neighbors at accumulating atoms and harnessing energy. But success breeds competition. Inevitable slight variations in the self-replicating cycle, triggered by the introduction of new molecular species or by differences in environment, initiated an era of increased competition. More efficient cycles flourished at the expense of the less efficient. Evolution by natural selection had begun on Earth.

The simplest self-replicating molecular systems are not alive. Given all the time in the world such systems won't evolve. A more interesting possibility is an environment in which two or more self-replicating suites of molecules compete. These dueling molecular networks vie for resources, mimicking life's unceasing struggle for survival. A stable system without competition, by contrast, has no incentive to evolve.

Two common processes – variation and selection – provide a powerful mechanism for self-replicating systems to evolve. For a system to evolve it must first display a range of variations. Natural systems display random variations through mutations, which are undirected changes in the chemical makeup of key biomolecules. Most variations harm the organism and are doomed to failure. Once in a while, however, a random mutation leads to an improved trait – a more efficient metabolism, better camouflage, swifter locomotion, or greater tolerance for extreme environmental conditions. Such beneficial variations are more likely to survive in the competitive natural world – such variations fuel the process of natural selection.

Competition drives the emergence of natural selection. Such behavior appears to be inevitable in any self-replicating chemical system in which resources are limited and some molecules have the ability to mutate. Over time, more efficient networks of autocatalytic molecules will increase in concentration at the expense of less efficient networks. In such a competitive milieu the emergence of increasing molecular complexity is inevitable; new chemical pathways overlay the old. So it is that life has continued to evolve over the past four billion years of Earth history.

2.5.1 Molecular evolution in the lab

Laboratory studies of molecular evolution have led to methods for the artificial evolution of molecules that accomplish specific chemical tasks. Harvard biologist Jack Szostak, for example, has explored this approach in attempts to design an evolving synthetic laboratory life form [117]. Szostak focuses on devising chains of RNA that perform one key task such as copying other RNA molecules [118].

Szostak and co-workers [119, 120, 121, 122] tackle RNA evolution by first generating a collection of trillions of random RNA sequences, each about 120 nucleotides (RNA "letters") long. The vast majority of these chains do nothing, but a few are able to perform a specific task, for example grabbing onto another molecule that has been attached to the sides of a glass beaker. Flushing out the beaker with water removes all of the unattached RNA strands and thus selects and concentrates RNA sequences bonded to the target. These select strands are then copied with a controlled rate of mutational errors to produce a new batch of trillions of RNA sequences, some of which are even better suited to attach to the beaker's sides. Repeating this process improves the speed and accuracy with which the RNA latches onto the target molecules.

An enticing target of molecular evolution research is self-replicating RNA – a big molecule that carries genetic information, catalyzes its own reproduction, and mutates and evolves as well. But no plausible geochemical environment could have fed such an unbound molecule, nor would it have survived long under most natural chemical conditions. Szostak's group ultimately hopes to encapsulate such a replicative RNA strand in a lipid vesicle to synthesize a self-replicating cell-like entity [117].

The synthesis of an "RNA organism" would provide a degree of credibility to the RNA world hypothesis, that a strand of RNA (or some chemically more stable precursor genetic molecule) formed the basis of the first evolving, self-replicating chemical system. But lab life will not have emerged spontaneously from chemical reactions among the simple molecular building blocks of the prebiotic Earth, nor would such an object be capable of reproduction without a steady supply of laboratory chemicals. Deep mysteries will remain.

2.6 Three scenarios for the origin of life

What do we know so far? Abundant organic molecules must have been synthesized and accumulated in a host of productive prebiotic environments. Subsequently, biomolecular systems including lipid membranes and genetic polymers might have formed through self-organization and by selection on mineral surfaces. As molecular complexity increased, it seems plausible that simple metabolic cycles of self-replicating molecules emerged, as did self-replicating genetic molecules.

The greatest gap in understanding life's origin lies in the transition from a more-or-less static geochemical world with an abundance of interesting organic molecules, to an evolving biochemical world. How that transition occurred may be reduced to a choice among three possible scenarios [7]:

1. *Life began with metabolism and genetic molecules were incorporated later.* Following Günter Wächtershäuser's hypothesis, life began autotrophically. Life's first building blocks were the simplest of molecules, while minerals provided chemical energy. In this scenario a self-replicating chemical cycle akin to the reverse citric acid cycle became established on a mineral surface. All subsequent chemical complexities, including genetic mechanisms and encapsulation into a cell-like structure, arose through natural selection as variants of the cycle competed for resources and the system became more efficient. In this scenario, life first emerged as an evolving chemical coating on rocks.
2. *Life began with self-replicating genetic molecules and metabolism was incorporated later.* Following the RNA world hypothesis, life began heterotrophically. Organic molecules in the prebiotic soup, perhaps aided by clays or some other template, self-organized into information-rich polymers. Eventually, one of these polymers, perhaps surrounded by a lipid membrane, acquired the ability to self-replicate. All subsequent chemical complexities, including metabolic cycles, arose through natural selection as variants of the genetic polymer became more efficient at self-replication. In this scenario, life emerged as an evolving polymer with a functional genetic sequence.
3. *Life began as a cooperative chemical phenomenon between metabolism and genetics.* A third possible scenario rests on the possibility that neither primitive metabolic cycles (which lack the means of faithful self-replication) nor primitive genetic molecules (which are not very stable and lack a reliable source of chemical energy) could have progressed far by themselves. If, however, a crudely self-replicating genetic molecule became attached to a crudely functioning surface-bound metabolic coating, then a kind of cooperative chemistry might have kicked in. The genetic molecule might have used chemical energy produced by metabolites to make copies of itself, while protecting itself by binding to the surface. Any subsequent variations of the genetic molecule that fortuitously offered protection for itself or for the metabolites, or improved the chemical efficiency of the system, would have been preserved preferentially. Gradually, both the genetic and metabolic components would have become more efficient and more interdependent.

Scientists do not yet know the answer, but they are poised to find out. Whatever the process, ultimately competition began to drive the emergence of ever more elaborate chemical cycles by the process of natural selection. Inexorably, life emerged, never to relinquish its foothold on Earth.

Acknowledgements

I thank Connie Bertka and three anonymous reviewers for valuable comments on the manuscript. This work was supported by the NASA Astrobiology Institute and the Carnegie Institution of Washington.

References

[1] H. J. Morowitz. *The Emergence of Everything* (New York: Oxford University Press, 2002).

[2] S. L. Miller. Production of amino acids under possible primitive earth conditions. *Science*, **17** (1953), 528–529.

[3] I. Prigogine. *Order out of Chaos: Man's New Dialogue with Nature* (Toronto: Bantam Books, 1984).

[4] G. Nicolis and I. Prigogine. *Exploring Complexity: An Introduction* (New York: W. H. Freeman and Company, 1989).

[5] J. H. Holland. *Hidden Order* (Reading, MA: Helix Books, 1995).

[6] J. H. Holland. *Emergence: From Chaos to Order* (Reading, MA: Helix Books, 1998).

[7] R. M. Hazen. *Genesis: The Scientific Quest for Life's Origin* (Washington, DC: Joseph Henry Press, 2005).

[8] R. A. Bagnold. *The Physics of Blown Sand and Desert Dunes* (London: Chapman and Hall, 1941).

[9] R. A. Bagnold. *The Physics of Sediment Transport by Wind and Water* (New York: American Society of Civil Engineers, 1988).

[10] S. Camazine, J. L. Deneubourg, N. R. Franks, J. Sneyd, G. Theraulaz, and E. Bonabeau. *Self-Organization in Biological Systems* (Princeton, NJ: Princeton University Press, 2001).

[11] R. Solé and B. Goodwin. *Signs of Life: How Complexity Pervades Biology* (New York: Basic Books, 2000).

[12] E. J. Chaisson. *Cosmic Evolution: The Rise of Complexity in Nature* (Cambridge, MA: Harvard University Press, 2001).

[13] M. A. Kessler and B. T. Werner. Self-organization of sorted patterned ground. *Science*, **299** (2003), 380–383.

[14] J. L. Hansen, M. Van Hecke, A. Haaning, *et al.* Instabilities in sand ripples. *Nature*, **410** (2001), 324.

[15] C. De Duve. *Vital Dust: Life as a Cosmic Imperative* (New York: Basic Books, 1995).

[16] C. Cleland and C. Chyba. Defining life. *Origins of Life and Evolution of the Biosphere*, **32** (2002), 387–393.

[17] R. M. Hazen, A. Steele, J. Toporski, G. D. Cody, M. Fogel, and W. T. Huntress Jr. Biosignatures and abiosignatures. *Astrobiology*, **2** (2002), 512–513.

[18] S. L. Miller. Production of some organic compounds under possible primitive earth conditions. *Journal of the American Chemical Society*, **77** (1955), 2351–2361.

[19] S. L. Miller and H. C. Urey. Organic compound synthesis on the primitive Earth. *Science*, **130** (1959), 245–251.

[20] S. J. Dick and J. E. Strick. *The Living Universe: NASA and the Development of Astrobiology* (New Brunswick, NJ: Rutgers University Press, 2004).

[21] I. Fry. *The Emergence of Life on Earth: A Historical and Scientific Overview* (New Brunswick, NJ: Rutgers University Press, 2000), pp. 318–320.

[22] J. E. Strick. *Sparks of Life: Darwinism and the Victorian Debate over Spontaneous Generation* (Cambridge, MA: Harvard University Press, 2000).

[23] A. I. Oparin. *Proiskhozhdenie Zhizny* (in Russian) (Moscow: Rabochii, 1924). (An English translation appears in: J. D. Bernal. *The Origin of Life* (London: Weidenfeld and Nicolson, 1967).)

[24] J. B. S. Haldane. The origin of life. *The Rationalist Annual* (1929), 3–10.

[25] C. Wills and J. L. Bada. *The Spark of Life: Darwin and the Primeval Soup* (Cambridge, MA: Perseus, 2000).

[26] J. Oró. Synthesis of adenine from ammonium cyanide. *Biochemical and Biophysical Communications*, **2** (1960), 407–412.

[27] J. Oró. Mechanism of synthesis of adenine from hydrogen cyanide under possible primitive earth conditions. *Nature*, **191** (1961), 1193–1194.

[28] R. Shapiro. Prebiotic ribose synthesis: a critical analysis. *Origins of Life and Evolution of the Biosphere*, **18** (1988), 71–85.

[29] C. F. Chyba and C. Sagan. Endogenous production, exogenous delivery, and impact-shock synthesis of organic molecules: an inventory for the origins of life. *Nature*, **355** (1992), 125–132.

[30] J. F. Kasting and J. L. Siefert. Life and the evolution of Earth's atmosphere. *Science*, **296** (2002), 1066–1068.

[31] J. F. Kasting. Methane-rich proterozoic atmosphere? *Geology*, **31** (2003), 87–90.

[32] F. Tian, O. B. Toon, A. A. Pavlov, and H. De Sterck. A hydrogen-rich early Earth atmosphere. *Science*, **308** (2005), 1014–1016.

[33] C. Chyba. Rethinking Earth's early atmosphere. *Science*, **308** (2005), 962–963.

[34] J. B. Corliss, J. Dymond, L. I. Gordon, *et al.* Submarine thermal springs on the Galapagos rift. *Science*, **203** (1979), 1073–1083.

[35] J. B. Corliss. J. A. Baross, and S. E. Hoffman. An hypothesis concerning the relationship between submarine hot springs and the origin of life on earth. In *Proceedings of the 26th International Geological Congress, Geology of the Oceans Symposium*, eds. X. Le Pichon, J. Debyser, and F. Vine (Paris 1980, X. *Oceanologica Acta*, **4** (supplement) 1981), pp. 59–69.

[36] J. A. Baross and S. E. Hoffman. Submarine hydrothermal vents and associated gradient environments as sites for the origin and evolution of life. *Origins of Life and Evolution of the Biosphere*, **15** (1985), 327–345.

[37] N. G. Holm. Why are hydrothermal systems proposed as plausible environments for the origin of life? *Origins of Life and Evolution of the Biosphere*, **22** (1992), 5–14.

[38] T. Gold. *The Deep Hot Biosphere* (New York: Copernicus, 1999).

[39] R. J. Parkes, B. A. Craig, S. J. Bale, *et al.* Deep bacterial biosphere in Pacific Ocean sediments. *Nature*, **371** (1993), 410–413.

[40] T. O. Stevens and J. P. McKinley. Lithoautotrophic microbial ecosystems in deep basalt aquifers. *Science*, **270** (1995), 450–454.

[41] J. K. Frederickson and T. C. Onstott. Microbes deep inside the Earth. *Scientific American*, **275**(4) (1996), 68–73.

[42] K. Pedersen. The deep subterranean biosphere. *Earth-Science Reviews*, **34** (1993), 243–260.

[43] M. T. Madigan and B. L. Marrs. Extremophiles. *Scientific American*, **276**(4) (1997), 82–87.

[44] G. Wächtershäuser. Before enzymes and templates: theory of surface metabolism. *Microbiology Review*, **52** (1988), 452–484.

[45] G. Wächtershäuser. Pyrite formation, the first energy source for life: a hypothesis. *Systematic Applied Microbiology*, **10** (1988), 207–210.

[46] G. Wächtershäuser. The case for the chemoautotrophic origin of life in an iron-sulfur world. *Origins of Life and Evolution of the Biosphere*, **20** (1990), 173–176.

[47] G. Wächtershäuser. Evolution of the first metabolic cycles. *Proceedings of the National Academy of Sciences USA*, **87** (1990), 200–204.

[48] G. Wächtershäuser. Groundworks for an evolutionary biochemistry: the iron-sulfur world. *Progress in Biophysics and Molecular Biology*, **58** (1992), 85–201.

[49] H. Beinert, R. H. Holm and E. Münck. Iron-sulfur clusters: nature's modular, multi-purpose structures. *Science*, **277** (1997), 653–659.

[50] E. Blöchl, M. Keller, G. Wächtershäuser, and K. O. Stetter. Reactions depending on iron sulfide and linking geochemistry with biochemistry. *Proceedings of the National Academy of Sciences USA*, **89** (1992), 8117–8120.

[51] W. Heinen and A. M. Lauwers. Organic sulfur compounds resulting from interaction of iron sulfide, hydrogen sulfide and carbon dioxide in an aerobic aqueous environment. *Origins of Life and Evolution of the Biosphere*, **26** (1996), 131–150.

[52] C. Huber and G. Wächtershäuser. Activated acetic acid by carbon fixation on (Fe,Ni)S under primordial conditions. *Science*, **276** (1997), 245–247.

[53] C. Huber and G. Wächtershäuser. Peptides by activation of amino acids with CO on (Ni,Fe)S surfaces: implications for the origin of life. *Science*, **281** (1998), 670–672.

[54] G. D. Cody, N. Z. Boctor, T. R. Filley, *et al.* Primordial carbonylated iron-sulfur compounds and the synthesis of pyruvate. *Science*, **289** (2000), 1337–1340.

[55] G. D. Cody, N. Z. Boctor, R. M. Hazen, J. A. Brandes, H. J. Morowitz, and H. S. Yoder, Jr. Geochemical roots of autotrophic carbon fixation: hydrothermal experiments in the system citric acid-H2O-(±FeS)-(±NiS). *Geochimica et Cosmochimica Acta*, **65** (2001), 3557–3576.

[56] G. D. Cody, N. Z. Boctor, J. A. Brandes, T. R. Filley, R. M. Hazen, and H. S. Yoder, Jr. Assaying the catalytic potential of transition metal sulfides for prebiotic carbon fixation. *Geochimica et Cosmochimica Acta*, **68** (2004), 2185–2196.

[57] J. L. Bada and A. Lazcano. Some like it hot, but not the first biomolecules. *Science*, **296** (2002), 1982–1983.

[58] J. L. Bada, S. L. Miller, and M. Zhao. The stability of amino acids at submarine hydrothermal vent temperatures. *Origins of Life and Evolution of the Biosphere*, **25** (1995), 111–118.

[59] D. L. Nelson and M. M. Cox. *Lehninger's Principles of Biochemistry*, 4th edn. (New York: Worth Publishers, 2004).

[60] P. L. Luisi and F. J. Varela. Self-replicating micelles: a chemical version of a minimal autopoietic system. *Origins of Life and Evolution of the Biosphere*, **19** (1989), 633–643.

[61] D. W. Deamer. The first living systems: a bioenergetic perspective. *Microbiology and Molecular Biology Review*, **61** (1997), 239–261.

[62] S. Segré, D. W. Deamer, and D. Lancet. The lipid world. *Origins of Life and Evolution of the Biosphere*, **31** (2001), 119–145.

[63] A. C. Lasaga, H. D. Holland, and M. J. Dwyer. Primordial oil slick. *Science*, **174** (1971), 53–55.

[64] N. Lahav, D. White, and S. Chang. Peptide formation in the prebiotic era: thermal condensation of glycine in fluctuating clay environments. *Science*, **201** (1978), 67–69.

[65] L. E. Orgel. Polymerization on the rocks: theoretical introduction. *Origins of Life and Evolution of the Biosphere*, **28** (1998), 227–234.

[66] J. V. Smith. Biochemical evolution. I. Polymerization on internal, organophilic silica surfaces of dealuminated zeolites and feldspars. *Proceedings of the National Academy of Sciences USA*, **95** (1998), 3370–3375.

[67] J. P. Ferris. Catalysis and prebiotic synthesis. *Origins of Life and Evolution of the Biosphere*, **23** (1993), 307–315.

[68] J. P. Ferris. Prebiotic synthesis on minerals: bridging the prebiotic and RNA worlds. *Biology Bulletin*, **196** (1999), 311–314.

[69] J. P. Ferris. Mineral catalysis and prebiotic synthesis: montmorillonite-catalyzed formation of RNA. *Elements*, **1** (2005), 145–149.

[70] S. Pitsch, A. Eschenmoser, B. Gedulin, S. Hui, and G. Arrhenius. Mineral induced formation of sugar phosphates. *Origins of Life and Evolution of the Biosphere*, **25** (1995), 297–334.

[71] A. G. Cairns-Smith. The origin of life and the nature of the primitive gene. *Journal of Theoretical Biology*, **10** (1968), 53–88.

[72] A. G. Cairns-Smith. *Genetic Takeover and the Mineral Origins of Life* (Cambridge: Cambridge University Press, 1982).

[73] A. G. Cairns-Smith. The first organisms. *Scientific American*, **252**(6) (1985), 90–100.

[74] A. G. Cairns-Smith. *Seven Clues to the Origin of Life* (Cambridge: Cambridge University Press, 1985).

[75] W. A. Bonner. The origin and amplification of biomolecular chirality. *Origins of Life and Evolution of the Biosphere*, **21** (1991), 59–111.

[76] W. A. Bonner. Chirality and life. *Origins of Life and Evolution of the Biosphere*, **25** (1995), 175–190.

[77] R. M. Hazen and D. S. Sholl. Chiral selection on inorganic crystalline surfaces. *Nature Materials*, **2** (2003), 367–374.

[78] J. R. Cronin and S. Pizzarello. Amino acids in meteorites. *Advances in Space Research*, **3** (1983), 5–18.

[79] M. H. Engel and S. A. Macko. Isotopic evidence for extraterrestrial non-racemic amino acids in the Murchison meteorite. *Nature*, **296** (1997), 837–840.

[80] S. Pizzarello and J. R. Cronin. Non-racemic amino acids in the Murray and Murchison meteorites. *Geochimica et Cosmochimica Acta*, **64** (2000), 329–338.

[81] S. Clark. Polarized starlight and the handedness of life. *American Scientist*, **87** (1999), 336–343.

[82] J. Bailey, A. Chrysostomou, J. H. Hough, *et al.* Circular polarization in star-formation regions: implications for biomolecular homochirality. *Science*, **281** (1998), 672–674.

[83] J. Podlech. New insight into the source of biomolecular homochirality: an extraterrestrial origin for molecules of life. *Angewandt Chemie International Edition English*, **38** (1999), 477–478.

[84] R. M. Hazen. Chiral crystal faces of common rock-forming minerals. In *Progress in Biological Chirality*, eds. G. Palyi, C. Zucchi and L. Caglioti (Oxford: Elsevier, 2004), pp. 137–151.

[85] R. M. Hazen. Mineral surfaces and the prebiotic selection and organization of biomolecules. *American Mineralogist*, **91** (2006), 1715–1729.

[86] R. M. Hazen, T. Filley, and G. A. Goodfriend. Selective adsorption of L- and D-amino acids on calcite: implications for biochemical homochirality. *Proceedings of the National Academy of Sciences, USA*, **98** (2001), 5487–5490.

[87] C. De Duve. The beginnings of life on Earth. *American Scientist*, **83** (1995), 428–437.

[88] F. Dyson. *Origins of Life* (Cambridge: Cambridge University Press, 1999).

[89] S. W. Fox. A theory of macromolecular and cellular origins. *Nature*, **205** (1965), 328–340.

[90] S. W. Fox. *The Emergence of Life: Darwinian Evolution from the Inside* (New York: Basic Books, 1988).

[91] L. E. Orgel. RNA catalysis and the origin of life. *Journal of Theoretical Biology*, **123** (1986), 127–149.

[92] H. J. Morowitz. *The Beginnings of Cellular Life: Metabolism Recapitulates Biogenesis* (New Haven, CT: Yale University Press, 1992).

[93] E. K. Wilson. Go forth and multiply. *Chemical and Engineering News*, **76** (December 7, 1998), 40–44.

[94] G. Von Kiedrowski. A self-replicating hexadeoxynucleotide. *Angewandt Chemie International Edition English*, **25** (1986), 932–935.

[95] D. Sievers and G. Von Kiedrowski. A self-replication of complementary nucleotide-based oligomers. *Nature*, **369** (1994), 221–224.

[96] J. Rebek Jr. Synthetic self-replicating molecules. *Scientific American*, **271**(1) (1994), 48–55.

[97] D. H. Lee, J. R. Granja, J. A. Martinez, K. Severin, and M. R. Ghadiri. A self-replicating peptide. *Nature*, **382** (1996), 525–528.

[98] S. Yao, I. Ghosh, R. Zutshi, and J. Chmielewski. A pH-modulated, self-replicating peptide. *Journal of the American Chemical Society*, **119** (1997), 10559–10560.

[99] M. Eigen and P. Schuster. *The Hypercycle: A Principle of Natural Self-Organization* (Berlin: Springer-Verlag, 1979).

[100] S. A. Kauffman. *The Origins of Order: Self-Organization and Selection in Evolution* (New York: Oxford University Press, 1993).

[101] C. Huber, W. Eisenreich, S. Hecht, and G. Wächtershäuser. A possible primordial peptide cycle. *Science*, **301** (2003), 938–940.

[102] M. J. Russell and A. J. Hall. The emergence of life from iron monosulphide bubbles at a submarine hydrothermal redox and pH front. *Journal of the Geological Society of London*, **154** (1997), 377–402.

[103] M. J. Russell and A. J. Hall. From geochemistry to biochemistry: chemiosmotic coupling and transition element clusters in the onset of life and photosynthesis. *The Geochemical News*, **113** (2002), 6–12.

[104] C. R. Woese. *The Genetic Code* (New York: Harper & Row, 1967).

[105] F. H. C. Crick. The origin of the genetic code. *Journal of Molecular Biology*, **38** (1968), 367–379.

[106] L. E. Orgel. Evolution of the genetic apparatus. *Journal of Molecular Biology*, **38** (1968), 381–393.

[107] K. Kruger, P. J. Grabowski, A. J. Zaug, J. Sands, D. E. Gottschling, and T. R. Cech. Self-splicing RNA: autoexcision and autocyclization of the ribosomal RNA intervening sequence of Tetrahymena. *Cell*, **31** (1982), 147–157.

[108] C. Guerrier-Takada, K. Gardiner, T. Marsh, N. Pace, and S. Altman. The RNA moiety of ribonuclease P is the catalytic subunit of the enzyme. *Cell*, **35** (1983), 849–857.

[109] A. J. Zaug and T. R. Cech. The intervening sequence RNA of Tetrahymena is an enzyme. *Science*, **231** (1986), 470–475.

[110] G. F. Joyce. RNA evolution and the origins of life. *Nature*, **338** (1989), 217–224.

[111] G. F. Joyce. The rise and fall of the RNA world. *New Biology*, **3** (1991), 399–407.

[112] G. F. Joyce and L. E. Orgel. Prospects for understanding the RNA world. In *The RNA World*, eds. R. F. Gesteland and J. F. Atkins (Cold Spring Harbor, NY: Cold Spring Harbor Laboratory Press, 1993), pp. 1–25.

[113] L. E. Orgel. Some consequences of the RNA world hypothesis. *Origins of Life and Evolution of the Biosphere*, **33** (2003), 211–218.

[114] A. Eschenmoser. Chemical etiology of nucleic acid structure. *Science*, **284** (1999), 2118–2124.

[115] A. Eschenmoser. The TNA-family of nucleic acid systems: properties and prospects. *Origins of Life and Evolution of the Biosphere*, **34** (2004), 277–306.

[116] P. E. Nielsen. Peptide nucleic acid (PNA): a model structure for the primordial genetic material? *Origins of Life and Evolution of the Biosphere*, **23** (1993), 323–327.

[117] J. W. Szostak, D. P. Bartel and P. L. Luisi. Synthesizing life. *Nature*, **409** (2001), 387–390.

[118] G. F. Joyce. Directed molecular evolution. *Scientific American*, **267**(6) (1992), 90–97.

[119] J. A. Doudna and J. W. Szostak. RNA catalyzed synthesis of complementary strand RNA. *Nature*, **339** (1998), 519–522.

[120] R. Green and J. W. Szostak. Selection of a ribozyme that functions as a superior template in a self-copying reaction. *Science*, **258** (1992), 1910–1915.

[121] D. P. Bartel and J. W. Szostak. Isolation of new ribozymes from a large pool of random sequences. *Science*, **261** (1993), 1411–1418.

[122] J. W. Szostak and A. D. Ellington. In vitro selection of functional RNA sequences. In *The RNA World*, eds. R. F. Gesteland and J. F. Atkins (Cold Spring Harbor, NY: Cold Spring Harbor Laboratory Press, 1993), pp. 511–533.

3

From Aristotle to Darwin, to Freeman Dyson: changing definitions of life viewed in a historical context

James E. Strick

3.1 Introduction

The origin of life has always been a topic charged with religious import. This chapter aims to survey briefly some origin-of-life ideas over time; first, if we look in detail at how the debate over life's origin played out between Darwin and his supporters, we will see a clear example of this. Note that I said Darwin's *supporters*, rather than his opponents. We expect a story about the new evolutionary science to include much heated objection from religious groups, for whom so many aspects of Darwin's theory produced problems. But a look at how divisive the issue was among the Darwinians themselves is an even more complex and enlightening story, perhaps of more relevance to scientists in their own work.

By the 1850s, William Benjamin Carpenter, Professor of physiology at the Royal Institution of London, believed, as did Herbert Spencer, that the vital force of living things was completely interconvertible with forces from non-living nature such as heat, electricity, kinetic energy, etc. However, also like Spencer and a number of others in the Darwinian camp, Carpenter believed that the conversion of heat or chemical energy into vital energy could *only* be accomplished through the agency of already living matter; in other words, that non-living matter could never organize itself into living matter capable of generating vital force [1]. This served as a convenient theoretical barrier separating modern physiology and its ally, the new evolution theory, from the bête noir of spontaneous generation. (Spontaneous generation meant life arising directly from non-living matter; the idea had become associated through Buffon, Lamarck, and others with the politics of revolutionary France. In Britain the most prominent spontaneous generation advocate was Charles Darwin's early mentor, the Lamarckian Robert Grant, also a supporter of radical political reforms.) Carpenter's spin on how, despite the conservation of energy principle,

Exploring the Origin, Extent, and Future of Life: Philosophical, Ethical, and Theological Perspectives, ed. Constance M. Bertka. Published by Cambridge University Press. © Cambridge University Press 2009.

there was still a qualitative difference between vital and non-living forces, did keep the politically unsavory doctrine of spontaneous generation at arm's length from the new professional scientists among whom Darwin was a leader. However, it simultaneously posed a new conundrum for the emerging life sciences centered around Darwinism: it seemed to prove an absolute, uncrossable divide between the living and non-living worlds, leaving no alternative but a supernatural explanation for the first origin of life on Earth.

That suited Carpenter, a very religious man, just fine; but others were worried about the contradiction. Many younger supporters of Darwin's theory were enthusiastic about its charter of seeming to banish supernatural explanations from life sciences and of establishing new proof for the unbroken continuity of nature. These included T. H. Huxley and his protégés, the next generation of rising Darwinian talent, prominent among them the physician and medical researcher Henry Charlton Bastian. Bastian was deeply interested in the problem of consciousness; and Bastian, more than any of the other young Darwinians, became fascinated with the origin of life itself and convinced that the question could be approached by experiment in a laboratory. While some Darwinians imagined even the simplest cells must require a long, gradual process of formation, Bastian described experiments that he thought showed cells emerging very rapidly, in a few days, from non-living starting materials. In hermetically sealed containers, Bastian boiled infusions of different materials, both organic and inorganic, then observed the cooled contents microscopically and saw a profusion of different living microorganisms. At this time many observers thought the term "spontaneous generation" to automatically imply such rapid formation. By 1870 he claimed experimental proof for spontaneous generation and argued this was important support for the Darwinian theory.[1]

Those liberal-minded Christians who sought some kind of accommodation between the new evolutionary science and their religious beliefs were an audience of great interest to Darwin. Hence, in *On the Origin of Species* [2], Darwin went out of his way, even misrepresenting his own views on life's origin, to use language that gave those readers some breathing room. He devoted only one sentence in the entire book to the origin of life, and there he said:

…I should infer from analogy that probably all the organic beings which have ever lived on this earth have descended from some one primordial form, into which life was first breathed (p. 484).

---

[1] Works by Henry Charlton Bastian include the following: Facts and reasonings concerning the heterogeneous evolution of living things. *Nature*, **2** (June 30, 1870), 170–177; (July 7), 193–201; and (July 14), 219–228; *Modes of Origin of Lowest Organisms* (London: Macmillan, 1871); *The Beginnings of Life*, 2 vols. (London: Macmillan, 1872); *Evolution and the Origin of Life* (London: Macmillan, 1874); Discussion on the germ theory of disease. *Transactions of the Pathological Society of London*, **26** (1875), 255–345; The Commission of the French Academy and the Pasteur–Bastian eperiments. *Nature*, **16** (1877), 276–279; On the conditions favouring fermentation and the appearance of bacilli, micrococci, and torulae in previously boiled fluids. *Journal of the Linnean Society of London*, **14** (1879), 1–94; *The Origin of Life* (London: G.P. Putnam's Sons, 1911).

Many of Darwin's friends criticized him for this dissembling, as did Richard Owen while touting an alternative evolution theory to Darwin's. In private Darwin repeatedly regretted having weakened the clarity of his position by using such Biblical language. As he put it, with sarcasm toward Owen's spontaneous generation concept, which likened snot to protoplasm:

It will be some time before we see "slime, snot, or protoplasm" ... generating a new animal. But I have long regretted that I truckled to public opinion, and used the Pentateuchal term of creation, by which I really meant "appeared" by some wholly unknown process. It is mere rubbish, thinking at present of the origin of life; one might as well think of the origin of matter [3].[2]

But clearly Darwin felt that, at least in the short run, it was unwise to alienate so large a pool of potential allies during the early years of controversy over his theory. An indicator that Darwin agrees with the criticism of his friends is that he dropped the entire phrase "into which life was first breathed by the Creator" from the third edition of *Origin*, which appeared in April 1861. But, tellingly, Darwin did not insert in its place a more forthright explanation of the kind he had given in private. He simply left out any comment on the matter of how that "one primordial form" had arisen. And despite much private speculation, he never made any further public comment on the topic to his dying day in 1882. He left the passage untouched in all remaining editions of the book (the last revision was the sixth edition of 1872).

I have demonstrated elsewhere that Bastian's experiments on the possibility of spontaneous generation were widely seen in the 1870s as relevant to much broader, socially charged questions such as the nature of consciousness; the interchangeability of "vital force," and especially "nerve force," with other physical and chemical forces in nature; the mechanism of heredity, and of memory generally; cell theory; and the origin of diseases, both contagious and degenerative [4].[3] Clearly, a lot was at stake, and some of it, most deeply of all, was inseparably tied to whether there was any need at all for a Creator God. To those who opposed Darwinian evolution on religious grounds, it was obvious that Darwin and Huxley implied a naturalistic origin of life, just as it was obvious to Bastian and his supporters [5, 6[4]]. But for religious critics of evolution this meant an even more important reason to oppose Darwinism.

Thus, in 1870 when Bastian began to publicly press the implications of Darwinism for a materialistic origin of life, he created great difficulty for many of

---

[2] See also I. Fry. *The Emergence of Life on Earth* (New Brunswick, NJ: Rutgers University Press, 2000), pp. 55–57. By the third edition (April 1861) of his *Origin*, Darwin had removed the words "into which life was first breathed."

[3] More primary materials on Victorian speculation about the problem of consciousness, with all its moral implications, can be found in *Embodied Selves: An Anthology of Psychological Texts, 1830–1890*, eds. J. Bourne Taylor and S. Shuttleworth (Oxford: Oxford University Press, 1998).

[4] It is this work by Stirling, attacking Huxley and Darwin, that Bastian criticizes in *Nature*, February 1870 in his article "Protoplasm."

those liberal Christians who had, up until that point, felt they could enthusiastically support Darwin.[5] In one public venue Frederick Barnard, president of Columbia University, wrote that he found Bastian's *Beginnings of Life* so convincing, that educated Christians must now feel forced to *choose* between Darwinism and a universe with no Creator God.[6] Surely Barnard was not the only such person, but his remarkable frankness bears quoting at some length, to show just why Darwin and Huxley tried to avoid pushing the issue as forcefully as Bastian. Barnard said:

We are told, indeed, that the acceptance of these views need not shake our faith in the existence of an Almighty Creator. It is beautifully explained to us how they ought to give us more elevated and more worthy conceptions of the modes by which He works His will in the visible creation. We learn that our complex organisms are none the less the work of His hands because they have been evolved by an infinite series of changes from microscopic gemmules, and that these gemmules themselves have taken on their forms under the influence of the physical forces of light and heat and attraction acting on brute mineral matter. Rather it should seem we are a good deal more so. This kind of teaching is heard in our day even from the theologians. ... It is indeed a grand conception which regards the Deity as conducting the work of His creation by means of those all-pervading influences which we call the forces of nature; but it leaves us profoundly at a loss to explain the wisdom or the benevolence which brings every day into life such myriads of sentient and intelligent beings, only that they may perish on the morrow of their birth. But this is not all. If these doctrines are true, all talk of creation or methods of creation becomes absurdity; for just as certainly as they are true, God himself is impossible. ... But we are told it is unphilosophical, in the pursuit of truth, to concern ourselves about consequences. ... To this canon I am willing to subscribe up to a certain point. But if, in my study of nature, I find the belief forced upon me that my own conscious spirit ... is but a mere vapor, which appeareth for a little time and then vanisheth away forever, that is a truth ... for which I shall never thank the science which has taught it me. Much as I love truth in the abstract, I love my hope of immortality still more; and if the final outcome of all the boasted discoveries of modern science is to disclose to men that they are more evanescent than the shadow of the swallow's wing upon the lake, ... give me then, I pray, no more science. Let me live on, in my simple ignorance, as my fathers lived before me, and when I shall at length be summoned to my final repose, let me still be able to fold the drapery of my couch about me, and lie down to pleasant, even though they be deceitful dreams [7].[7]

Spontaneous generation seemed to bring home to many evolutionists the stark materialist implications of what Huxley had called "the physical basis of life."

---

[5] The exception, so uncommon that he proves the rule, was spontaneous generation advocate and Darwinian, Rev. Thomas Stebbing. See his *Essays on Darwinism* (London: Longmans, 1871).

[6] Barnard himself may have already had problems with Darwinism, though this is not clear from the existing record. See W. Chute. *Damn Yankee! The First Career of Frederick A. P. Barnard* (Port Washington, NY: National University Publications, 1978), p. 69.

[7] It is interesting that this passage was one of very few excerpted from a bowdlerized version of Barnard's essay published in *Medical America in the Nineteenth Century*, ed. G. Brieger (Baltimore, OH: Johns Hopkins, 1972), pp. 278–292; passage expurgated on p. 286.

Among the younger Darwinians, Alfred Russel Wallace came out early and very strongly in favor of Bastian's arguments, in one of the first reviews to appear of *The Beginnings of Life* [8]. The medical journals also applauded the strengthening of Darwinism by Bastian's experiments, emphasizing that he extended the logical "continuity in nature" so important to the new scientific naturalism, bridging the arbitrary gap between living and non-living matter.

At this, Huxley and his colleague, physicist John Tyndall, began to campaign more vigorously for a unified front among Darwin supporters. They thought Bastian must be a careless experimenter (though on many of the most important points they were wrong); thus they adopted a strategy of severing the linkage between evolution and present-day spontaneous generation, and they declared Bastian and all his supporters, de facto, "non-Darwinians." Huxley in his famous BAAS address [9], "Biogenesis and Abiogenesis," first publicly declared that Darwinian science did indeed imply a naturalistic origin of life; he called this "abiogenesis" to make clear that it differed from spontaneous generation in that it could only have occurred in the Earth's distant past.[8] It is from this time that Darwin wrote privately to Hooker:

It is often said that all the conditions for the first production of living organisms are now present, which could ever have been present. But if (and oh what a big if) one could conceive in some warm little pond with the right amounts of ammonia and phosphoric salts, – light, heat, electricity, etc. present, thus a protein compound was chemically formed, ready to undergo itself such complex changes, at the present day such matter would be instantly devoured or absorbed, which would not have been the case before living creatures had formed [10].

Clearly, Darwin's private views were highly sympathetic to a naturalistic origin of life, at least on the primitive Earth. And he continued to follow the debates between Bastian and Huxley with great interest. But he certainly never judged it appropriate to make his own views public, even when Bastian's career was nearly ruined by Huxley's campaign to undermine his scientific reputation. The reaction of Barnard and others like him can only have confirmed for Darwin how important his original choice had been to keep silent, or even at first throw a bone to Christians on the origin-of-life issue. Thus until at least the 1870s, the scientists themselves stepped very carefully around religious implications of their work.

William Thomson (later Lord Kelvin) produced his theory of meteoritic import of life to Earth in his 1871 Presidential Address to the BAAS as a clear rebuttal to

---

[8] Huxley, ironically, hijacked Bastian's own term "biogenesis" without attribution and gave it the opposite meaning, i.e. life only from already existing life. Huxley's rhetorical success in making his terms stick (and thereby distancing proper evolutionary science from the older doctrine of "spontaneous generation" and its unsavory associations, was no small part of his ability to vanquish Bastian. So much so that most textbooks today include "biogenesis" among terms Huxley is credited with coining.

Huxley's position on abiogenesis from 1870 [11]. Thomson was a deeply religious man and a committed opponent of Darwinism. Not surprisingly, in their personal correspondence the Darwinians made fun of Thomson's recourse to such an unscientific hypothesis. From their point of view, he was abandoning his usual scientific critical faculties because he was so desperate to find a way to avoid the implication of abiogenesis. In this early incarnation, Huxley thought Thomson's version of "panspermia" merely pushed the problem back to another planet but was no real solution to the origin-of-life question.[9]

As Iris Fry has pointed out, the Pope, even in his 1997 statement declaring that evolution was supported by enough facts that it was now "more than a hypothesis," still made it clear that he was not ready to accept a theory of evolution based on a materialist interpretation, but only one that was spiritualist in its philosophy (i.e. one that acknowledges that if a human body takes its form from pre-existing living matter, the soul was immediately created by God). Fry's book *The Emergence of Life on Earth* [12] offers an excellent summary of how varied Christian theological positions could be over time, however. Early Church doctrine was dominated for hundreds of years by the thinking of St. Augustine, for example. Quoting Fry,

According to Augustine, continuing spontaneous generation is made possible by a divine decree, issued at the moment of creation and active forevermore, under whose power the generation of living creatures from the earth and various organic materials persists. Augustine referred to biblical commands such as "Let the land produce vegetation," "Let the water teem with living creatures," and "Let the land produce living creatures according to their kinds." At the first event of creation, he said, God formed the "seed-principles" of all the species, which are the potentials of all living creatures to come. These potentials materialize throughout the history of the Earth according to a preordained plan, in both sexual and spontaneous generation, when the circumstances are favorable [12] (p. 18).

Fry points out that, even when Thomas Aquinas modified this doctrine by combining Augustine's "seed-principles" with Aristotelian philosophy, arguing for much more active, direct intervention by God, still, under both Augustine's and Aquinas's doctrines

and for centuries thereafter, no doubt was raised as to the possibility of spontaneous generation. This harmony between the belief in spontaneous generation and various Christian doctrines is noteworthy, especially since it tends to be forgotten following the radical change in the church's attitude toward this issue that took place at the end of the seventeenth

---

[9] See, for example, Huxley to Hooker, August 11, 1871, Joseph Dalton Hooker Papers, Royal Botanical Gardens, Kew, tss. pp. 101–102; Hooker to Huxley, August 19, 1871, Hooker papers tss. pp. 109–110; Hooker to Darwin, August 5, 1871, DCC #7896, *Life and Letters of Sir Joseph Dalton Hooker*, vol. 2: 126–127; Tyndall to Bence Jones, August 13, 1871, John Tyndall papers, Royal Institution, London (microfilm copies of all correspondence at Martin Science Library, Franklin and Marshall College), tss. p. 805.

century. Until that change there was no reason – neither in the way organisms were thought of, nor in the predominant philosophy of nature – that would prevent the church from supporting the widespread belief in spontaneous generation [12] (p. 19).

Both Augustine's and Aquinas's positions are explored in much more detail by Ernan McMullin in Chapter 5.

3.2 Changing concepts of life

Talking about the origin of life, then, is still potentially theologically charged, based upon how one defines life. Now for our (by no means exhaustive) sampling of concepts of what life actually is. Here I hope to show that broader historical context (than religious implications) can be just as informative about trends in defining life. A useful survey of numerous definitions of life up until the end of the nineteenth century can be found in the opening pages of the 1899 F. J. Allen article "What is life?"[13].

Louis Pasteur argued as early as 1860 that optical activity is "the only distinct line of demarcation that we can show today between dead and living matter." (He believed that "cosmic asymmetric forces" were responsible for this.) One of Allen's most useful observations is that most of the attempted definitions have some merit, but they all tried to find a single key formulation to capture an extremely complex, many-sided phenomenon. "Life," he says, "is too complex to be described in a concise aphorism" [13] (p. 45). Pasteur was surely pointing out one of the most crucial features of living systems, but as the 1997 work of Cronin and Pizzarello on non 50:50 enantiomers (the opposite right- and left-handed mirror image versions in which many organic molecules occur) in extraterrestrial amino acids from meteorites has shown, even that formerly sharp line of demarcation has begun to blur into a spectrum [14].

Allen then plunges into a review of what is known about the biochemistry of metabolism and a very modern-sounding discussion of the central role of "energy-traffic" in cells. He is impressed with the 1875 theory of the German physiologist Eduard Pflüger, that cyanogen is the defining unstable compound of living reactions, and, ironically enough, notwithstanding his opening caveat, Allen soon arrives at his own pet, too-concise formulation, which he calls a law: that "every vital action involves the passage of oxygen either to or from nitrogen" [13] (p. 52).[10] He cites another German, Oscar Loew, who has recently suggested that the basis

---

[10] The original is E. Pflüger. Über die physiologische Verbrennung in den lebendigen Organismen. *Arch. Gesam. Physiol.* (*Pflügers Archiv*) **10** (1875), 251–278. Pflüger's concepts were still influential, more recently, in the ideas on biogenesis developed by Wilhelm Reich. See, for example, Reich. *The Cancer Biopathy* [1948] (New York: Farrar, Straus and Giroux, 1973), p. 9; also S. Kramer. quoted in *Origins of Life II*, ed. L. Margulis (New York: Gordon and Breach, 1971), pp. 8–13.

of life is continuous atomic motion within large unstable organic molecules. Allen commends the theory for explaining some phenomena, but thinks "Dr. Loew does not lay enough stress on the functions of nitrogen, to which I personally trace so many of the phenomena of life.... It is chiefly owing to the instability of nitrogen compounds that continuous atomic motion is possible" [13] (p. 55). He gives priority to N, among the CHON.

Here we are seeing a transition to a new, biochemical way of talking about the most basic characteristics of life, both structural and metabolic. But Allen's attempt to hold on to some single *sine qua non* of life represents the deep grip of a late nineteenth century scientific gestalt that we need to look at more closely, in order to understand why biochemical ideas that in some ways sound so modern are still shaped by it. From the 1860s through the first decade of the twentieth century theories abounded as to the simplest "living unit." From Thomas Graham's colloid, containing "energia," to Huxley's "protoplasm," to Darwin's "gemmules" of pangenesis, to Hering and Haeckel's "plastidules," to Nägeli's "micelles," to Verworn's "biogen," in many ways Allen's focus on nitrogen as *the* crucial "living element" is in this tradition. These theories reflect the fact that from 1860 to perhaps as late as 1910, "explaining life" was interpreted by most investigators to require addressing all or most of the parts of a large problem complex. This included generation, development, inheritance, sensation or irritability, evolution, and in some cases the postulate of a specific biological energy.[11] The gradual teasing apart of the strands of this once-unified problem complex, for instance the separation, from about 1910–1920 of transmission genetics from embryology/development and its more inclusive notion of "inheritance," greatly altered discussions of the origin of life.

The effect of the new discipline of transmission genetics, most of all due to the work of the T. H. Morgan *Drosophila* school, began to be clearly seen in the August 1926 paper of H. J. Muller "The gene as the basis of life"[12] and the 1929 paper of J. B. S. Haldane "The origin of life" [15]. In Muller's paper, as John Farley has pointed out, defining the sudden appearance of a gene molecule as the sharp point at which life emerged has the paradoxical effect of reviving the supposedly most outdated feature of the old "spontaneous generation" doctrine: namely the idea that anything as complex as life could possibly arise in a sudden rapid event. In A. I. Oparin's writings from the 1920s, a gradual chemical evolutionary process is put forward as much more plausible; i.e. the idea of the *emergence* of life through a

---

[11] For the postulate of a specific biological energy, see B. Moore. *The Origin and Nature of Life* (London: Williams and Norgate, 1913), reprint in 1921, pp. 157, 225–226; also P. Kammerer. *Allgemeine Biologie* (Berlin and Leipzig: Deutsche Verlags-Anstalt Stuttgart, 1915), p. 8. Kammerer seems to have changed his mind or at least removed this wording, by the second edition in 1920. But it is from Kammerer that Wilhelm Reich appears to have drawn most directly in positing a specific biological energy.

[12] H. J. Muller. The gene as the basis of life. In *Proceedings of the 4th International Congress of Plant Biology*, vol. 1 (Menasha, WI, 1929), pp. 897–921; the talk was originally given in 1926.

series of transitional stages. While Haldane was interested in the latest ideas about genes, he leaned more toward the Oparin approach:

In the present state of our ignorance we may regard the gene either as a tiny organism which can divide in the environment provided by the rest of the cell; or as a bit of machinery which the "living" cell copies at each division. The truth is probably somewhere in between these two hypotheses. ... Unless a living creature is a piece of dead matter plus a soul (a view which finds little support in modern biology) something of the following kind must be true. A simple organism must consist of parts A, B, C, D, and so on, each of which can multiply only in the presence of all, or almost all, of the others [15] (p. 6).

Here Haldane makes an early but cogent statement on what are now called "emergent properties."

A brief look, now, at two recent books will quickly bring out numerous points about trends in defining life in the past 40–50 years and the historical context in which these trends developed. The books are Freeman Dyson's *Origins of Life* and Maynard Smith and Szathmáry's *The Origins of Life: from the Birth of Life to the Origin of Language* [16, 17].[13]

Being a physicist allows Dyson to see [16] (pp. 6–7) the extent to which much thinking among biologists is predisposed by their own philosophical assumptions; as for instance, why such an overwhelming majority of life scientists raised since Watson and Crick believe information-carrying molecules more fundamental to life than biochemical metabolism. This, despite the fact that, ever since researchers have seen the origin of life to be predicated upon the origin of DNA, RNA, or some other more primitive information-carrying molecule, the result has been the chicken–egg problem because of the interdependence of proteins and nucleic acids in extant cells. Dyson is less aware, or at least does not stop to comment on, the degree to which his own reasoning [16] (pp. 7–8) is being guided just as forcefully by notions about "hardware" and "software" inherited from the culture of computer technology. This is not to imply that use of these analogies in thinking about living systems is *necessarily* faulty. But rather, just as the dominance of machines in scientific cultures cannot be said to be historically unrelated to the growth of the mechanistic view of life from 1850–1950, one ought at least to note that the ideas of hardware and software are not merely disconnected intellectual "ideas" floating around, but also fundamentally *cultural* resources, being drawn upon here by scientists. Thus, it is worth asking the question: do these ideas come into the scientific arena freighted with any other interesting cultural or philosophical baggage? More on this later.

---

[13] I survey more broadly the discussions of a definition of life and research into its origins in S. Dick and J. Strick. *The Living Universe: NASA and the Development of Astrobiology* (New Brunswick, NJ: Rutgers University Press, 2004), especially chapters 3–5.

Dyson emphasizes the need to distinguish between replication and reproduction, in order to break the logical "catch-22" deadlock that results when one considers DNA- or RNA-centered systems to be the *sine qua non* of life. Dyson gives Von Neumann credit for emphasizing the distinction between replication and metabolism. This is the most significant distinction Dyson rightly emphasizes in his book. But one can only wonder why he lauds Schrödinger and Von Neumann's early and vague approaches to this distinction, while studiously avoiding mention of Sidney Fox and his school – those who first made the issue of "proteins first" vs. what they called "the nucleic acid monopoly" central in the origin-of-life debate. Of course, the two big name physicists have become revered in science (and Dyson himself is a physicist), while Fox was a protein chemist who eventually became marginalized by the mainstream origin-of-life community.[14] So if one is constructing a "forerunners" pedigree for one's most important idea, perhaps the temptation is overwhelming to attribute that idea to winners and silently pass over losers, especially if one at the outset intends to write a highly condensed narrative that disclaims any attempt at comprehensiveness. From a historian's point of view, this practice is in itself an object of study.

Dyson points out that Schrödinger saw biology through the eyes of the physicist-turned-biologist Max Delbrück and historians have elaborated at some length on the construction of a master narrative of the history of molecular biology that emphasizes only the line from Schrödinger, Delbrück, and Salvador Luria (the team which in the early 1940s famously began experiments on bacteriophage genetics) to Watson and Crick. Dyson says that thus Delbrück's focus on replication (and later on nucleic acids) as the central feature of the origin of life gained undue prominence in the field and came to dominate the mindset of most researchers. This insight into "paradigms" and their control of thinking in the field has been stated repeatedly over the last twenty-five years.[15]

Maynard Smith and Szathmáry's *The Origins of Life* sets out to describe and explain what they plausibly argue are the eight major qualitative transitions that have occurred in the history of life since the origin of replicating molecules.[16] The book is an eloquent and very illuminating analysis of these transitions and of some very important parallel trends among them [17] (pp. 19–25). But as a result of such breadth of conceptual reach, it manages to survey only somewhat

---

[14] See W. Hagan. Review of Fox's *The Emergence of Life. Isis*, **80** (1989), 162–163. Also, A. Brack. Review of chemical evolution: physics of the origins and evolution of life. *Origins of Life and Evolution of the Biosphere*, **29** (1999), 109–112, 110. Fox's work, as well as the process of marginalizing him, are discussed further in S. Dick and J. Strick. *The Living Universe: NASA and the Development of Astrobiology* (New Brunswick, NJ: Rutgers University Press, 2004), chapter 2.

[15] See for example, S. Fox. The proteinoid theory of the origin of life and competing ideas. *American Biology Teacher*, **36** (1974), 161–172, 181.

[16] Note that, since the mid-1960s most researchers have preferred to use the plural, *origins* of life, to emphasize their belief that the process may have occurred multiple times, or in multiple steps.

superficially the origin of life *per se*. The major transitions they address are: replicating molecules→populations of molecules in compartments; independent replicators→chromosomes; RNA as gene and enzyme→DNA and protein; prokaryote→eukaryote; asexual clones→sexual populations; protists→animals, plants, and fungi; solitary individuals→colonies; primate societies→human societies and the origin of language.

In *The Origins of Life*, immediately from the first page, the focus is on *information*. The question of metabolism being of equal importance, let alone first in time (as in Dyson), is very briefly raised [17] (p. 12), only to be dismissed or minimized: their overall usage betrays a strong bias toward an "information-first" view of life. Their approach clearly assumes that life is synonymous with replication.[17]

There is no more *historical* a phenomenon in modern biology than the dialectically related rise of information theory and computers, and the simultaneous importation of such analysis into biological thinking, beginning no later than Schrödinger's 1944 work *What Is Life?*[18] Maynard Smith and Szathmáry tackle this strikingly parallel development of concepts right away. It would seem strange or incomprehensible to Darwin, they say [17] (p. 2), that template reproduction allows transmission of instructions in a homogeneous-looking, as yet unformed egg or zygote. The idea is much less strange to us because "we are familiar with the idea that patterns of magnetism on a magnetic tape can carry the instructions for producing a symphony." Indeed, they close their book with a tantalizing guess that the move to transmitting information in electronic form may be potentially a transition on the scale of the other major transitions around which the book is framed. It is astute of the authors to recognize how much our cultural experience *enables* our view, especially on questions of such fundamental importance as "what is life?" However, being more or less complete advocates of the information-first approach to conceptualizing life, they seem to miss the other implication of the power of historical context. If our cultural experience enables our view, it also simultaneously *constrains* it. The primacy of computers and electronic information in our lives makes images of "programming" of instincts, "hard-wiring" of certain traits, etc., highly compelling metaphors for how we think about "life" in late twentieth century, high-tech Western society. But these metaphors tend to channel one's thinking strongly, above and beyond the actual experimental evidence, as in the nature–nurture debate where "master molecule" and "inborn hard-wiring" metaphors have

---

[17] See for example, p. 35 and p. 37 where they say "catch-22 of the origin of life," but they mean the origin of replication.

[18] Erwin Schrödinger. *What Is Life?* (Cambridge: Cambridge University Press, 1945), originally given as a lecture in 1944. E. Fox Keller's *Refiguring Life* (New York: Columbia University Press, 1995) is one of several recent analyses that begin to examine this topic, as is L. Kay's *Who Wrote The Book of Life?* (Palo Alto, CA: Stanford University Press, 2000). Donald Fleming also raised this issue in his Émigré physicists and the biological revolution. In *The Intellectual Migration*, eds. D. Fleming and B. Bailyn (Cambridge, MA: Harvard University Press, 1969).

boosted the stock of biological determinism far above even the rapidly growing knowledge-base of molecular genetics. We may well reflect on the dominance of such models, when they say [17] (p. 13) "a living being resembles a computer, rather than just a program, although it has its own program as subsystem." The irony of the back-and-forth relations between culture and nature is never more provocative than in this passage [17] (pp. 12–13), where computers and computer "viruses" are used as the standard against which to evaluate whether *biological* viruses should be thought of as truly alive. Is this not putting the cart before the horse in some fundamental ontological sense? One is reminded of N. W. Pirie's 1937 remark:

When one is asked whether a filter-passing virus is living or dead the only sensible answer is: "I don't know; we know a number of things it will do and a number of things it won't, and if some commission will define the word 'living' I will try to see how the virus fits into the definition." His answer as a rule does not satisfy the questioner, who generally has strong but unformulated opinions about what he means by the words living and dead. The virus worker may then be told, to take one example out of many equally foolish, that a certain virus cannot be living because its physical properties enable us to investigate it in some detail [18] (p. 12).

As we shall see in origin-of-life debates, the possibly crucial question that gets drowned out by talk of the primacy of information (and thus, of nucleic acids) is: can there be any other central characteristic of living systems as fundamental as, or perhaps even more fundamental than, information? Granted, Maynard Smith and Szathmáry give a brilliant and powerful analysis of events *since* the evolution of information-carrying molecules. But their bias leaves us with the chicken–egg problem: if metabolism is dominated mostly by proteins, but is a prerequisite for the functioning of nucleic acid information molecules, how can a system like our current living cell, even the simplest prokaryote, with each of these two parts totally dependent upon the other, ever have evolved in the first place? (In this way, giving primacy to nucleic acids/information plays into the hands of creationists.) This is the issue upon which Dyson's book is so focused.

Not that Dyson is the first to raise this issue. Ever since Harvard biochemist Leonard Troland's 1914 [19] paper emphasizing autocatalytic enzymes and Muller's 1926 [20] gene-first response, this tension has been a central focus of debate and discussion in the origin-of-life literature. Noted advocates toward Muller's end of the spectrum have included Norman Horowitz and Carl Sagan. Toward the opposite end have been A. I. Oparin, J. D. Bernal, N. W. Pirie, and Sidney Fox. The boom of interest in ribozymes and the "RNA world," beginning with Thomas Cech's 1982 discovery of catalytic RNA molecules, was precisely because it was hoped this phenomenon would finally offer a way out of the impasse that dominated much of twentieth century discussion. If the simplest nucleic acid information molecules can also simultaneously perform the enzyme role, previously thought only to be a property of proteins, then catalytic RNA molecules could be the "missing link" bridging the gap between these two now separate but interdependent functions. Maynard Smith and Szathmáry clearly hope [17] ribozymes offered the solution to the catch-22. But this now seems to have been excessively optimistic [21].[19] For, while RNA does seem to have the dual capabilities to bridge the gap, its monomers are so difficult to form spontaneously, and are so short-lived under primitive Earth conditions, that the question of how to get from an abiotic world to the RNA world is not much easier to solve than before the RNA world transitional stage was known.[20] Information is surely one of the fundamental issues for living systems as we know them, but we are still unclear as to whether it came first or is the *sine qua non* of the first life or its molecular predecessors.

In conclusion, I hope this chapter has demonstrated the need to look at the religious/theological context as well as the broader historical/cultural context in order to begin to understand how ideas about defining life and life's origins have changed over time. Since the theory was first discussed, common cultural assumptions about the nature of life and/or common religious assumptions about its origin have been prominent forces shaping the conceptions of the scientific community.

References

[1] W. B. Carpenter. On the mutual relations of the vital and physical forces. *Philosophical Transactions of the Royal Society (London)*, **140** (1850), 727–757.

[2] C. Darwin. *On the Origin of Species* (London: John Murray, 1859).

[3] Darwin to Joseph Hooker, March 29, 1863, published in *Correspondence of Darwin*, vol. 11, pp. 277–278 (*Darwin Correspondence Calendar* – hereafter DCC, #4065); expurgated version in Darwin's *Autobiography*, p. 272 and in *Life and Letters of Charles Darwin*, vol. 3, pp. 17–18.

[4] J. Strick. *Sparks of Life* (Cambridge, MA: Harvard University Press, 2000), especially chapters 2, 5, and 6.

[5] L. Beale. Protoplasm and living matter. *Monthly Microscopical Journal*, **1** (1869), 277–288.

[6] J. H. Stirling. *As Regards Protoplasm, in Relation to Professor Huxley's Essay on the Physical Basis of Life* (Edinburgh: Blackwood and Sons, 1869), 2nd edn., 1872.

---

[19] See also S. Freeland, R. Knight, and L. Landweber. Do proteins predate DNA? *Science, 286* (1999), 690–692; G. F. Joyce. The rise and fall of the RNA world. *New Biologist, 3* (1991), 399–407. This, along with many other critical papers in the field, are reprinted in an invaluable collection, D. Deamer and G. Fleischaker (eds.). *Origins of Life: the Central Concepts* (Boston, MA: Jones and Bartlett, 1994).

[20] The field underwent a similar burst of growth following the Miller–Urey experiment of 1953. But it was found that, here too, one of the most basic quandaries proved more elusive than expected. In 1981 researchers realized that the Earth's early atmosphere was most likely *not* a chemically reducing one. It was probably neutral or even mildly oxidizing; under which conditions Miller–Urey-type syntheses do not produce any significant amounts of organic precursors of biomolecules. See R. Kerr. Origin of life: new ingredients suggested. *Science*, **210** (October 3, 1980), 42–43; also J. Levine, T. Augustsson, and M. Natarajan. The prebiological paleoatmosphere: stability and composition. *Origins of Life, 12* (1982), 245–259, first presented in October 1981, included as pp. 97–111 in D. Deamer and G. Fleischaker (eds.). *Origins of Life: the Central Concepts* (Boston, MA: Jones and Bartlett, 1994).

[7] F. Barnard. The germ theory of disease and its relations to hygiene. *Public Health Reports and Papers*, **1** (1873), 79–80.

[8] A. R. Wallace. Review of *The Beginnings of Life*. *Nature*, **6** (August 8 and 15, 1872), 284–287, 299–303.

[9] T. H. Huxley. Biogenesis and Abiogenesis. *Nature*, **2** (September 15, 1870), 400–406.

[10] Darwin to Joseph Hooker, February 1, 1871, DCC# 7471; Darwin papers, American Philosophical Society Library, Philadelphia.

[11] W. Thomson. Presidential address to the BAAS. *Nature*, **4** (August 3, 1871), 262–270.

[12] I. Fry. *The Emergence of Life on Earth* (New Brunswick, NJ: Rutgers University Press, 2000).

[13] F. J. Allen. What is life? *Proceedings of Birmingham Natural Historical and Philosophical Society*, **11** (1899), 44–67.

[14] J. Cronin and S. Pizzarello. Enantiomeric excesses in meteoritic amino acids. *Science*, **275** (February 14, 1997), 951–955.

[15] J. B. S. Haldane. The origin of life. *Rationalist Annual* (1929), 3–10.

[16] F. Dyson. *Origins of Life*, 2nd edn. (Cambridge: Cambridge University Press, 1999).

[17] J. M. Smith and E. Szathmáry. *The Origins of Life: from the Birth of Life to the Origin of Language* (Oxford: Oxford University Press, 1999).

[18] N. W. Pirie. The meaninglessness of the terms "life" and "living." In *Perspectives in Biochemistry*, ed. Joseph Needham (Cambridge: Cambridge University Press, 1937), pp. 11–22.

[19] L. Troland. The chemical origin and regulation of life. *Monist*, **24** (1914), 92–133.

[20] H. J. Muller. The gene as the basis of life. In *Proceedings of the 4th International Congress of Plant Biology*, vol. 1 (Menasha, WI, 1929), 897–921.

[21] L. Orgel. The origin of life: a review of facts and speculations. *Trends in Biochemical Science,* **23** (1998), 491–495.

4

Philosophical aspects of the origin-of-life problem: the emergence of life and the nature of science

Iris Fry

4.1 Introduction

At a public lecture at the Sorbonne in 1864, the renowned French scientist Louis Pasteur raised the fundamental philosophical question pertaining to the origin of life: can matter organize itself to form a living system? Pasteur answered this question with a decisive no. He was referring both to the emergence of life on the primordial Earth and to the possibility of "spontaneous generation" – the formation of living organisms out of matter here and now. Life, Pasteur believed, was originally created by God and organisms are subsequently born only from parents.[1]

Toward the end of the eighteenth century, the German philosopher Immanuel Kant had also pondered the question of the nature of biological organization and its emergence from matter. Unlike other physical objects, Kant noted, an organism is an interactive system in which parts and whole are reciprocally dependent.[2] This circular, self-reproducing nature defines a living system and the critical question is whether we can understand the production of such an organized, functional whole in causal, materialistic terms [1] (77: 293).

For Kant, the idea of physical principles of self-organization, working mechanistically without any guiding plan and yet capable of producing an organized whole, was inconceivable. Hence, the very possibility of the emergence of life from matter seemed to him absurd [1] (80:305, note 5; 81:311). Although not sharing Kant's

---

[1] Several historical studies reveal that Pasteur's position on the relationship between matter and life was far more complex than apparent from the standard Pasteur's scientific biography. See G.L. Geison. *The Private Science of Louis Pasteur* (Princeton, NJ: Princeton University Press, 1995), pp. 133–138.

[2] This unique nature of organisms was evident to any keen observer of the living world long before the establishment of modern biology. The most notable manifestation of this awareness is to be found in the fourth century BCE in the biological treatises of Aristotle. See M.C. Nussbaum. Aristotle on teleological explanations. In *Aristotle's De Motu Animalium: Text with Translation, Commentary, and Interpretive Essays* (Princeton, NJ: Princeton University Press, 1978), pp. 76–80.

Exploring the Origin, Extent, and Future of Life: Philosophical, Ethical, and Theological Perspectives, ed. Constance M. Bertka. Published by Cambridge University Press. © Cambridge University Press 2009.

and Pasteur's convictions, most scientists toward the end of the nineteenth century nevertheless preferred not to deal with the origin-of-life problem. Following the growth of cytological knowledge and the rise of Biochemistry and Genetics, the realization of the enormous complexity of the cell seemed to have put a ban on research of this subject, turning it into a scientific taboo [2] (p. 494).

A serious change in attitude emerged in the twentieth century. It was expressed in the formulation of new theories on the emergence of life, and by the middle of the century, in the establishment of a scientific discipline devoted to the subject. These developments reflected the accumulation of new scientific data in biology and chemistry and in many other scientific fields, such as astronomy and geology. However, this breakthrough could not have been made without the radical change in the conception of the living world produced by Darwin's theory of evolution and coming into maturity in the twentieth century. In *On the Origin of Species* [3], published in 1859, Darwin did not discuss the question of the origin of life. In his private letters he expressed the feeling that contemporary scientific knowledge was far from sufficient to deal with this difficult problem [4]. He assumed that a single form, or a few original living forms, existed on the primordial Earth and explained the evolution of the whole tree of life from "so simple a beginning" [3] (p. 470).[3] Yet, it was evident to Darwin's supporters and opponents alike that a consistent theory of evolution implied that life had formed in the ancient past from inorganic matter through natural processes. The naturalistic character of Darwin's theory stemmed mainly from his major explanation of the phenomena of evolution, the mechanism of natural selection. This and other evolutionary mechanisms were thought to be the result of natural, material processes with no recourse to a supernatural, purposive power.

The general philosophical postulate asserting that organized living systems emerged on Earth through physical and chemical mechanisms is supported by science today and is a crucial part of the evolutionary worldview. This worldview emphasizes with equal force both continuity and novelty. It is distinct from the creationist view, which maintains an unbridgeable gap between matter and life that is overcome only by an intervening transcendent divine agent. It is also different from earlier simplistic mechanistic concepts that ignored the unique features of living systems. Origin-of-life research focuses on the physical and chemical factors that could have constituted the necessary infrastructure for the emergence of life, and thus emphasizes the continuous nature of the process. At the same time, researchers suggest theoretical models to account for the emergence of the new, unique features characteristic of biological organization, already displayed by the first organized systems.

---

[3] Not in public, however, Darwin did contemplate the possibility of a primordial material organization out of simple chemical compounds in a "warm little pond" under the influence of various energy sources, to form a living entity. See F. Darwin. *The Life and Letters of Charles Darwin*, 3 vols. (New York: Johnson Reprint Corporation, 1969), pp. 3, 18 (originally published London: Murray, 1887).

Although there is as yet no consensus on the possible mechanisms of the emergence of life, and though this process has not yet been simulated in the laboratory, scientists are confident that a primitive living system emerged on Earth a few billion years ago by natural means. This conviction is based on the achievements of the natural sciences. Above all it reflects the robust status of the theory of evolution that underlies the evolutionary worldview. Like any scientific conviction, it stems from the interaction among the empirical, theoretical, and philosophical aspects of science, being shaped by the social, political, and cultural context.[4] This conviction also serves as a motivation to the continuation of research.

The following discussion will examine two main themes that are closely interwoven. First, the principles by which theories of the origin of life face the challenge of accounting for the emergence of biological organization will be considered. Here, the role played by physical and chemical constraints on the primordial Earth and the crucial function of natural selection in the process will be analyzed. Second, the investigation of the origin-of-life problem is presented as a case study of the nature of science. Here, the need to acknowledge the interaction between the empirical and the philosophical in the working of science and the assault on the evolutionary worldview by the Intelligent-Design movement will be discussed.

4.2 The arch and the scaffolding

As noted above, the core problem of the origin of life is the emergence of the unique functional organization of the living system – the interdependence among its various components and their dependence on the system as a whole. Echoing other scientists and philosophers, the Scottish chemist Graham Cairns-Smith referred to life as appearing to have been designed [5] (p. 2). Not surprisingly, natural theology traditionally regarded this appearance as the strongest evidence of God's purposeful design of nature [6]. Present-day creationists claim that signs of "intelligent design" revealed at the molecular level are even more pronounced than ever [7]. Scientists, particularly evolutionary biologists, on the other hand, rely on material evolutionary processes, especially on the mechanism of natural selection, to explain the apparent design of biological organization. It is commonly accepted by science that the last common ancestor at the root of the evolutionary tree was already an evolved cell, containing the basic biochemical machinery and manifesting apparent design [8]. It was on the basis of these features that it could continue to evolve by natural selection, consequently giving rise throughout the history of the Earth to enormously diverse and adaptive forms of life.

While evolutionary biology deals with the evolution of the last common ancestor onward, the focus of origin-of-life research is on the natural, gradual processes that

---

[4] Due to space limits and the nature of my argument, these influential factors will not be discussed here.

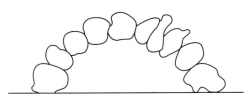

Figure 4.1 An arch of stones in which each stone is held in place by its neighbors. © Cambridge University Press 1985. Printed with permission of Cambridge University Press.

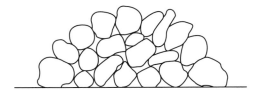

Figure 4.2 The gradual building of the "paradoxical structure" of the arch is possible by the construction of supporting scaffolding. © Cambridge University Press 1985. Printed with permission of Cambridge University Press.

led to the emergence of the last common ancestor out of a lifeless environment. To illuminate both the problem involved in the emergence of biological organization and its naturalistic solution, Cairns-Smith introduced a highly suggestive image representing the closely integrated biological organization – an arch of stones in which each stone is held in place by its neighbors (see Figure 4.1). An arch could not arise gradually, one stone at a time, hanging in midair [5]. Similarly, how could a step-by-step evolution lead to a system in which everything depends on everything else [5] (p. 39)?[5] The arch-like architecture of the cell is manifested in numerous cyclic, interlocked subsystems, crucial to the functioning of the organism. The most noticeable subsystem is the interdependent relationship between nucleic acids and proteins, giving rise to the chicken–egg problem: how could nucleic acids have emerged without proteins or proteins without nucleic acids?

Cairns-Smith pointed out that the answer to the puzzle, the gradual building of the "paradoxical structure" of the arch, is possible by the construction of supporting scaffolding: building a stone wall gradually and then removing the stones to leave the arch standing (see Figure 4.2) [5] (pp. 59–61). For Cairns-Smith, the idea of a scaffold was more than a figurative device offering a naturalistic alternative to the design notion. In his scenario for the emergence of life he suggested

---

[5] One of the proponents of the Intelligent Design movement, the biochemist Michael Behe, referred to this question as "the problem of irreducible complexity." Behe contended that "An irreducibly complex system cannot be produced ... by slight, successive modifications of a precursor system, because any precursor system ... that is missing a part is by definition nonfunctional" (M. J. Behe. *Darwin's Black Box* (New York: Free Press, 1996), p. 39). We'll come back to this claim later.

that organisms based on organic chemistry developed at a later evolutionary stage on pre-existing clay scaffolding. The first "organisms" capable of undergoing replication, mutation, and evolution through natural selection were made of mineral crystals of various clays [5] (pp. 98–106). Inorganic scaffolding made of materials ubiquitous on the primordial Earth was simple enough to evolve gradually and could have given rise to a more complex organic chemistry. At a certain historical point, when organic compounds were already built on the scaffolding and interconnected to form organic arch-like systems, there occurred a "genetic takeover" and the mineral scaffolding was then discarded [5] (pp. 107–113).

It should be clarified that the purposeful (teleological) language that is used to describe Cairns-Smith's arch-and-scaffolding scenario is a kind of a shorthand substitute for more elaborate causal statements describing mechanistic, selective processes. Just as evolution through natural selection accounts for the apparent design of organisms, no purpose or plan is supposed to have been involved in the construction of the scaffolding and the arch at the origin-of-life stage.[6]

Cairns-Smith's suggestion of removable mineral scaffolding was his answer to what he saw as the limitation of prebiotic chemistry to produce sophisticated biologically relevant organic compounds without such a device [5] (pp. 4–6). The chemical discontinuity involved in this scenario has been questioned by several critics claiming that it challenges a basic "continuity conception" that is guiding research into the origin and evolution of life [9, 10]. Researchers point to the existence in present organisms of "molecular fossils" – structures or functions that arose early in the history of life. Being highly beneficial, they were "frozen in time," since any change would have entailed many additional changes and the risk of the demise of such systems [11]. For this reason many scientists believe that the primordial organization of life was similar to the present biochemical system. Most origin-of-life theories therefore postulate various prebiotic stages, gradually increasing in organization and complexity that were at least partly incorporated into the first living systems rather than discarded. Christian de Duve suggests that "protometabolism," the complex set of chemical reactions in which organic building blocks were engaged prior to the emergence of RNA, had to be "congruent" with metabolism into which it grew. The two had to follow similar pathways [12].

Despite this basic difference between Cairns-Smith's and other theories, I suggest that the arch-and-scaffolding model can serve as a common, unifying theme of all origin-of-life theories. First, all theories grapple with the fundamental problem of the construction of a living arch – a system, separated from its environment, in

---

[6] At the same time, the teleological terminology inherent in biological discussions is not fortuitous. It reflects the fact that organisms are being naturally selected based on criteria of functionality and adaptability.

which both replication based on nucleic acids and metabolism operated by protein enzymes were interconnected. The figure of the "arch" seems to demand, even point to, the role of a "scaffold." Second, the concept of the scaffolding should direct our attention to the role of physical and chemical selection mechanisms and other channeling factors active in the prebiotic environment that laid the ground for the working of natural selection. As will be pointed out, with the recent emphasis in origin-of-life research on specific locales on the primordial Earth where life could have emerged, the image of scaffolding acquires a most tangible, physical meaning. It stresses the continuity between the primordial environment and emerging life.[7] Third, philosophically, the application of this image will allow us to combat the creationist contention that life is based on "irreducible complex systems" the emergence of which cannot be accounted for by causal, natural processes. Lastly, the arch-and-scaffolding tool enables comparative analysis of the various scenarios and theories formulated in the field, noting both their common denominator and the different "stone walls" suggested by each [13] (pp. 185–193).

4.3 On the role of selection mechanisms in the emergence of life

Biologists and philosophers of biology are engaged in an ongoing controversy on the adequate definition of life (and in a debate whether an exhaustive definition is at all possible) [9 (pp. 4–9), 14, 15]. Nevertheless, it is generally accepted that any definition of life must refer to the evolution of life's adaptive complexity through the mechanism of natural selection [13 (pp. 238–242), 16]. Indeed, origin-of-life researchers, despite the differences among their theories, aim at explaining the emergence of systems capable of undergoing biological evolution.

An examination of the various scaffolds suggested with this purpose in mind reveals that natural selection functions in all of them as an essential agent. Are we not involved here in a vicious circle? How could natural selection have operated prior to the existence of living systems capable of reproducing, mutating, and competing for resources? The first experiments demonstrating *in vitro* Darwinian evolution in the laboratory were conducted in the 1960s and 1970s. These experiments revealed that populations of molecules capable of replicating, mutating, and reproducing these mutations responded to applied selective pressures by evolving and adapting to new "environmental conditions" [17, 18]. It became clear that any group of entities, not only living systems, could evolve through natural selection provided it conformed to a set of specific conditions: reproduction, variation, inheritance

---

[7] On the continuous emergence of levels growing in complexity, see H. J. Morowitz. A theory of biochemical organization, metabolic pathways, and evolution. *Complexity*, **4**(6) (1999), 39–53; R. M. Hazen. *Genesis: The Scientific Quest for Life's Origin* (Washington, DC: Joseph Henry Press, 2005).

of these variations, relative advantage conferred by some of these variations, and competition [5] (pp. 2–3). This awareness made it possible to explore the role of natural selection in the origin of life itself. Though it was clear that molecular evolution achieved in the first *in vitro* experiments depended on the participation of a highly complex replicating enzyme, these results nevertheless led to a search for primitive entities that could have emerged on the primordial Earth and could have fulfilled the required conditions.

4.3.1 The metabolic and genetic traditions

As one reviews the origin-of-life field, it is possible to classify theories according to various criteria.[8] Historically, the investigation of the origin of life was established early in the twentieth century both by biochemists and geneticists who brought to the subject their particular preconceptions, methods, and terminologies [19].[9] Not surprisingly, the biochemical and genetic research traditions also involved different conceptions of the defining traits of life. While the biochemical tradition, from its inception to the present, characterizes life, primarily, as an integrated multi-molecular entity sustaining its homeostasis through metabolic processes, the genetic tradition focuses on the nature of life as a replicating system [20]. Chemically, the "metabolists" emphasized the emergence and function of proteins, while the "geneticists" regarded a self-replicating genetic molecule as the *sine qua non* of the emergence of life. More recently, the "chemical division" is less definitive;[10] however, the difference in emphasis is still between an early or late appearance of genetic material ("gene-first" and "gene-second" theories) [13] (pp. 135–178).

Clearly, these two traditions, or "camps," attack the chicken–egg problem from different directions, facing the challenge of accounting for the emergence and evolution of the "living arch" by different means. In order to examine the role allotted by each group to natural selection and to the means that enable its operation, a brief analysis of metabolic and genetic scaffolding is in order.

---

[8] See for example, H. Kamminga. Historical perspective: the problem of the origin of life in the context of developments in biology. *Origins of Life and Evolution of the Biosphere*, **18** (1988), 1–11; W. L. Davis and C. P. McKay. Origins of life: a comparison of theories and application to Mars. *Origins of Life and Evolution of the Biosphere*, **26** (1996), 61–73; J. Pereto. Controversies on the origin of life. *International Microbiology*, **8** (2005), 23–31.

[9] These traditions were initially personified by the figures of the Russian biochemist Alexander Oparin and the British geneticist J. B. S. Haldane. The American physicist Leonard Troland and the American geneticist Herman Muller were also important contributors to the genetic tradition in the first decades of the twentieth century (I. Fry. The origins of research into the origins of life. *Endeavour*, **3** (2006), 24–29).

[10] For examples of these trends, see I. Fry. *The Emergence of Life on Earth: A Historical and Scientific Overview* (New Brunswick, NJ: Rutgers University Press, 2000), pp. 177–178, 190–193. See also M. M. Hanczyc, S. M. Fujikawa, and J. W. Szostak. Experimental models of primitive cellular compartments: encapsulation, growth, and division. *Science*, **302** (2003), 618–622.

4.3.2 Genetic scaffolding

The most important genetic conception addressing the chicken–egg problem is the RNA-world theory. Suggested in the late 1960s as a tentative idea it acquired empirical support with the discovery of naturally occurring ribozymes in the 1980s and the continuous isolation of additional ribozymes in molecular-evolution experiments [13] (pp. 136–142). This theory constitutes the basic genetic scaffolding concept. The strong appeal of any "gene-first" theory stems from the fact that evolution through natural selection is best and most easily explained by the mechanism in extant cells of replication of nucleic acids, or nucleic-acid-like molecules. The discovery that RNA molecules can function as both information carriers and enzymes strengthens this appeal. It is thus clear that natural selection is the central element in genetic scaffolding. However, extensive work in prebiotic chemistry has revealed huge difficulties interfering with the synthesis of RNA and its constituents under primordial conditions [21]. The realization that even more primitive scaffolding was needed for the emergence of a self-replicating RNA system led to the search for simpler genetic systems [22, 23]. It also directed the attention of origin-of-life "geneticists" to the necessary physical and chemical prebiotic factors without which natural selection could not have operated.

This is a distinct shift from early "genetic" studies that paid little attention to the constraining effects of prebiotic chemistry on the emergence of RNA and self-replication [18, 24]. A common element now in all versions of genetic scaffolding is the extensive role of catalysts – metal ions, mineral surfaces, and peptides – both in the synthesis of building blocks and polymers as well as in the various template-directed reactions [25, 26]. Catalysis acts as a channeling factor, enhancing a limited number of reactions from the wide spectrum of theoretically possible ones and enabling the synthesis of polymers of a more specific structure. Some researchers experiment with alternative replication mechanisms that are less demanding chemically than those active in current cells and thus more in line with the "strictures of prebiotic chemistry" [27]. They also consider how mixed populations of RNA sequences, varying in physical and chemical properties, could have evolved. Out of these heterogeneous aggregates of polymers, more homogeneous and efficient sequences may have been selected, eventually giving rise to ribozymes and the RNA world.[11] Understanding the narrow constraints imposed on emerging life by the prebiotic environment has also led to a more open attitude toward non-genetic

---

[11] The selection of RNA sequences, homogeneous in terms of their constitution and chirality, from a mixed population depended on differences in stability and in the efficiency of binding. It was thus chemical selection that made it possible, first through the mechanism of template-directed ligation, and later through the template-directed-elongation mechanism, for natural selection to become active, see A. Kanavarioti. Template-directed chemistry and the origins of the RNA world. *Origins of Life and Evolution of the Biosphere*, **24** (1994), 489–492; K.D. James and A.D. Ellington. A search for missing links between self-replicating nucleic acids and the RNA world. *Origins of Life and Evolution of the Biosphere*, **25** (1995), 528.

models. Leading researchers of the chemistry of RNA consider the possible role of metabolic, non-genetic systems in changing the chemical environment and thereby influencing the chances of the emergence of a genetic system [28, 29].

4.3.3 Metabolic scaffolding

Early genetic studies regarded the synthesis of a few self-replicating sequences as enough to start the emergence of life through the action of natural selection. They tended to disregard the role played by the prebiotic environment in building the infrastructure needed for natural selection to work. Early metabolic theories, on the contrary, focused on the channeling effects of the prebiotic environment leading to the establishment of peptide catalysts and metabolic cycles and were not very concerned with the mechanism of evolution and the role of natural selection [30, 31]. Most metabolic theories, old and new, in focusing on the integrated organization of the living system as its characteristic trait, postulate the emergence of a protocellular unit separated from its environment as an early, crucial step in scaffolding construction. The postulated protocells are claimed to have emerged as a result of physical and chemical prebiotic processes, especially physical and chemical selection processes based on thermodynamic and kinetic criteria. Material self-organization of available prebiotic chemical compounds is a fundamental principle in all metabolic scaffolding.[12]

The protocellular character of emerging metabolic entities was associated from the start with their tendency to grow and divide. This was viewed as an unsophisticated means of heredity, assumed to form the basis of a primitive process of natural selection and evolution. In distinction to the predominant "genetic" conception asserting that evolution through natural selection is possible only through the replication of a genetic molecule, "metabolists" claim that reproduction of protocells at the origin of life could have been independent of replication [20] (pp. 5–9). More recent metabolic theories attempt to provide a "metabolic" mechanism for natural selection and evolution based on the emergence of autocatalytic reaction cycles within the protocellular units. Unlike the autocatalysis of a single genetic polymer, autocatalysis of a whole cycle is achieved by the mutual catalysis of its chemical components when a certain level of chemical complexity of the system is reached [9 (pp. 153–154), 32, 33, 34, 35]. All these models also grapple with the question of the emergence of metabolic novelties, branching from the autocatalytic cycles that

---

[12] See, for example, A. I. Oparin. *Origin of Life*, trans. S. Morgulis (New York: Dover Publications, 1953 [1936]); S. A. Kauffman. *The Origins of Order: Self-Organization and Selection in Evolution* (New York: Oxford University Press, 1993); H.J. Morowitz. *Beginnings of Cellular Life* (New Haven, CT: Yale University Press, 1992); D. W. Deamer. The first living systems: a bioenergetic perspective. *Microbiology and Molecular Biology Review*, **61**(2) (1997), 239–261; R. M. Hazen. *Genesis: The Scientific Quest for Life's Origin* (Washington, DC: Joseph Henry Press, 2005).

on rare, though finite, occasions can give rise to a variant autocatalytic cycle, thus providing a basis for selection and evolution.[13]

Recently, there has been a growing tendency of origin-of-life research to focus on the question of specific geophysical and geochemical settings in which life could have emerged. The notion that life might have emerged at the bottom of the sea near hydrothermal vents is gaining wide support [36].[14] The autotrophic metabolic theory of the organic chemist, Günter Wächtershäuser, postulates the emergence of a "surface metabolist," consisting of a surface of the mineral pyrite (FeS_2) and bonded organic compounds, at such submarine sites. The scaffold as a system within which processes of chemical selection based on differences in chemical affinity and stability set the stage for the action of natural selection is prominent in Wächtershäuser's theory [33].[15]

4.4 The arch-and-scaffolding image: the philosophical angle

We have observed how origin-of-life theories, both genetic and metabolic, suggest various modes of scaffolding all aimed at answering the fundamental question, how did matter organize itself to form life? These include both theoretical constructs, e.g. the RNA world, and actual physical structures, e.g. a pyrite surface or a lipid vesicle. Extensive research and the acquisition of data in the fields of prebiotic chemistry, geophysics, biochemistry, and molecular biology, make the notion of scaffolding necessary and cogent. This idea of scaffolding may also shed light on some philosophical dilemmas that traditionally accompany the discussion of the emergence of life.

As pointed out, although the construction of scaffolding and the emergence of an arch may appear to be a purposeful process, it can in fact be accounted

---

[13] Instead of identifying biological information with the "digital" machinery consisting of the "letters" in biopolymers, these theories postulate a "compositional" mode of information that can be stored in and propagated by mutually catalytic networks, see D. Segré, D. Ben-Eli, and D. Lancet. Compositional genomes: prebiotic information transfer in mutually catalytic noncovalent assemblies. *Proceedings of the National Academy of Sciences*, **97**(8) (2000), 4112–4117.

[14] The submarine scenario is also attracting fierce criticism, see M. Levy and S. L. Miller. The stability of the RNA bases: implications for the origin of life. *Proceedings of the National Academy of Sciences. USA*, **95** (1998), 7933–7937; J.L. Bada and A. Lazcano. Some like it hot, but not the first molecules. *Science*, **296** (2002), 1982–1983; D. Penny. An interpretive review of the origin of life research. *Biology and Philosophy*, **20** (2005), 633–671.

[15] The focus on a tangible, actual scaffolding, continuous with its geochemical setting, necessary for the emergence of the living arch is also prominent in the metabolic theories of the Scottish geologist Michael Russell and the physical chemist Harold Morowitz. See W. Martin and M.J. Russell. On the origins of cells: a hypothesis for the evolutionary transitions from abiotic geochemistry to chemoautotrophic prokaryotes, and from prokaryotes to nucleated cells. *Philosophical Transactions of the Royal Society of London. Series B, Biological Sciences*, **358** (2003), 59–85 (available online at: http://www.pubs.royalsoc.ac.uk/phi_bio/news/martin.htm); H. J. Morowitz. A theory of biochemical organization, metabolic pathways, and evolution. *Complexity*, **4**(6) (1999), 39–53.

for by material processes. Crucial to the evolutionary worldview, the natural self-organization of prebiotic chemical building blocks to form a living system suggests an alternative to the need for a teleological divine intervention in the process. The arch-and-scaffolding image also negates the idea that the emergence of life was the result of an improbable chance event [10]. It is now unimaginable to assume, as was the case at the earlier stages of origin-of-life research, that a "happy accident," producing a molecule capable of self-replication, was responsible for the emergence of life [37, 38, 39]. Obviously, in distinction to the biochemical events in extant cells, the chemical processes involved in the construction of scaffolding were not directed by genetic information, and in that sense were random [40] (p. 141). However, due to physical and chemical constraints these processes were not statistically random – not all chemical options had the same chance to materialize, hence the development of life took the more probable channels [39 (pp. 22–30), 40 (pp. 213–217)].

In a captivating metaphor, the Princeton biologist Harold Blum compared evolution to a great tapestry. The basic chemical properties of the environment constitute the strong warp into which mutation and natural selection introduce the multicolored woof. Blum, however, indicated that through the restrictions exerted on mutation by the physical nature of the environment, the warp influenced the pattern itself: "Purely physical factors … must have imposed such restrictions from the beginning of life, since that beginning was inextricably tangled up with the physical nature of the then existing earth" [41].

Does the emphasis on physical and chemical continuity mean that the origin-of-life process can be described as "reductionistic"? Among the various versions of "reductionism" contemplated by philosophers of science, the one relevant to our discussion is "theoretical reductionism" – the claim that theoretical concepts and explanations in biology can be reduced, in principle, to those in chemistry [42, 43].

According to the physical chemist Manfred Eigen, the two philosophical options for the explanation of the emergence of genetic information (that he identifies with the emergence of life) are reduction, or the assumption of *vis vitalis* (vital force) [39] (p. 122). At the same time, Eigen does claim that genetic information represents a quality that "far transcends chemistry" (p. 124). In his discussion of evolution, the chemist Harold Morowitz also speaks about "a deep logic of evolution within the reductionistic roots of quantum mechanics, the periodic table, and the boundary conditions on the planet Earth" [44]. The theme of reductionism is obviously connected to the question of the "autonomy of biology," i.e. the relationship between biology on the one hand and physics and chemistry on the other. Traditionally, supporters of the notion of the autonomy of biology adopted a skeptical attitude toward the possibility of solving the origin-of-life problem

[43 (p. 584), 45].[16] The scaffolding approach, embodying the emergence of life as part of the evolutionary process, transcends the erroneously presumed dichotomy between biology and physico-chemistry and between reductionistic and organismic attitudes to life and its emergence. By insisting on both continuity and the emergence of more complex levels of organization, on physical and chemical factors as conditions necessary for the emergence of biological properties non-existent before, the arch-and-scaffolding image is a synthesis of both.

4.5 The emergence of life and the nature of science

A current wave of new creationism, in the USA, and also elsewhere, challenges the reasoning underlying the arch-and-scaffolding image, attacking the very possibility of the natural emergence and evolution of biological organization. Using molecular terminology, the new creationists claim that "irreducible complex systems" [7] could not have emerged and evolved *via* a gradual selective process. Rather, they suggest a supernatural Intelligent Design as the only explanation of the origin of life and important stages in evolution [7] (pp. 187–205) (see also [46]). The Intelligent Design (ID) proponents dismiss the role of prebiotic chemistry and the application of natural selection to the origin of life and insist that nothing but a direct intervention by a purposeful, supernatural agent could have brought about the appearance of life on Earth [47].

Since most scientists identify the empirical procedures of science as its most distinct characteristic, they believe that the conflict with the creationists is basically empirical rather than philosophical. Biochemists and evolutionary biologists therefore respond to the Intelligent-Design challenge by demonstrating theoretically and empirically how highly complex anatomical and biochemical systems could have evolved naturally [48, 49, 50]. This exclusive focus on the empirical aspects of science is especially significant when confronting the ID movement on the origin-of-life question. Since there is still no agreed-upon scientific solution to the problem, it is felt that ID claims in this domain cannot be fully rejected at the moment. Indeed, origin-of-life research is often viewed as a problem in the fight against creationism – as the "soft underbelly of evolutionary biology" [51].

The philosopher of biology, Michael Ruse believes that the status of origin-of-life study differs from that of Darwinism.[17] He thus finds it advisable, facing a

---

[16] The evolutionary biologist Ernst Mayr later changed his opinion, considering the origin of life to be a probable process. See E. Mayr. The search for extraterrestrial intelligence. In *Extraterrestrials: Where Are They?*, 2nd edn., eds. B. Zuckerman and M. H. Hart (Cambridge: Cambridge University Press, 1995), p. 152.

[17] See M. Ruse. Prologue: a philosopher's day in court. In *But is it Science?*, ed. M. Ruse (Amherst, NY: Prometheus Books, 1996), p. 27; but see the change in Ruse's more recent books. He now describes the origin of life as an "extension of Darwinism, for scientists are now starting to think in terms of the ways in which selection might have taken over the process" (M. Ruse. *Can a Darwinian be a Christian?*

believer who asserts the special intervention by God in the creation of life, "to let it be" [52]. The biochemist, Christian de Duve points out that as far as science is concerned a supernatural intervention is unnecessary, "as well as unlikely, in the light of present knowledge." Yet, he feels that since the science of today "cannot prove wrong" those who believe in divine intervention, "it would be a mistake to make this point into an issue" [53].[18] Apparently, according to this view, science could provide facts "to demonstrate an entirely material origin of life" [54]. Thus, the problem with the scientific status of origin-of-life research is its present inability to provide such a demonstration – the problem is with the lack of "proof," or "evidence."

The history of the attempts by philosophers of science to determine the essential characteristics of science and demarcate it from non-science is notoriously controversial. A notable relevant episode took place following a legal battle with the creationists in the early 1980s. In his testimony in this trial, Michael Ruse described "creation science" as a religion and a non-science. Ruse distinguished "creation science" from the theory of evolution, which he characterized as science by its empirical methodology [55]. A heated dispute among philosophers followed in which other philosophers of science pointed out to Ruse that some of the hypotheses of "creation science" (e.g. the young age of Earth) are empirical and testable (which in fact made it possible to refute them). The best way, they suggested, to deal with the creationists is thus piecemeal, by refuting empirically each of their contentions [56, 57].

Interestingly, both parties to this controversy seemed to disregard the philosophical dividing line between science and creationism. The following discussion will aim to point out how this dividing line is associated with the question of the nature of science. The core procedures of science – putting hypotheses to the test and establishing empirical evidence – are based on philosophical presuppositions that provide necessary guidelines and make possible the work of science. Crucial to the philosophical underpinning of science is the presupposition that natural phenomena are to be explained on the basis of natural causes obeying the laws of nature. This presupposition, whose implications lie at the heart of the conflict with creationism, rules out any explanation based on purposeful intervention and is often described as "methodological naturalism." As put by the philosopher of science, Robert Pennock, "lawful regularity is at the very heart of the naturalistic worldview and to say that some power is supernatural is, by definition, to say that it can violate natural laws" [58]. Whether the naturalistic worldview also involves a

---

(Cambridge: Cambridge University Press, 2001), p. 65; M. Ruse. *Darwinism and its Discontents* (Cambridge: Cambridge University Press, 2006), p. 53).

[18] For a similar sentiment, see K.R. Miller. *Finding Darwin's God* (New York: Cliff Street Books, 1999), p. 276.

"metaphysical" or "ontological naturalism," underlying an agnostic or an atheistic point of view, is a controversial issue among philosophers and scientists.[19]

The interaction between the empirical and philosophical aspects of science can be fully appreciated when examined in its historical development. During the last few hundred years, the naturalistic philosophical presupposition guiding research and the empirical results of this research continuously fed into each other. As a result, the more successful empirical science became, the stronger this philosophical presupposition grew. It should be clarified: philosophical ideas are not "caused" by empirical evidence. Similarly, philosophical ideas do not "cause" new theories in science. Nevertheless, both empirical results and philosophical concepts contribute to changes in the intellectual climate, allowing new scientific questions to be raised and making new scientific solutions possible. This dynamic interchange, always taking place within specific cultural and political contexts, has given rise during the last few hundred years to the naturalistic worldview whose implications clearly transcend questions of methodology.

The naturalistic conception, relating to the realm of physical phenomena, was developed as part of the scientific revolution of the sixteenth and seventeenth centuries. For a while, particularly until the middle of the eighteenth century, the new Copernican cosmology promoted a revival of natural theology, especially in England. The discoveries of science were seen as the strongest indication of the existence and properties of God. However, by the end of the eighteenth century, the establishment of cosmology and astronomy based on Newtonian principles replaced the previous explanation of nature through purposeful divine intervention.

The removal of God from the direct explanation of natural phenomena and the rejection of intention and purpose as causes in the working of nature – intertwining elements of the new cosmological worldview – came into their own following the rise in the nineteenth century of the evolutionary conception of nature. The revolution precipitated by Darwin's theory of evolution depended on these previous developments and carried them even further. When the "hand of God" as the cause of the adaptive complexity of organisms was replaced by the material mechanisms of evolutionary biology, the remaining stronghold of natural theology in the realm of biology came down.

Historians differ in their interpretation of the changes that led to the decreasing relevance of an intervening God for the explanation of natural phenomena. They point to diverse processes – political, economical, and cultural – as crucial in the

---

[19] For support of the "methodological" position see M. Ruse. *Can a Darwinian be a Christian?* (Cambridge: Cambridge University Press, 2001), pp. 100–102; R.T. Pennock. *The Tower of Babel* (Cambridge, MA: The MIT Press, 2000), pp. 189–214. Among proponents of the "metaphysical" stance see R. Dawkins. Obscurantism to the rescue. *Quarterly Review of Biology*, **72**(4) (1997), 397–399; D. Dennett. *Darwin's Dangerous Idea* (New York: Simon and Schuster, 1995); S. Weinberg. *Dreams of a Final Theory* (New York: Vintage Books, 1994), pp. 241–261.

development of a new naturalistic worldview.[20] However, the major influence of the growth of science on this development cannot be denied.

The interaction between empirical and theoretical scientific achievements and a philosophical standpoint is evident also in the development of the study of the origin of life. Not until the theory of evolution proved itself strong enough theoretically and empirically could the question of the origin of life be incorporated into the study of evolution. As already noted, in the early twentieth century, when the mechanism of natural selection was held in much doubt, many scientists still considered the origin of life a taboo, better left untouched [2, 59]. This attitude changed toward the middle of the century when the "neo-Darwinian synthesis" established natural selection as the major evolutionary mechanism and rejected all purposive explanations of evolution previously entertained [60]. Devising experiments that showed the power of natural selection at the molecular level allowed scientists to formulate various theoretical prebiotic scenarios. Data gathered by astronomy, geology, chemistry, and biology provided stronger constraints that limited the number of possible theoretical origin scenarios.

Despite the greater knowledge and sophistication in evaluating the problem, it is not solved yet. Researchers are still debating the mechanisms and processes that could have brought about the emergence of the first living systems. No empirical simulation in the laboratory of a possible hypothesis was yet achieved. Following the above brief historical examination of the interdependence between empirical evidence and the naturalistic worldview, the question of the scientific status of origin-of-life study can now be reconsidered. It can be confidently stated that this status does not stand or fall on the merit of empirical evidence alone. Despite the difference in the empirical evidence for evolution and the origin of life, the distinction between unproven-as-yet origin of life and evolution-as-a-fact is not justified. The question whether life emerged naturally is not put to the test afresh in every new experiment. Scientists feel confident about this question on the basis of a robust worldview drawing its strength from the historical record of science.

Indeed, science is not a body of demonstrated truths and scientific activity does rely on belief where we cannot provide proof [61]. Naturalism as a system of beliefs is clearly of a different status than empirical data and like all philosophical claims cannot be confirmed or refuted by facts. There are no facts to "demonstrate an entirely material origin of life [54]", nor facts to refute it. Scientific belief, however, is not detached from empirical facts and can be strengthened or weakened by the accumulation of observations and experiments. In this respect, the confidence of origin-of-life scientists in the eventual solution of the problem is different from a dogmatic belief.

---

[20] See, for example, J. H. Brooke. *Science and Religion: Some Historical Perspectives* (Cambridge: Cambridge University Press, 1991), pp. 372–380.

It is on this crucial interaction between the empirical and philosophical aspects of science that ID promoters focus their attack. Phillip Johnson, the leader and initiator of the ID movement, argues that instead of practicing "empirical science," which should avoid "all philosophical prejudice and follow the evidence wherever it leads," evolutionary biologists in fact practice science "as applied materialist philosophy" [62]. In the same vein, the study of the origin of life is presented as "metaphysical," failing to be based on empirical evidence [62] (see also [63], p. 197).

In their struggle against science, ID promoters contend that unlike the study of evolution, their own "science" is based on "experience" [63] (p. 177). Realizing obviously that there can be no experience of a supernatural designer, ID theorists move instead to discuss human designers, basing their argument on the analogy between human and divine intelligence. However, this is a faulty analogy: the causal connection between human designers and their organized, functional products can be explored, tested, and verified. On the other hand, a supernatural intelligent agent and his "design activity" cannot undergo the same scrutiny. Instead of distinguishing between natural and supernatural causes, the new creationists differentiate between "natural" and "intelligent" causes. Relying on a general, generic sense of the term "cause," they move freely between intelligent human causation open to observation and experimentation and the alleged analogical supernatural intelligent causation [13] (pp. 204–206).

Presenting ID ideas as "scientific" is thus based on changing the concept of science altogether. In order to allow their "science" to detect the "footprints" ("specified complexity," "irreducible complexity," etc.) of a purposeful agent, ID theorists have to insist that intention and supernatural agents should be part and parcel of science (see [64] (especially pp. 197–201) and [65]).

Interestingly, some scientists and philosophers, ignoring the philosophical dimension of science, seem also to disregard the philosophical implications of the new creationism for the nature of science. In a recent volume titled *Why Intelligent Design Fails*, one of the editors contends that "ID is a very ambitious claim and ... it must produce strong evidence before scientists go along with the proposed revolution" [66].[21] Urging ID promoters to "go scientific" by producing testable hypotheses and, eventually, tested evidence is a contradiction in terms: no observation or experiment, no empirical measure whatsoever, can test supernatural intention and purpose supposedly intervening in nature and affecting natural phenomena.[22] The

---

[21] Similarly, the philosopher, Niall Shanks in a critical review of one of Johnson's books says that scientists do not currently support ID claims for a supernatural creation, "but that is because these are hypotheses for which there is not currently any high-quality evidence" (N. Shanks. Creationism, evolution and baloney. *Meta Science*, **9**(1) (2000), 90).

[22] Howard Van Till, emeritus professor of astronomy and physics at Calvin College, points out that an intelligent designer has to perform two tasks: "Mind action has to be followed by hand action ... to be intelligently

supporters of science should realize that the conflict between science and ID is not one between experimental evidence and the lack of it. The conflict is between those who are capable of studying nature and indeed do so and those who cannot and hence do not.

To recapitulate, the study of the origin of life is an instructive case study of the nature of science: the very fact that researchers still struggle to find a specific scenario that could have led to the emergence of life, while not doubting the in-principle natural character of such a scenario, reveals the complex interaction characterizing science between empirical evidence, theory, and the naturalistic worldview.

References

[1] I. Kant. *Critique of Judgment. Part 2. Critique of Teleological Judgement*, trans. W. S. Pluhar (Indianapolis, IN: Hackett Publishing Company, 1987 [1790]).

[2] H. J. Muller. The gene material as the initiator and the organizing basis of life. *American Naturalist*, **100**, no. 915 (1966), 493–517.

[3] C. Darwin. *On the Origin of Species* (New York: Washington Square Press, 1963 [1859]).

[4] F. Darwin. *The Life of Charles Darwin* (London: Senate, 1995), p. 257. (Originally published London: Murray, 1902.)

[5] A. G. Cairns-Smith. *Seven Clues to the Origin of Life* (Cambridge: Cambridge University Press, 1985).

[6] W. Paley. *Natural Theology: or, Evidence of the Existence and Attributes of the Deity, Collected from the Appearances of Nature* (Farnborough, UK: Gregg, 1970 [1802]).

[7] M. J. Behe. *Darwin's Black Box* (New York: Free Press, 1996), pp. 39, 110–111, 230.

[8] C. De Duve. *Vital Dust* (New York: Basic Books, 1995), pp. 112–117.

[9] H. J. Morowitz. *Beginnings of Cellular Life* (New Haven, CT: Yale University Press, 1992), p. 27.

[10] I. Fry. Are the different hypotheses on the origin of life as different as they seem? *Biology & Philosophy*, **10** (1995), 389–393.

[11] A. M. Weiner and N. Meizels. The genomic tag model for the origin of protein synthesis. In *Evolution of Life*, eds. S. Osawa and T. Honjo (Tokyo: Springer-Verlag, 1991), p. 53.

[12] C. De Duve. *Singularities: Landmarks on the Pathways of Life* (Cambridge, New York: Cambridge University Press, 2005), pp. 17–21.

[13] I. Fry. *The Emergence of Life on Earth: A Historical and Scientific Overview* (New Brunswick, NJ: Rutgers University Press, 2000).

[14] N. Lahav. *Biogenesis: Theories of Life's Origin* (New York: Oxford University Press, 1999), pp. 111–113.

[15] P. L. Luisi. About various definitions of life. *Origins of Life and Evolution of the Biosphere*, **28** (1998), 613–622.

[16] R. Dawkins. Universal biology. *Nature*, **360** (1992), 25–26.

designed is … to be *both* conceptualized for a purpose *and* assembled by the action of an extranatural agent" (H.J. Van Till. Does "intelligent design" have a chance? *Zygon*, **34**(4) (1999), 670). Not by chance, as noted by Van Till, proponents of the ID movement fail to say so candidly. They cannot of course specify how the designer intervenes in nature – his "hand action" – and hence they cannot, in principle, suggest any means to test this intervention.

[17] R. Safhill, H. Schneider-Bernloehr, L. E. Orgel, and S. Speigelman. In vitro selection of bacteriophage Qß variants resistant to enthidium bromide. *Journal of Molecular Biology*, **51** (1970), 531.

[18] M. Eigen, W. Gardiner, P. Schuster, and R. Winkler-Oswatitsch. The origin of genetic information. *Scientific American*, **244**(4) (1981), 82–83.

[19] I. Fry. The origins of research into the origins of life. *Endeavour*, **3**(1) (2006), 24–29.

[20] F. Dyson. *Origins of Life* (Cambridge: Cambridge University Press, 1985).

[21] G. F. Joyce and L. E. Orgel. Prospects for understanding the origin of the RNA world. In *The RNA World*, eds. R. F. Gesteland, T. Cech, and J. F. Atkins (Plainview, NY: Cold Spring Harbor Laboratory Press, 1998), pp. 49–77.

[22] L. E. Orgel. The origin of life on the earth. *Scientific American*, October (1994), 60–61.

[23] B. C. F. Chu and L. E. Orgel. Peptide-formation on cysteine-containing peptide scaffolds. *Origins of Life and Evolution of the Biosphere*, **29** (1999), 441–449.

[24] L. E. Orgel. *The Origins of Life* (New York: John Wiley & Sons, 1973).

[25] G. Ertem and P. Ferris. Synthesis of RNA oligomers on heterogeneous templates. *Nature*, **379** (1996), 238–240.

[26] A. Kanavarioti. Template-directed chemistry and the origins of the RNA world. *Origins of Life and Evolution of the Biosphere*, **24** (1994), 489–492.

[27] K. D. James and A. D. Ellington. A search for missing links between self-replicating nucleic acids and the RNA world. *Origins of Life and Evolution of the Biosphere*, **25** (1995), 528.

[28] G. F. Joyce. RNA evolution and the origins of life. *Nature*, **338** (1989), 222.

[29] A. Eschenmoser. Chemistry of potentially prebiological natural products. *Origins of Life and Evolution of the Biosphere*, **24** (1994), 393–394.

[30] A. I. Oparin. *Origin of Life*, trans. S. Morgulis (New York: Dover Publications, 1953 [1936]).

[31] S. W. Fox and K. Dose. *Molecular Evolution and the Origin of Life* (San Francisco, CA: W. H. Freeman, 1972).

[32] S. A. Kauffman. *The Origins of Order: Self-organization and Selection in Evolution* (New York: Oxford University Press, 1993).

[33] G. Wächtershäuser. Groundwork for an evolutionary biochemistry: the iron-sulfur world. *Progress in Biophysics and Molecular Biology*, **58** (1992), 110–111.

[34] B. Shenhav, D. Segré, and D. Lancet. Mesobiotic emergence: molecular and ensemble complexity in early evolution. *Advances in Complex Systems*, **6**(1) (2003), 15–35.

[35] E. Smith and H. J. Morowitz. Universality in intermediary metabolism. *Proceedings of the National Academy of Sciences* **101**(36) (2003), 13168–13173.

[36] R. M. Hazen. *Genesis: The Scientific Quest for Life's Origin* (Washington, DC: Joseph Henry Press, 2005), pp. 95–119.

[37] J. Monod. *Chance and Necessity* (Glasgow: Collins Fontana Books, 1974), p. 137.

[38] F. Crick. *Life Itself* (New York: Simon and Schuster, 1981), pp. 39, 88.

[39] M. Eigen. *Steps Towards Life* (Oxford: Oxford University Press, 1992).

[40] C. De Duve. *Blueprint for a Cell* (Burlington, NC: Neil Patterson Publishers, 1991).

[41] H. F. Blum. *Time's Arrow and Evolution* (New York: Harper & Brothers, 1962 [1951]), pp. 208–209.

[42] F. J. Ayala. Biological reductionism: the problems and some answers. In *Self-Organizing Systems: The Emergence of Order*, ed. F. E. Yates (New York: Plenum, 1987), pp. 315–324.

[43] E. Mayr. *The Growth of Biological Thought* (Cambridge, MA: Harvard University Press, 1982), pp. 59–63.

[44] H. J. Morowitz. A theory of biochemical organization, metabolic pathways, and evolution. *Complexity*, **4**(6) (1999), p. 39.

[45] K. Popper. Reduction and the incompleteness of science. In *Studies in the Philosophy of Biology*, eds. F. Ayala and T. Dobzhansky (Berkeley, CA: University of California Press, 1974), p. 270.

[46] B. C. Thaxton, W. L. Bradley, and R. L. Olsen. *The Mystery of Life's Origin* (New York: Philosophical Library, 1994), pp. 200–217.

[47] P. E. Johnson. *The Wedge of Truth: Splitting the Foundations of Naturalism* (Downers Grove, IL: InterVarsity Press, 2000), pp. 48–62.

[48] A. D. Gishlick. Evolutionary path to irreducible systems: the avian flight apparatus. In *Why Intelligent Design Fails*, eds. M. Young and T. Edis (New Brunswick, NJ: Rutgers University Press, 2004), pp. 58–71.

[49] I. Musgrave. Evolution of the bacterial flagellum. In *Why Intelligent Design Fails*, eds. M. Young and T. Edis (New Brunswick, NJ: Rutgers University Press, 2004), pp. 72–84.

[50] K. R. Miller. *Finding Darwin's God* (New York: Cliff Street Books, 1999), pp. 129–161.

[51] E. C. Scott. Creationism, ideology, and science. *Annals of the New York Academy of Sciences*, **775** (1996), 515–516.

[52] M. Ruse. *Can a Darwinian be a Christian?* (Cambridge: Cambridge University Press, 2001), p. 65.

[53] C. De Duve. *Life Evolving* (New York: Oxford University Press, 2002), p. 288.

[54] G. Easterbrook. Science and God: a warming trend? *Science*, **277** (1997), 893.

[55] M. Ruse. Witness testimony sheet *McLean v. Arkansas*. In *But is it Science?* ed. M. Ruse (Amherst, NY: Prometheus Books, 1996), p. 318.

[56] L. Laudan. Science at the bar – cause for concern. In *But is it Science?* ed. M. Ruse (Amherst, NY: Prometheus Books, 1996), p. 354.

[57] E. Sober. *Philosophy of Biology* (Boulder, CO: Westview, 1993), pp. 46–53.

[58] R. T. Pennock. Naturalism, evidence and creationism: the case of Phillip Johnson. *Biology and Philosophy*, **11** (1996), 552.

[59] I. Fry. On the biological significance of the properties of matter: L. J. Henderson's theory of the fitness of the environment. *Journal of the History of Biology*, **29** (1996), 177–188.

[60] W. B. Provine. Progress in evolution and meaning in life. In *Evolutionary Progress*, ed. M. H. Niteckie (Chicago, IL: Chicago University Press, 1988), pp. 58–62.

[61] P. Kitcher. *Abusing Science* (Cambridge, MA: The MIT Press, 1982), p. 34.

[62] P. E. Johnson. Author's response. *Meta Science*, **9**(1) (2006), 106.

[63] W. L. Bradley and C. B. Thaxton. Information & the origin of life. In *The Creation Hypothesis: Scientific Evidence for an Intelligent Designer*, ed. J. P. Moreland (Downers Grove, IL: InterVarsity Press, 1994).

[64] P. E. Johnson. *Darwin on Trial* (Downers Grove, IL: InterVarsity Press, 1993).

[65] A. W. Dembski. *Intelligent Design* (Downers Gove, IL: InterVarsity Press, 1999), pp. 122–152.

[66] T. Edis. Grand themes, narrow constituency. In *Why Intelligent Design Fails*, eds. M. Young and T. Edis (New Brunswick, NJ: Rutgers University Press, 2004), p. 17.

5

The origin of terrestrial life:
a Christian perspective

Ernan McMullin

When the "origin-of-life" issue is debated nowadays, the question most often asked is: How did the first cell originate? What sequence of processes could have given rise to the extraordinary complexity of even the most rudimentary living cell? Once such a cell was in place, the assumption is that metabolism and descent with modification could begin and the Darwinian selection process could get under way, generating over the course of aeons the vast profusion of natural kinds, living and extinct, that we know. But what sort of selection principles, working on what sort of materials, could have sufficed in the first place to build up the kind of intricate structure that even the smallest functioning cell requires? In short, how did the living come from the non-living in the first place?

In earlier centuries, the transition from non-life to life would have seemed unproblematic, indeed entirely commonplace. It would have appeared obvious that living comes from non-living in the world of nature all the time: maggots develop in decaying flesh, tiny worms appear in rotting fruit, and so on. At the lowest levels of living complexity, matter (it seemed) could generate life unaided. Only with the advent of the microscope and finally the experiments of Pasteur was it shown that this kind of spontaneous generation was only apparent: no real transition from non-living to living actually occurred in it. So how *did* life begin? With the discovery in recent decades of just how biochemically complex the cell is, the transition has become even more of a puzzle. As other essays in this collection will show, important advances have been made and hints of possible paths have appeared. But it would seem fair to say that an even moderately plausible reconstruction in scientific terms of how the first functioning cells came to be is still lacking.

5.1 A Christian perspective?

One recent response to this situation goes under the title of "Intelligent Design" (ID). Reminiscent of the natural theology of the seventeenth and eighteenth centuries,

Exploring the Origin, Extent, and Future of Life: Philosophical, Ethical, and Theological Perspectives, ed. Constance M. Bertka. Published by Cambridge University Press. © Cambridge University Press 2009.

it postulates the "special" action on the part of a creating intelligence as the most plausible, some would argue the only plausible, explanation of how the first cells were formed. More broadly, it points to various features of organisms that would likewise, at a later evolutionary stage, have required the distinctive action of a shaping intelligence to bring their first appearance about. The "irreducible complexity" of such features as the mammalian blood-clotting mechanism, so it is argued, would exclude a gradualist account of their first coming-to-be [1]. Unlike the "creation-science" of an earlier generation of Christian critics of Darwin, the proponents of ID tend to limit themselves to what they regard as the key transitions in the evolutionary story, among which the origin of the first terrestrial living things holds pride of place.

The ID argument can take two rather different logical forms, as already implied. One, which might be called "strong" ID, maintains that these transitions cannot, as a matter of principle, be explained in the gradualist terms of neo-Darwinism or of conventional natural science generally. The other, "weak" ID, emphasizes the current lack of satisfactory accounts in evolutionary terms of how those transitions occurred and concludes that an explanation in terms of a Designer's action outside the ordinary course of Nature is, at present, at least, clearly the preferable explanation, notably for those who already accept the existence of a Creator on other grounds. (For others, the epistemic situation would, of course, be different, though proponents of the weaker form of ID argument would argue that it gives a motive for belief in such a Creator, the strength of which would depend on how far from satisfactory formulation the current scientific attempts to explain the designated transitions are.)

The title of this chapter, "The origin of terrestrial life: a Christian perspective," might lead the reader to expect that the ID position, whether in its stronger or weaker form, would be the likely choice for such a perspective. Strong ID is, however, open to obvious challenge: it would require one to know somehow or other in advance what the limits are of the intrinsic transformative capacities of matter. Claims that conventional science can *never*, in principle, explain the first appearance of some feature of the living world, including that of life itself, seem presumptuous in the light of the achievements of the sciences of recent centuries. If, on the other hand, the argument were to depend primarily on the fact that scientists have encountered far more checks than anticipated in their search for a "natural" origin of life and at the moment might be thought to be at a loss (weak ID), an obvious response would be that the resources of the scientific imagination, backed with careful experimental trial, are far from exhausted and pointers are beginning to appear. After all, only half a century has elapsed since serious research on this issue first got under way.

I will, however, leave aside any further assessment of the merits and demerits of the ID approach; that has been amply done elsewhere. I prefer to focus here in quite another direction, leaving it to the reader to judge whether or not it offers a more plausible expectation, from the Christian point of view, of the sort of explanation that should, even in advance, be expected of how life made its first appearance on Earth. What gives this alternative a special plausibility for the Christian is that it is rooted in the earliest ages of Christian theology, long before Darwin was heard of. It cannot, then, be dismissed as a merely defensive ploy to cover the retreat of Christian belief in the face of triumphant scientific advance. Though its relevance for our purposes is to the origin of terrestrial life, its broader relevance is to the evolutionary account generally.

5.2 Two sources

The theologians of the early Christian church relied on two major sources as they pondered the question of cosmic origins. The first of those was, of course, Genesis, the book that serves as solemn introduction to the Bible. But there was an immediate problem. The first two chapters of Genesis set forth two rather different accounts of origins. The first chapter describes the creation as spread out over six days. First come the heavens and the Earth, then in succession the creation of light, of the firmament dividing the waters above from those beneath, of vegetation and trees, of the Sun and Moon, of fish and birds, of land animals, and finally of the first human beings, male and female (Genesis 1 to 2:4). The second account tells the story rather differently: the Earth is already in place, but barren and empty. After a flood rises from the Earth and waters it, God fashions a man from the dust of the Earth and plants a garden containing every sort of tree; then God adds animals and birds, drawing them "from the earth," and finally makes a woman to be man's helpmate (Genesis 2:5 to 2:25).

The sequence of events does not appear to be the same in the two accounts nor is the focus the same. The first emphasizes God as the author of all that is; the second concerns itself with the creation of man and woman and with their relation to the rest of creation. Since Genesis was assumed to have had Moses as single author, attentive readers would ask themselves: What could Moses have had in mind by telling the story of creation in two installments so different from one another? The first chapters of Genesis were the source of choice for Christian theologians of the early and the medieval church who concerned themselves with the elucidation of cosmic origins. And this would leave them with a question: How are the first chapters of Genesis to be made consistent with one another?

The second resource to which these theologians could turn was the works of the philosophers, Greek and Roman, the secular learning of that day. In this case, the appropriation was cautious and usually only implicit. These were, after all, pagan sources as far as the theologians were concerned; the philosophers lacked the sort of privileged access to truth claimed for the Bible. Nevertheless, they had to be reckoned with and might be availed of when they could help to illuminate some theological issue or, as in this case, a puzzling biblical text.

Only three of the earlier philosophical cosmogonies seemed to have the potential to be of service to the theologians. The first was that of Plato, the most influential of the earlier Greek philosophers in those formative years of Christian theology. His *Timaeus* had been expressly conceived as a cosmogony, mythic in form but intended as a "likely story." Plato begins from the evidences of the work of intelligence that he sees everywhere in the physical world. How are they to be accounted for? One way, the one to which he devotes most attention, is to postulate a "Demiourgos" or Craftsman, a God who through Reason brought order from an original disorder of matter in chaotic motion, a matter not itself of the Demiurge's making. Matter, the realm of necessity, is resistant; the order imposed by Reason is never perfect, never permanent. In this cosmogony, there is a Maker and a beginning.

Very different was the Stoic tradition, which was destined to take a firm hold in the Roman world of the first two centuries of the Christian era. For the Stoic, physics and ethics were inseparable: the order discovered in the one becomes the ideal to be achieved in the other. God is the power that makes all things be what they are and holds them together in harmony. God and the universe are one: all change is immanent in God and ordered by God's *logos spermatikos* (seedlike word). The operation of this *logos* is sometimes described as a creative fire, sometimes as a tension, depending on whether its capacity to make things new or its capacity to maintain order is in question. The eclectic cosmogony of Plotinus, which gained in popularity in the Roman world as Stoicism declined, drew on a variety of sources, on Platonism mainly but also on other Greek philosophies as well as on Oriental religious thought. Plotinus represents creation as an emanation from the One. The One first gives rise to *Nous* (Intellect) from which in turn comes Soul. Then Soul continues this downward process of emanation, making use of the productive power of the ideas present within it to form "seedlike principles" (*logoi spermatikoi* in the original Greek, *rationes seminales* in Latin translation). Somewhat after the fashion of Aristotle's substantial forms, these combine with matter, the indeterminate, to form the familiar bodies of the sensible world. They are, as it were, the last flicker of the One, removed as they are from the source of good and intelligibility from which the timeless process of emanation begins.

In the monistic system of Stoicism, there is no qualitative difference between God and the universe: the *logos spermatikos* is, effectively, God. And the physical universe is the embodiment of the order produced by that *logos*, the model indeed for the sort of order to which the ethical life should aspire. Whereas for the neo-Platonism of Plotinus, the One and matter, the indeterminate, are as far removed from one another as conceivable: indeed, it is unclear how matter, so understood, can have a place in a fully emanationist system. Here the *logoi spermatikoi* are plural and the world they inform is touched with evil by the mere fact of being in part material.

When the theologians of that day looked for aid in understanding the sometimes obscure or laconic text of the Scriptures when it touched on matters that were also of broader philosophic concern, it was to resources such as these that they turned, just as their modern counterparts often draw on the philosophies and sciences of our own time. Preeminent among those theologians was Augustine, bishop of Hippo in North Africa, whose work was to have such a profound influence on the Christian theology of later ages.

5.3 Augustine on creation

As a young man, Augustine was attracted to the teachings of Mani, then widely disseminated in the Mediterranean world. Born in Babylon around AD 216, Mani created an eclectic religion with elements drawn from practically every religious source of the day: from Zoroastrianism, Judaism, Buddhism, and Christianity. His teaching featured an uncompromising dualism between polar opposites, portrayed as warfare between Light and Darkness, Good and Evil. The dramatic cosmogony he proposed described a primal struggle between two forces that existed independently of one another. So strongly dualist a cosmogony posed a challenge for the cosmogony of Genesis, with its single all-powerful Creator of all that is and its repetition of "and God saw that it was good" at each stage of the creation. The Manicheans, not surprisingly, subjected to unsparing attack the six-day account of the creation, so different from their own version of cosmic origins, treating it as childish and incoherent, unable to account for the pervasive presence of evil in the world.

When Augustine returned to the Christian faith, he set his impressive talents to the task of developing and deepening the theological doctrines that were coming to define the still-new Christian church [2] (pp. 11–16). It was not surprising that meeting the Manichean challenge to Genesis would rank high in his priorities. Two early commentaries on Genesis, the second unfinished, left him dissatisfied. Finally, in AD 401, he began the composition of what would become one of his major works, the *De Genesi ad litteram*, which would eventually run to twelve books and would occupy him on and off for fourteen years.

The work was intended to be a "literal" commentary, not in the modern sense of that term – it frequently treats as "literal" what we today would describe as metaphor – but, rather, in contrast to "allegorical" (i.e. regarding the Old Testament text as foreshadowing later Christian themes). "Literal" for Augustine in practice meant something like: "in the primary sense intended by the author." A decision that he had to make right away was how to reconcile the striking differences between the two accounts of creation in Genesis 1 and 2. His solution would seem, to the modern reader at least, quite a radical one. He was evidently deploying the resources of the Stoic and neo-Platonic philosophies of the day to arrive at a more coherent, if unexpected, reading of the six-day account in chapter 1, one that departed almost entirely from the literal in our sense of that term but one that (more or less) enabled the two chapters to constitute a single story.

He proposes first that the "days" of which the first account of the creation speaks cannot be understood in the everyday sense of that term "day" here:

It is not to be taken in the sense of our day, which we reckon by the course of the sun; but it must have another meaning, applicable to the three days mentioned before the creation of the heavenly bodies. This special meaning of "day" must not be maintained just for the first three days, with the understanding that after the third day [i.e. when the Sun makes its first appearance] we take the word "day" in its ordinary sense. But we must keep the same meaning even to the sixth and seventh days [3] (par. 43).

The creation was not, from the human perspective, a series of discrete acts on God's part spread out over six days or even six indefinitely long periods of time. Augustine lays much weight on a single text here: "He made all things together" (Sirach 18:1).[1] The "days" of creation reduce to the single moment of cosmic initiation when all the natural kinds are already somehow prefigured. The six-fold distinction in Genesis 1 must, then, refer to something other than time. He could only speculate as to what this might be.[2] This idea of a creation in which "all things" are somehow already present from the beginning owed something to neo-Platonism for its philosophic inspiration. It had already been broached by Basil and by Gregory of Nyssa, a Greek theologian of the Alexandrine school. Here is how Gregory puts it:

The sources, causes, and potencies of all things were collectively sent forth in an instant, and in this first impulse of the Divine Will, the essences of all things assembled together: heaven, aether, star, fire, air, sea, earth, animal, plant – all beheld by the eye of God. … There followed a certain necessary series according to a certain order … as nature, the maker, required … appearing not by chance … but because the necessary arrangement of nature required succession in the things coming into being [4].

Augustine developed this notion much further than his predecessors had done. From God's timeless perspective, he maintained, there is but a single act of creation: "one

---

[1] Ironically, the crucial "together" ("*simul*") in this passage now appears to have been a mistranslation of the original Hebrew in the Old Latin translation Augustine was using.

[2] He suggests that it could be two successive stages in the angelic knowledge of the "ordered arrangement according to causal connection" [3], LMG, Book IV, par. 33) implicit in the creation of the material universe, but he emphasizes how tentative that suggestion must remain.

can speak of 'before' and 'after' in the relationship of creatures, although all is simultaneous in the creative act of God" [3] (par. 56). But from *our* perspective, the perspective of the timebound creature, God's creative act has two "moments," two phases:

one in the original creation when God made all creatures before resting from all His works on the seventh day, and the other in the administration of creatures by which He works even now. In the first instance God made all things together without any moments of time intervening, but now He works within the course of time, by which we see the stars move from their rising to their setting … [5] (par. 27).

He stresses the distinctness of the two phases from *our* point of view:

We should not think of [the first creation of things] as if it were the same as His working now in time. Rather, He made that which gave time its beginning as He made all things together, disposing them in an order based not on intervals of time but on causal connections … [5] (par. 12).

Later theologians would mark the distinction by means of different names, "creation" and "conservation," even though, to repeat, both are understood as aspects of the single creative act from the Creator's perspective. (The unavoidable duality of perspective is a major source of ambiguity in all this.) Augustine is careful to stress that the Creator's involvement does not end with the originating bringing-to-be: "For the power and might of the Creator … makes every creature abide; and if this power ever ceased to govern creatures, their essences would pass away and all nature would perish" [3] (par. 22).

5.4 Augustine's seedlike principles

But now, in what sense can one say that all things *were* created simultaneously in the first moment of the universe's existence? Recall that this is what Augustine requires in order to have an instantaneous act cover all the multiple contents of the universe described in Genesis 1. It is at this point that he makes his most distinctive contribution, appropriating from the Stoics and neo-Platonists the notion of seedlike principles (*rationes seminales*).[3] But these principles function rather differently for him than did those of his predecessors. The Creator, he proposes, implanted in the matter or "earth" that came to be in the first moment of creation, the seedlike principles of each kind that would later make its appearance, each in its own time, when the conditions would be favorable.

---

[3] It is difficult to find an adequate translation here for "*ratio*" or "*logos*." Since the *rationes* are realized in the primal matter somewhat after the fashion of Aristotelian forms, the literal translation, "reason," seems hardly appropriate. The term "principle" adopted here is intended to convey the notion of a really existing causal factor.

The Greek theologians had already noted the significant role played by earth in all the makings described in Genesis 2. In contrast to the summoning into existence by an act of the Divine will described in the first chapter, the "makings" mentioned in the second chapter (excepting that of woman) are said to be from the earth, an earth that had just been watered and made fruitful. Augustine draws attention to the essential role attributed here to earth and water in the first appearance of each natural kind at its appropriate moment. His proposal is that potencies, causal possibilities, must have been implanted in the matter of the first creation, potencies that would in due time lead to the later appearance of each kind, here more exactly described as a fashioning from materials already at hand, unlike the coming-to-be from nothing prior, characteristic of the first appearance of a source of those materials.

In the Old Latin translation of the text that Augustine used, the second account of creation begins: "This is the book of the creation of heaven and earth. When day was made, God made heaven and earth and every green thing of the field before it appeared above the earth." [4] Augustine gives an elaborate construal of this passage, noting different possible ways to read the text. The phrase that particularly caught his attention was the reference to plants that existed within the earth before making their appearance above ground. It provided him with a ready analogy on which he could draw to support his more general thesis about the role of potential existence within the earth prior to appearance in mature form when conditions were right.

How literally were analogies of this kind to be taken here? The seedlike principles are clearly not seeds in any ordinary sense. Augustine readily admits that they are difficult for us to imagine since they are of their nature hidden from view. Yet, as he reminds the reader, this is no different from the principle or cause that brings about our own growing old. Such principles lie within us; though they are inaccessible to our senses, "by another kind of knowledge we conclude that there is in nature some hidden force by which latent forms are brought into view" [6] (par. 27). Analogies can help: "There is, indeed, in seeds some likeness to what I am describing because of the future developments stored up in them. Indeed, it is the seedlike principle that is the more basic of the two, since it comes before the familiar seeds we know" [6] (par. 11). He goes further:

In the seed, then, there was invisibly present all that would develop in time into a tree. And in the same way we must picture the world, when God made all things together, as having had all things that were made in it and with it when day was made. This includes not only heaven with sun, moon and stars ... but also the beings which earth produced in potency and in their causes before they came forth in the course of time [5] (par. 45).

[4] Later translators have agreed in substituting for "before it appeared" something like "there was as yet no vegetation upon the earth."

The force of this analogy lies in the notion of potentiality: the original creation contained within it the potentialities for all the living kinds that would later appear. The seedlike principles were conceptually distinct from one another, but Augustine did not think of them as distinct physical bodies that lay somehow embedded, after the fashion of ordinary seeds, within the primal matter. They were real, they were physical, but they did not have to occupy a specific location, as a seed would [7].[5] To assert that the seedlike principle for a particular natural kind lay within the earth meant no more (and no less) than that the earth had conferred upon it what it would take for that natural kind to develop eventually within it in a natural way: "All things were created by God in the beginning in a kind of blending of the elements, but they could not develop and appear until the circumstances were favorable" [8].

It sounds as though the seedlike principle was enough of itself, once the environment was right, to produce the new kind in a natural way, that is, by virtue of the "causal connections," as Augustine calls them, that the Creator implanted from the beginning within matter generally. Is this, then, what is meant by the "making" attributed to the Creator in the second story of creation? Granted that no new act of creation, of radical bringing-to-be, is needed, is no further, lesser, supplementation of the causal capacities of matter on the part of the Creator required?

At times, Augustine hesitates to say so, in regard to some comings-to-be, at least. To assert this would raise a troublesome issue: does the new kind of animal come to be as infant or as adult? If infant, how does it survive? If as adult, the ordinary laws of nature would not suffice to bring it about, and a further miraculous intervention on God's part would be needed. Augustine leaves the matter open:

In either case, whichever way God made [Adam's body] He did what was in accordance with His almighty power and wisdom. God has established in the temporal order fixed laws governing the production of kinds of beings and qualities of beings and bringing them forth from a hidden state in full view, but His will is supreme overall. By His power He has given numbers to his creation, but He has not bound His power by these numbers [6] (par. 13; p. 194).

The "numbers" referred to here are the laws of nature. Augustine often drew upon a sentence from Wisdom (11:21): "Thou hast ordered all things according to measure, number, and weight," to describe the kind of order that the Creator imparted to the activity of the physical world. In these chapters of the *De Genesi* he goes out of his way, as he frequently does elsewhere in his theological works, to convey that God is not bound by this order. He can depart from it by way of miracle, and this too is

---

[5] McKeough cites numerous Catholic writers, beginning with Edouard Thamiry (*De rationibus seminalibus*, Paris, 1905) who debated the meaning and significance of the doctrine of seminal principles in the early part of the twentieth century.

"natural" in the broader sense that openness to miracle is a basic trait of the natures that the Creator brought to be.[6] But Augustine makes it clear that miracle would not ordinarily be involved for the seedlike principle to give rise to the appropriate natural kind when "the conditions are right." What the Creator's "making" would amount to in these cases would be the regular conservation in being afforded to all creatures and their actions, ordered as they are by the "numbers" by which they have been constituted. Still, it should be kept in mind that Augustine *does* allow that this may not be sufficient when something apparently impossible is in question, like the production directly from the earth of the adult animal body by means of the multiple causalities conferred on matter in the beginning [2] (pp. 13–15).

It is possible now to see how the postulation of seedlike principles that enables one to speak without contradiction of an initial simultaneous "creation of all things" goes some way towards resolving the troublesome tension between the Genesis chapters. Chapter 1 can be taken to describe the instantaneous bringing into existence of the primal matter, containing within it the causal resources for all the kinds that would later appear and therefore in a real sense already present within it. Chapter 2 follows the subsequent history of the working out in time of these causal possibilities and the actual first appearances "from the earth" of plants, wild beasts, and birds, a "making" that may or may not require a supplementation of the natural potentialities of matter on the Creator's part. There are loose ends, of course, as Augustine was the first to admit, notably in regard to the appearance of the animal, and especially the human, body. If the seminal principles were not sufficient of themselves to bring that appearance about, so that some sort of supplementary action on God's part was required, a "making" in a stronger sense then, could one properly hold that the matter of the initial creation already in some sense contained "all things"? Augustine tries to meet this objection by allowing a second extended sense of "seedlike principle," but it is at best a strained response. In what follows, the term will be used in Augustine's primary sense to designate a feature of the regular causal order of the physical world, capable of bringing about the appearance of a new physical kind when the conditions are propitious.

5.5 Genesis and *On the Origin of Species*

To trace the fortunes of Augustine's account of cosmic origins through the centuries that followed would lead far afield. A few brief remarks will have to suffice. Medieval theologians, with whom Augustine carried more authority than any source other

---

[6] The term, "natural," is used in this essay, however, in the more usual sense where God's "special" action is not involved.

than the Bible, were aware of his foray in cosmogony. But for those who sought a story of origins that all could understand, Genesis 1, read literally, carried undeniable attraction. Hexaemeral ("Six Day") literature flourished, embellishing the biblical story in imaginative ways and making it the introduction to a general account of the natural world for the Christians of that time.

For those among them who took Aristotle as their guide, there were serious reasons to doubt the Augustinian version of origins. Aquinas, who revered Augustine, did his best to present the two alternatives, literal-biblical and Augustinian, as fairly as possible [2] (pp. 16–21).[7] But for an Aristotelian, the Augustinian proposal would have held little attraction. The notion of living kinds making their first appearance in a quasi-natural way, impelled by causal powers present since the first creation, ran contrary to the fundamental principles of Aristotle's physics. No such powers had ever been discovered actually at work. Easier for a Christian Aristotelian to call on a six-day succession of miracles in the first place: the very notion of a cosmic beginning itself had to lie outside the bounds of nature as Aristotle understood the natural world.

The advent centuries later of the Protestant Reformation led to a heightened emphasis on literal interpretation of the Bible among Protestants and Catholics alike. Genesis was a favored source in cosmology. Tensions between it and Aristotle's physics were much debated. An interpretation that departed as far from the literal as Augustine's reading of Genesis 1 was not likely to win much support. Jumping far ahead once again, this was to change, to some modest extent at least, in the last decades of the nineteenth century. After centuries in the shade, Augustine's vision of cosmogony would once more gain a modest degree of notice.

The reason, of course, was simple. In the wake of Darwin's *On the Origin of Species*, Augustine's account offered an authentic Christian alternative to the literalist reading of Genesis that Darwin's work had so dramatically, and so effectively, called into question. It demonstrated that the greatest theologian of the early Christian church did not regard the literalist reading of the six-day story of creation as plausible, let alone obligatory. Furthermore, and even more important, the alternative reading he suggested opened the way to a gradualist account of cosmic origins. Calling on Augustine for support seems to have been particularly appealing to a small number of Catholic writers[8] who were sympathetic to evolutionary ideas as well as to patristic precedent and who were wrestling with the

---

[7] The relevant passages of the *Summa Theologica* are to be found in Part I, questions 70–74.

[8] It is quite striking that the Augustinian notion of seedlike reasons was rarely, if ever, cited among the far more numerous Protestant writers who, in this early period, defended a broadly Darwinian, gradualist, account of the origin of living things. See, for example, J.H. Roberts. *Darwinism and the Divine in America: Protestant Intellectuals and Organic Evolution 1859–1900* (Notre Dame, IN: University of Notre Dame Press, 2001); D.N. Livingstone. *Darwin's Forgotten Defenders: The Encounter between Evangelical Theology and Evolutionary Thought* (Grand Rapids, MI: Eerdmans, 1987).

prevailing literalism of the approach to Genesis among Catholic scholars. These authors clearly expected the appeal to the Fathers of the Church to carry particular weight.

As early as 1871, a distinguished English biologist, St. George Mivart, a defender of evolution (though departing from Darwin's account in significant ways), pointed to the relevance of Augustine's treatment of the origin of natural kinds as opening the way, theologically speaking, for an evolutionary approach [9]. In the US, John Zahm at the University of Notre Dame, a leading figure in the contemporary debates about the relations of science and religion, likewise underlined the Augustinian precedent as a theological warrant for those who would see evolution as God's way of bringing the natural kinds to be [10]. In 1909, a noted geologist, Henri de Dorlodot, representing the Catholic University of Louvain, offered a tribute at a celebration in Cambridge of the fiftieth anniversary of the publication of *On the Origin of Species*, and went on to develop over some years a theological defense of Darwin that eventually appeared in book form in 1921. He analyzed a series of patristic texts in some detail, focusing on Gregory of Nyssa and Augustine and concluding that "the authority of the Fathers would lead us to accept the theory of absolute natural evolution – a theory which was formally professed, as we have shown, by St. Gregory of Nyssa and St. Augustine" [11]. And again: [St. Augustine held] "that the theory of the absolute natural evolution of living beings is a necessary deduction from the infallible teaching of Holy Writ" [11] (pp. 87, 149). De Dorlodot's English translator, Ernest Messenger, went on later to publish a work of his own carrying a similar, though more cautiously worded, message, which had a very wide circulation in Britain and the USA [12].

The initiatives on the part of these writers, arguing for the Augustinian credentials of the notion of organic and perhaps pre-organic evolution, met with opposition, occasionally strenuous opposition on the part of Church authorities, notably in Zahm's case. Part of the problem was the ambiguity in what was being claimed. When these writers claimed Augustine's sanction for "natural evolution," did this mean *Darwinian* evolution? Specifically, did it extend to "transformism," as the view that species evolved by the transformation of earlier species was being called? Surely, the critics argued, Augustine could not have had in mind this specifically Darwinian innovation, involving natural selection as its agent. Augustine would have assumed that each seedlike principle was linked one-to-one with a specific natural kind. It would hardly have occurred to him that one kind might over the course of time be transformed into another. With his background in Greek philosophy, he would assuredly never have countenanced this and could call on both Plato and Aristotle for assurance in that regard. So it would certainly not be correct to suggest that Augustine's theology of seedlike principles supports Darwin's *specific* version of how organic evolution might come about.

Though this is correct, a weaker claim is still significant. Part of the problem here is the ambiguity of the term "evolution." Since Darwin's time, the term has meant only one thing for most people: evolution is what Darwin proposed and argued for in *On the Origin of Species*. When the writers mentioned above recalled Augustine's *De Genesi* as a theological precedent for "natural evolution," what their readers tended to take from this, even when Darwin was not specifically mentioned, was that Augustine had already proposed a quasi-Darwinian account of the origins of living things. This Augustine did not do, as we have seen. But what the authors meant (or at least *should* have meant!) was something much broader but still significant: that Augustine defended a gradualist account of origins, where causal resources were already implanted in the original creation that would be sufficient, under favorable environmental circumstances, to allow all of the profusion of plant and animal life to make its appearance later. How this was to be achieved, how the seedlike principles would operate, Augustine quite frankly admitted that he did not know. Darwin has filled in that picture in ways that Augustine would have marveled at.

But would he have welcomed it? Some might argue that his seed analogy is much too teleological in implication to be even compatible with the Darwinian picture. The seed, after all, must take a predetermined path as it develops into the mature organism. There is a preset goal that will be reached if the conditions are favorable. The zigzag path followed by natural selection could hardly be more different.

Here, once again, a reminder is needed that the seed analogy is not supposed to be taken literally, as Augustine himself so strongly emphasized. It is no more than a convenient comparison that works in some respects but not in others. As with any analogy, one must ask what it was intended to bring out. The point Augustine wanted to convey was clearly retrospective, not prospective, in intention. That is, his point was that the causal resources, whatever they might have been, were there right from the beginning to explain the later appearance of the different natural kinds. That was sufficient to enable him to hold that those kinds had themselves, in a limited but real sense, been present (i.e. in potency) from the first moment of cosmic beginning.

This is something that Darwinians could agree with and, indeed, in some contexts, would themselves want to emphasize. Taking the analogy in a prospective sense was not what Augustine had in mind, however, as though he maintained that one could predict from a knowledge of the seedlike principles what the course of later development would be:

The principle which makes this development possible is hidden to the eyes but not to the mind. But whether such a development must necessarily come about is completely unknown to us. We know that the principle that makes it possible is in the very nature of [the] body. But there is no clear evidence in that body that there is a principle by which it must necessarily take place [6] (par. 16).

Concluding, then, there was a firm precedent in the early Christian tradition for the view that the six-day Genesis account of cosmic origins was not to be taken literally and that the most satisfactory reading of it – satisfactory from both the theological and the philosophical standpoints – was to suppose that the Creator implanted in the original cosmic material the wherewithal for the diversity of living things that over the ages would later make its appearance.[9] So that sanctioning evolution, broadly understood as gradual development, in this way cannot be dismissed as no more than a defense measure introduced post-Darwin to fend off criticism inspired by *On the Origin of Species*. It has an honorable ancient lineage.

5.6 The origin of terrestrial life

It is not difficult to see what the implication of the Augustinian vision is for the multitude of researchers engaged in solving the puzzle of how the first cells might have come to be. It would be an emphatic: keep at it! The seedlike principles were almost surely there from the very beginning, even though it may not be within our reach to establish the specific causal sequence to which they gave rise in the earth and water of long, long ago. The philosophical/scientific grounds for this confidence would not be the same today as they were for Augustine. But the theological grounds would be in important respects the same.

What has, however, encouraged some to take this occasion to promote the special-creation alternative once again, what has, in consequence made the proper theological response to the origin-of-life issue once more a matter of debate, has been the problem of envisioning in scientific terms what the path might conceivably have been from the play of non-living molecules to the tight organization of even the simplest cell. This is what led Mivart and de Dorlodot, champions otherwise of Augustine's seedlike principles, to deny in the case of the former, and to doubt in the case of the latter, that such a path could exist. For a time after Miller's discovery in 1952 of amino-acids in the residue of a simple mixing of inorganic molecules supposedly mimicking the constitution of the primitive atmosphere of Earth, it seemed to many that a giant step had been made along that path. But that enthusiasm gradually evaporated as the biochemical complexity of even the most basic cell became more and more apparent. Great strides have been made in recent years, but significant obstacles remain. ...

An alternative, of course, would be to suppose that the Creator supplemented the normal operations of nature in this case to initiate a new stage in cosmic development. For those, however, who do not believe in the existence of a Creator, there

---

[9] Howard van Till develops what he calls the "functional integrity" of the creation along these same lines in Basil, Augustine, and the doctrine of the Creation's functional integrity. *Science and Christian Belief*, **8** (1996), 21–38.

obviously *has* to have been a causal path directly from the non-living to the living, no matter how far from being understood in terms of currently known causal possibilities this may be. For them there is no other option. Less combatively, others would insist, on methodological grounds, that the naturalist option should continue to be energetically pursued. And here is where debate can arise, as indeed it has: one side maintaining that the difficulty in conceiving a "natural" path in this case implies the need for a designer's intervention and thus may even become an independent argument for the existence of such a designer, the other side vigorously questioning the force of this appeal to a designer, on either metaphysical or methodological grounds.

In the context of this contemporary debate, Augustine's doctrine of seedlike principles takes on a real relevance. It might have seemed in advance somewhat strained to hark back to an intellectual context so remote from ours to illuminate a theological option that believers in a Creator, be they Christian, Jewish, Moslem, or other, might want to call on when faced with this dispute about the origin of life. Despite the evident difference of context, however, the theological principles that underlay Augustine's doctrine can be disentangled from what is no longer relevant to them and put to good use once again today.

These principles might fairly be summarized as follows. From the Augustinian perspective, it seems unlikely that there would be *no* possible way in which a transcendent Creator could impart to the original created world the capacities that would have made possible the natural transition from non-life to life on Earth. From this it would plausibly follow, first, that the Creator would have made those seedlike principles part of the original cosmic endowment. And second, it would also follow that it is appropriate for the scientist to proceed on the expectation that life would have come to be in a manner, in principle at least, accessible to human science. To your average scientist these might seem pretty minimal conclusions! But in the light of recent controversy about the appropriate Christian approach to the issue of the origin of the first life on Earth, it is not a small matter if one can show how they fit quite comfortably into the first great theology of creation.

References

[1] M. Behe. *Darwin's Black Box: The Biochemical Challenge to Evolution* (New York: Free Press, 1996), chapter 4.
[2] E. McMullin. Introduction: evolution and creation. In *Evolution and Creation*, ed. E. McMullin (Notre Dame, IN: University of Notre Dame Press, 1985), pp. 1–56.
[3] Augustine. *The Literal Meaning of Genesis*, 2 vols., trans. J. H. Taylor (New York: Newman, 1982), Book IV.
[4] Gregory of Nyssa. Apologetic Treatise on the Hexaemeron. In *Patrologia Graeca*, ed. J. P. Migne, 44, col. 72, trans. E. C. Messenger in his *Evolution and Theology* (New York: Macmillan, 1932), p. 24; translation slightly modified.

[5] Augustine. *The Literal Meaning of Genesis*, 2 vols., trans. J. H. Taylor (New York: Newman, 1982), Book V.

[6] Augustine. *The Literal Meaning of Genesis*, 2 vols., trans. J. H. Taylor (New York: Newman, 1982), Book VI.

[7] M. J. McKeough. *The Meaning of Rationes Seminales in St. Augustine* (Washington, DC: Catholic University of America, 1926), pp. 33–35.

[8] Augustine. *De Trinitate*, Book III, par. 9.

[9] St. George Mivart. *The Genesis of Species* (New York: Appleton, 1871).

[10] J. Zahm. *Evolution and Dogma* (Chicago, IL: McBride, 1896).

[11] H. de Dorlodot, *Le Darwinisme au point de vue de l'orthodoxie catholique* (Brussels: Lovanium, 1921), trans. E. C. Messenger as *Darwinism and Catholic Thought*, vol. 1, *The Origin of Species* (New York: Benziger, 1922).

[12] E. C. Messenger. *Evolution and Theology: The Problem of Man's Origins* (New York: Macmillan, 1932).

Note added in proof: Besides his discussion in the *Summa Theologica* of Augustine's interpretation of *Genesis* on cosmic creation, Aquinas also discusses this in an earlier work, his *Commentary on the Sentences* (II, 12, 1, 2), where he describes this interpretation as "the more reasonable one." See Norman Kretzmann. *The Metaphysics of Theism: Natural Theology in the Summa Contra Gentiles* (Oxford: Oxford University Press, 1997) pp. 190–193.

6

The alpha and the omega: reflections on the origin and future of life from the perspective of Christian theology and ethics

Celia Deane-Drummond

Astrobiology, encompassing the search for life on other planets, laboratory studies of the origin of early life forms from precursor materials, and prospects for the discovery of microbial life on other planets reflects outcomes at the cutting edge of science and technology. Yet the issues that such investigations raise are profound, for not only do they bear on our own sense of self in relation to the cosmos, but they also raise deep philosophical and religious issues concerned with purpose, meaning, and human identity. Given these profound challenges, the ethical and moral frameworks within which such developments take place need to be carefully considered, for outcomes of such deliberations have public and social importance alongside a potential scientific gain. The intention of this chapter is to:

1. analyze those philosophical and theological themes that arise in the context of the origin of life, focusing particularly on the intersection of physical and evolutionary parameters in the interplay of chance and necessity, alongside debates around purpose and design;
2. consider some of the ethical issues associated with both the origin and future of life from a Christian ethical perspective, including responsibility for future generations;
3. argue for the recovery of a sense of wisdom, from both a theological perspective and through *phronesis* or practical wisdom.

It would be impossible to do justice to the full range of possible positions that a theologian might take in relation to theological and ethical issues raised by astrobiology. For this study, I have drawn on the work of contemporary Protestant writers, such as Arthur Peacocke, as he has been very influential in the science and religion field. I am also convinced, however, that the medieval theologian Thomas Aquinas, especially in his understanding of prudence, has something useful to contribute to contemporary ethical debates. Arguments for using this method, rather

Exploring the Origin, Extent, and Future of Life: Philosophical, Ethical, and Theological Perspectives, ed. Constance M. Bertka. Published by Cambridge University Press. © Cambridge University Press 2009.

than, for example, engagement with Biblical or other traditional resources, such as Roman Catholic social teaching, are outside the scope of this chapter.

6.1 What is life?

One of the greatest challenges facing those scientists who are brave enough to enter the difficult waters of trying to recreate the early conditions of the solar system before life appeared on Earth is how to move from physical molecular processes to what might be broadly defined as "life." The definition of life will clearly influence those claims for or against having created "life" in a laboratory. The definition of life is also influenced by prior commitments to particular philosophies. A material-ist, for example, will have a very different definition compared with someone who believes that life is energized by spiritual forces. Those who adhere to mechanistic and reductionistic philosophies believe that complex activities can be explained by reference to molecules and their mutual interactions. In contrast, there are those who argue that the materialist explanation is insufficient to explain emergent prop-erties, that instead such properties become unique to the level of complexity under study. More holistic interpretations resist mechanistic definitions in favor of more complex functions as a requirement for life, and may even be associated with the ancient concept of vitalism, or a "life force," energizing life into being. While there are some religious believers who adhere to the concept of a "life force," this is rejected by biologists and physicists alike, as well as those theologians who are prepared to take the results of science seriously.

Some astrobiologists have suggested that we need to look for broad features of life, in order to look for other non-Earth "biosignatures," rather than associate life's definition with that life found on Earth [1] (p. 9). In this Popa goes further than most astrobiologists, who seek Earth-biosignatures understood as those external condi-tions that make life as we know it possible on Earth. I suggest that if very different forms of "life" were to be discovered on other planets then there would be a need to use very different terminology that reflects its different origin and function. John Maynard Smith and Eörs Szathmáry are, I think, correct to state that "if we are to distinguish between a living being and an artefact, it can only be by knowing its history" [2]. Life requires information flow, hence it needs to store and use that information, so that in the context of laboratory experiments to create life, it can be distinguished from non-life by viewing that information as superimposed on certain states of matter or energy [3].

6.2 Chance and necessity in the origin of life

Although the chance of fully functional life emerging spontaneously on Earth is infinitely small, most scientists argue that life began through small steps that were gradually built up over a very long time. Debates exist as to which process originated first. Was it primitive metabolic pathways? Perhaps building blocks subsequently formed more complex molecules that enabled a membrane to encase particular reaction centers, eventually leading to the formation of ribonucleic acid [4]. An alternative theory is that ribozymes were the most important, in which case ribonucleic acid also acted as enzymes for catalytic reactions [5]. The difficulty in this case is how to arrive at complex molecules from inorganic sources. The early experiments conducted by Stanley Miller at the University of Chicago in the 1950s showed how it was possible to make a wide range of amino acids as well as hydrogen cyanide when an electric spark was added to a reducing atmospheric mix of methane, ammonia, hydrogen, and water vapor [6]. Such research confirmed previous speculations by the Russian chemist Aleksandr Ivanovich Oparin (1894–1980) and the British geneticist John Haldane (1892–1964). Subsequently experimenters found that adenine and guanine could form from hydrogen cyanide, given the right conditions. RNA can replicate itself in the right conditions [7]. The difficulty, of course, is how to derive complex molecules from these relatively simple substrates, although authors like Thomas Cech believe that these details will be discovered in time; more importantly, perhaps, Cech believes that life originated relatively easily and spontaneously as a result of the chemistry of the early Earth. He also suggests that different forms of life with different chemistry would have arisen, and then been extinguished, but he claims "I am much less sure that each of these origins would have been chemically identical.... We have in our natural history one solution that persisted" [8]. Other scientists are more in favor of life emerging on some form of solid surface, such as that of iron-sulfur pyrite. According to this hypothesis, organic compounds attached to pyrite surfaces and then reacted to form cellular aggregates.[1]

Cech's suggestion that one form of chemical life persisted is illustrative of the way the evolutionary paradigm has reached down to a molecular level of explanation at the beginning of the emergence of life. He even believes that original new forms of life are still attempting to emerge even in the contemporary context, billions of years after life first began on Earth. Hence, he suggests that life is an *intrinsic* property of small organic molecules, emerging through self-assembly, eventually giving rise to complex forms. The difficulty, of course, is the directionality of that self-assembly towards greater and greater complexity. For example, while self-organizing properties are a function of matter, what makes such spontaneous self-organization shift towards greater and greater complexity? What is the source of such tendency towards evolutionary complexity? Radu Popa suggests that seeming purposeful behavior in the evolution of life is derived from the enhancement

---

[1] See P. Ulmschneider. *From Common Origins to the Future of Humanity* (Berlin: Springer-Verlag, 2003), p. 98.

of energy flows that are possible in more complex systems [1] (pp. 134–135). He strongly resists any suggestion that purposefulness is divine in origin, rather that over time the system stabilized in such a way as to have a vectorial tendency. This implies, of course, that life is a natural consequence of the restraints set up by physical laws, and these drive the origin of life. However, this seems to be a tautology. It stands to reason that complex life apparently goes against the second law of thermodynamics towards greater entropy, but in order to obey that physical law it must, of necessity, have enhanced energy flows. It seems unconvincing that energy flows in themselves are responsible for shifts towards greater complexity, for each level of organization will have its own properties distinctive for that level.

Debates about the origin of life are fascinating as they are situated at the borderland between discussions of cosmological and evolutionary theories. Physicists have often been more prepared to bring in theistic language in order to explain the remarkable set of coincidences that seem to have come together in order to make the Earth a planet fit for life, one where life is physically possible at all. For the Earth to be able to sustain life it needs to be close enough to an energy source so that there is sufficient heat, but not so far removed that it is locked in a frozen state. Overall, the combination of conditions found are appropriate for life as we know it, including the range of temperatures needed for liquid water to be present. The Earth, unsurprisingly, is situated in a "habitable zone," and other physical properties such as gravity, the distance of the Moon from the Earth, the lifetime of the Sun, and numerous other properties are "just right" for life to emerge. Physicists have called such a tendency the "anthropic principle," though more accurately it should be named the "life principle," for it specifies the conditions right for life to emerge. The question remains, why did such laws emerge in the Universe in the way they did, given the variety of other possible physical laws?

6.3 Is God a designer?

Given the multiple physical coincidences required to make life feasible at all, some physicists who are theists have returned to the idea of God as master Designer. Such ideas have been discredited historically, especially in the biological sphere, because processes of natural selection could explain those complex functions in biology that seemed to point to design [9]. Where design arguments are used as evidence for the existence of God the territory becomes even shakier, for once scientists find a biological or physical explanation, God seems superfluous to requirements. This is exactly what happened in the aftermath of Darwin to vestiges of natural theology remaining from the previous centuries. What about the origin of life itself? Does this suggest a master Designer in terms of molecules that are just right for life to emerge? Or perhaps it is even illicit to make such attempts to synthesize life, as these amount to forms of "playing God" with the order in the Universe, dabbling with that momentous change towards life from non-life? Debates still exist, of course, regarding whether life emerging from inorganic origins was a relatively "easy" process or highly improbable, given the particular physical conditions at the time. It is important to scrutinize the evidence so far and ask how far are such combinations of molecular mixes likely to have been around in the sequence suggested. Some biologists believe that the combination of events needed is sufficiently unlikely to be highly improbable, amounting almost to a "miraculous" understanding of life's origins, and in this sense, not really definitive as far as science is concerned [10]. Other scientists working in the field are more likely to be confident that a solution will be reached, and that over time they will be able to fill in the gaps in the knowledge of how life began. All scientists aim to arrive at a naturalistic explanation, but such a search should not be threatening to theists. In fact, one might argue that if a full explanation was reached, then this would in some ways serve a theological agenda, as it would effectively remove any sense that God intervenes at the start of the process (non-life to life) and then leaves it to organize itself to greater complexity through the processes of natural evolution. The latter view amounts to Deism, that is, God is simply one who starts off the process in the beginning.

How can we envisage God acting in such circumstances? Perhaps the most helpful way of viewing God is to envisage God as the foundation of all Being, including the foundation of physical and biological laws. Arthur Peacocke suggests that a rewriting of the book of Genesis needs to include the idea of the origin of life, so that following a survey of the way the Earth has come into being according to cosmology he suggests that "in some wet deep crevice, or pool, or deep in the sea, just over three billion years ago some molecules became large and complex enough to make copies of themselves and become the first specks of life" [11] (p. 1). Is this way of re-reading Genesis necessarily helpful, except in a metaphorical sense by contrast with the original text? This first book of the Bible was never intended to be a story about early cosmology, but rather a story written in order to affirm Israelite faith in a God who acts decisively in history in contrast to the prevalent Babylonian creation myths whose gods struggled with evil in the creation of the world. The God as envisaged in the Biblical account is a God who affirms the goodness of the material world and is the author of all that is. Belief in God as Creator was assumed by the early Israelites; hence the creation story in Genesis reminded believers that God was Lord of their history, a history that was rooted in the beginning of the creation of the world. The Genesis text speaks of life as a gift of God to all of creation, and to humanity in particular, being made in the image of God. While the title of Peacocke's book might imply that he supports natural theology, i.e. arguments for God's existence from reflection on the natural world, he is ready to admit that any hints at God's presence as discerned in science are only partial and fallible [11] (p. 127). The best we might say, perhaps, is that such scientific discoveries clear the way for the acceptance of theism, rather than point to theism itself.

Peacocke is correct, in my view, to resist any notion of "Intelligent Design," that is, the view that given the complexity of molecular pathways in the emergence of life, it must have been created all at once [11] (p. 69). These notions of special creation hark back to the earlier natural theologies in unhelpful ways, resurrecting a concept that is both naïve theologically and scientifically. In such circumstances even contemporary natural theologies of Intelligent Design become matters for historical curiosity as to their social and public significance, rather than their intellectual achievement. The reaction against such a move is to ignore theological engagement with the natural world altogether, but such problems can be avoided by reinstating the notion of a theology of nature that views any notions of "design" as a sequel to faith, rather than its anticipation. Perhaps Peacocke is a little too confident that all the answers to the early origin of life have been discovered through principles of self-organization. It seems to me that there are far more gaps in scientific understanding than he implies, but this does not open the window for more spurious notions of Intelligent Design.

The question that needs to be addressed in such a context is how God can be envisaged as continually active in the cosmos without intervening in a way that contravenes natural physical laws as understood according to contemporary science. Keith Ward believes that evolution by natural selection is too precarious a process to account for its directionality [12]. What about the biological phenomenon of convergence? According to this view different species that have developed along very different evolutionary paths have seemingly come up with the same functional solution to a particular problem, even though the genetic basis may be very different [13]. Such convergence also suggests constraints within the biological processes that are not amenable to purely physical explanations, though precisely how these restraints are derived is a matter of further research. To envisage God as simply the creator of these physical or biological laws seems too paltry an explanation, rather God could be envisaged as resonating with life, luring life towards a particular future as affirmed by God. This resonance is perhaps on a different plane of reality than that discovered by science. This is not vitalism, as the latter assumes that a "life force" is *required* in the biological explanation of reality. Rather, this is viewing life from a theological as well as a biological point of view. I have suggested that this resonance of God with the created world might be named as God's Wisdom.[2]

---

[2] Arguments for this position are outside the scope of this chapter. For detailed theological justification see C. Deane-Drummond. *Creation Through Wisdom* (Edinburgh: T & T Clark, 2000).

Wolfhart Pannenberg argues in *Towards a Theology of Nature* that a theology of nature needs to relate nature in its entirety to God, which includes a scientific understanding of natural processes [14] (p. 73). Note that he used the term *theology of nature*, i.e. a theological reflection on the significance of nature, unlike *natural theology*, i.e. an argument for God from contemplation of nature, often associated with God as designer. Contemporary theologians, reluctant to "burn their fingers," have avoided dealing with the subject of nature. Pannenberg views any concept of ordered fixed regularity of the Universe as resting on mistaken Greek notions of the cosmos. Instead he suggests that the Hebrew notion of God is one that stresses the contingency of divine will. He asks if contingent occurrences in some sense also disclose regularity. The "laws" of physics, he suggests, need to be considered under the category of contingency, for "only in this way would it be convincing that the order of the laws of nature on its part also is comprehended by the thought of creation and is not opposed to it" [14] (p. 79). In other words, the historical experience of Israel is of a God who acts powerfully in the midst of contingent events, so that while connections in occurrences arise, these only become visible from the end. Hence, he excludes the idea of purposefulness that directs everything *from the beginning*, for he believes that such forms of purposefulness amount to a loss of contingency. At the same time he does allow for "partial development tendencies within the total process" [14] (p. 83). Of course, his position means that in some sense the new enters *from ahead*, rather than from the past, so that forms are "overformed" by the new, rather than broken by them. This view is also implicit in his suggestion that the direction in evolution towards greater complexity comes through "field effects," rather than being implicit within the evolving species themselves [14] (p. 47). The concept of field effects seem to be taken from scientists who are on the more speculative end of the spectrum, and in the biological sphere at least, are not very convincing.

6.4 Ethical issues in the origin and future of life

What are the ethical issues associated with the origin-of-life experiments? So far the attempts to recreate life in the laboratory have been a painstaking affair, with initial concerns about contamination from existing microbial life forms interfering with the results. It therefore seems highly improbable that any fragile life so created would itself serve as a source of contamination for existing life, though presumably if a highly robust form of self-replicating molecules was created, then this would generate some ethical concerns. Nanotechnology, which takes advantage of the molecular level of chemical interactions, could potentially produce such autocatalytic self-replicating systems that would not be classified as "life" by most definitions, but might serve to generate environmental or health risks in the wrong contexts, or even possibly be used inappropriately in terrorist attacks. Hence, if the process could lead to a disruption of the natural world in which we find ourselves, then it needs to be monitored very carefully, especially if that multiplication cannot be adequately controlled.

I have argued above that attempts to create life should not be seen as a threat, rather they need to be welcomed as a way of removing the last traces of either a false deism, where God simply starts the process off and leaves it to evolve itself, or a false argument from design, where the intricacy of life encourages a belief that the first creative act in the evolution of life was a miraculous intervention by God. Both of these generate forms of theology that are naïve. Nonetheless, if humanity does discover a way of creating life, such a discovery needs to be used responsibly, with due account of the application of any such research. In addition, the overall budget expenditure on such research needs to be appropriate, and reflect both the gain in knowledge that this will bring, and the practical advantages. All aspects of research are now heavily competitive, and it is often tempting for scientists who are committed to a particular research project to exaggerate or even distort the benefits for humanity in order to win public support and funding. This tendency is ultimately a disservice for the science community, for it ultimately leads to mistrust when the purported benefits are not forthcoming in the way originally projected.

The above also relates to the importance of intention when discerning whether an action is ethical or not. Hence, it is not enough just to decide whether a particular action is technically feasible, but why do the scientists want to do the research in the first place? Is it going to serve not only their career interests, or satisfy curiosity, but is it also going to contribute to the public good in some way? Of course, there needs to be a place for more theoretical research that has, at first sight, little or no practical application. Yet where public money is being used in order to fund a major project, then that funding needs to be publicly accountable in the light of expenditure on other areas of concern to the human community. In addition, the outcome in terms of public good needs to be inclusive of environmental values that should inform the way research is conducted. It is, I suggest, inappropriate and irresponsible to litter stellar space or other planets with debris from failed attempts at space exploration. Hence, suggestions to set up conservation areas within planetary space are to be welcomed [15]. Perhaps one of the drives towards terraforming Mars and other planets is less about the very long-term future of millions of years before the Earth ceases to be habitable, and more about trying to find a technological solution to runaway exploitation of the Earth's resources. Searching for ways of terraforming other planets should not, in other words, serve to weaken human ethical responsibility for *this* Earth, and attempts to conserve biodiversity on Earth, as well as other natural resources.

In this I would support ideas that argue for the intrinsic goodness for the Earth and the Universe as such, rather than confining the concept of the good to the human community alone. Such an affirmation follows from Christian belief in God as Creator of the Universe, so that there is a sense that all that is, is good. It might be tempting for theologians to drop the notion of God as Creator, given the naturalistic explanations of the origin of the Universe, alongside pseudo-scientific assertions by Creationists and those who adhere to Intelligent Design. I suggest that the Genesis account is not so much a story about how the Earth and the Universe came into being, but ways of valuing that Universe in the light of the involvement of God in the human story. The material world is not evil, as the Manicheans thought, who supposed that spiritual Enlightenment comes through detachment from the created world in which we live. Rather, Christian belief in the incarnation of God in the material world in Christ reinforces the Hebrew notion of the world declared good in the beginning of the creation of the world. Given the goodness of the created world, it follows that actions that serve to destroy or deface it are to be resisted. Would introducing life on Mars change the goodness of Mars as a pristine environment where no life has previously existed? Does a lifeless world have value?

Those who argue for the intrinsic worth of those outside the human community have sometimes been accused of giving equal worth to all aspects of the natural order, and of not being able to discriminate between them. I would suggest that it is possible to hold to the notion of intrinsic value, while also being able to discriminate between different forms of life and non-life in terms of their worth. Only those who hold that the entire world is sacred would suggest that it is untouchable, and cannot be changed at all from its original ordering. A more helpful theological approach is to understand the world as sacrament, that is, one that points to God for those of faith. In this scenario, effecting change in the world is permitted, but not indiscriminately. Dilemmas about how far to change planetary space need to be considered in the light of the experience of human intervention on the Earth. Many of the changes introduced on the Earth had unforeseen consequences, even though they were done with the best of intentions at the time. Some of these consequences were accidental in this sense, though others have been carried out even while knowing what the environmental consequences might be.

According to the Christian tradition, it would be impossible to eradicate the human propensity for sin; hence the possibility of deliberate acts of environmental destruction on other planets cannot be ruled out. The experience of how the Earth has been treated, in spite of the knowledge of what one should do needs to be taken into account. Mary Midgley argues that our lack of responsibility for the Earth in spite of knowing what to do arises from an individualistic and competitive philosophy, fueled by notions arising from socio-biology that seem to support an ethic based on self-interest [16]. Much of the futurist projections by those writing about terraforming other planets fail to take this into account, and hence project an unrealistic picture of harmony and cooperation between peoples in the process of colonization. Given the timescale that is available for exploring space, according to estimates of the lifespan of the Earth, and the human propensity for mistakes, either deliberate or unintended, it would make sense to make any such changes very carefully and slowly. Midgley argues that we need to regain a vision of the Earth as the community in which humanity is situated, and learn the lesson of cooperation by reflection on the planet in terms of James Lovelock's Gaia hypothesis. This hypothesis argues that the planet is only fit for life because of the activity of the biota, hence it is not simply a matter of life adapting to the Earth, but life changing the environmental conditions to make it fit for life. Moving living things into inhabitable planets would take this one step further, as it would amount to shifting a planet that is dead to one that is "living." Of course, there is some debate as to how far Gaia is operative as a living organism, and I would resist ideas that imply Gaia has any teleology or consciousness in the way that is sometimes suggested. In addition, far from giving us an example of cooperation, Gaia could be interpreted as fostering the heightened value of microbial life, as these organisms are primarily responsible for atmospheric stabilization. I am less convinced than Midgley appears to be that a vision of Gaia is of any assistance in formulating an ethical basis for cooperative behavior. I will argue below that the virtue of prudence, which looks to the common good, is required for this kind of complex decision making where there are multiple and competing interests.

Given that these changes will need to be made slowly and over a long timescale, a further ethical consideration that needs to be borne in mind is our responsibility for future generations. Rachel Muers has discussed the question of responsibility towards future generations in a helpful article in connection with current environmental problems [17]. Some of this discussion is relevant to particular questions about environmental issues in terraforming other planets and solar systems. Discussion of the responsibility to future generations in a modern context is related to our ability to change the conditions possible for human existence by our impacts on the external environment. Does exploitation of other planets and creation of new human communities represent a radical break with the past, a moving into a new human future that is unprecedented, in such a way that those future persons will barely recognize themselves in their historical antecedents? Or is it a continuum with respect to human history as humanity has always and will continuously press towards a goal of progress? Is it a destabilization of what it means to be human or not? The conditions for social existence are likely to break down where there is environmental collapse on Earth, but where other planets become our home there is the additional risk of alienation of humanity from itself. Moreover, any extra-planetary environmental pollution event would be much harder to contain if the environmental conditions were themselves dependent on artificial life support systems.

While there is an intuitive sense that we have responsibilities towards future generations, this clashes with the difficulty of giving any such future persons rights, or other contractual means of having a relationship with those existing in the present. It is one area where rights language comes up against its own limits, reciprocity is impossible, and we reach a point where most philosophers would argue that it makes little sense in speaking about "possible" persons as being objects of ethical concern as expressed in the language of rights. Of course, in reproductive biology "possible persons" as embryos have become objects of rights language, but the two situations are different, for, short of freezing such embryos for another generation, discussion about future generations represents persons that have not yet even begun to come into existence. Does a communitarian philosophy fare any better in helping us decide about ethical responsibilities towards future generations? Muers argues that it does not, for a community is itself bounded by particular limitations, including its propensity to break down and its inability to secure indefinite continuity [17]. But the intuition remains that there is a sense that we need some way of taking into account the needs of future generations in arriving at ethical decisions that are likely to be long term. Muers argues that secular philosophy has failed to address ethical responsibility in this respect; it has reached its "limit." Theological ethics, by introducing the notion of the transcendent, offers a way of reflecting on timeless values and shared meanings that actually makes consideration about future generations more relevant for ethical concern.

6.5 Astrobiology and the search for wisdom

I will argue in this section that theology has a distinctive voice in these debates, not only by recovering a sense of wisdom, but also in practical decision making through the notion of prudence, or practical wisdom [18]. Wisdom as a virtue in the classic theological tradition is distinguished from prudence, for wisdom is also one of the intellectual virtues of theoretical reason, along with understanding and science. While prudence is wisdom in human affairs, wisdom pure and simple surpasses this as it aims directly at God, at knowledge of divine things. It has the facility to act as "judge" over all the sciences, both in the conclusions and premises. For the secular scientist the failure to know is a simple result of the failure of human reason. For Thomas Aquinas, while we can know the relationship between things and our minds, we can never penetrate fully into the mind of God, for "it is too rich to be assimilated completely, it eludes the effort to comprehend it" [19]. Hence the unknowability of creation stems from the fact that God creates it. As such creation only gives an imperfect picture of what God is like, we cannot simply read off God from the world in the manner implied by much contemporary science and religion writing. In addition, our minds are too crude an instrument to fathom the true nature of things; hence it is also because human beings are created that they cannot know fully the divine mind in things. Such a view does not lead to agnosticism, because it presupposes the idea of a Creator, but an awareness that we are always "on the way." Nor does it lead to a rejection of science, for even in its imperfect form creation shows the light of God as if it were in a mirror. How can we best describe this search for knowledge that is also a search into the mind of God? I suggest that such a search can be described in terms of *wisdom*. For wisdom is also an expression of the eternal mind of God, while at the same time affirming what can be known in creaturely existence. For Aquinas science is distinct from the gift of wisdom in that science is about the certitude of judgment that arises out of human or created things, while wisdom is judgment about the highest cause, namely God.

Prudence, or practical wisdom for Aquinas is the "mother" of all the other cardinal virtues. It might be hard to imagine that prudence is in any sense a pre-requisite to goodness, since prudence in colloquial use implies a hesitation, or even small-minded concerns about oneself. In contrast, the classical approach links pru-dence specifically with goodness, moreover there is no justice or courage without the virtue of prudence. Instinctive inclinations towards goodness become trans-formed through prudence, so that prudence gives rise to a perfected ability to make choices. Hence the free activity of humanity is good in so far as it corresponds to the pattern of prudence. Drawing on memory and experience, prudence has the facets of taking counsel, judging correctly and then moving to act in a certain way. Prudence also includes *practical execution* of what has been decided upon. It seems to me that the holistic task of prudence that serves to inform moral behavior is vitally important, for it invites self-reflection *at every stage* in the overall process from deliberation through to action. This applies in a particularly poignant way in astrobiology, where discussions about what should be done are taking place even prior to the possibility of their actual technical achievement. There are also no absolutes in prudence, so that there will always be a small measure of risk in any decision making, even if this decision is a prudent one.

Prudence demands openness to others in a way that prevents it from being indi-vidualistic, for taking counsel is the first part in the overall process of prudential activity. Those who are prudent take counsel and are "well advised." Such a char-acteristic is relevant in the present context, which is seeking to integrate different ways and approaches and arrive at provisional conclusions that are of fundamental significance for the future of life. In a similar way Aquinas believes that sound judgment is a subsidiary or "allied" virtue to prudence [20] (p. 51.3). In the context of astrobiology, we need to ask, what might "sound judgment" mean? The gift of science can be related to the judging activity in prudence, just as the deliberating activity is related to the gift of counsel. In as far as a person comes to know God through created things, this is the office of the gift of science, but the judgment of created things in the light of divine things is the gift of wisdom. This distinction is of great interest in the light of attempts by scientists to find God through cosmology. Thomas would suggest that such intimations of divine reality are the result of the gift of science, but while science might give knowledge as to what one might believe, wisdom affords a deeper relationship with God in a way that science cannot. Hence, he is able to qualify both the provision and the limits of scientific knowledge.

I suggest that accurate cognition at the first stage of prudential activity is not easily acquired, but comes gradually through patient effort in experience, one that cannot be short-circuited by an appeal to "faith" or a particular "philosophical" view. This silent contemplation of reality that takes counsel in all kinds of ways also includes the ability to be true to memory, meaning not just the recollection of the past, but one that is true to the nature of things. Aquinas lists the various components of prudence as memory, insight or intelligence, teachableness, acumen (*solertia*), reasoned judgment, foresight, circumspection and caution [20] (p. 49). Beyond this Aquinas distinguishes between individual prudence, aimed at the good of the individual, domestic prudence for family life and political prudence, aimed at the common good (p. 47.11). Relating the three aspects of prudence, individual, domestic, and political, is particularly significant in an ethics of astrobiology that includes decision making at all levels, from the individual decision of particular scientists, through to political decision making about funding and so on.

Aquinas distinguishes between imprudence, negligence, and sham prudence. Different types of imprudence are a result of imperfections at any stage in the process of deliberation, judgment, and action. If there is a failure to take counsel, this leads to foolish haste, if there is a failure in judgment this leads to thoughtlessness, if there is a failure to carry out an action this leads to inconstancy. Flaws at any stage of taking counsel, judging, and action lead to analogous vices. Imperfections of prudence at the first stage of cognition come about through a failure to be still in order to perceive reality. At the root of all these vices lies unchastity, that is the failure to be detached from the seduction of pleasures of sense [20] (p. 53.6). Prudence as a virtue overcomes this vice in allowing reasoning to order the moral life. Aquinas associated the vice of *curiositas* with magic, an addiction to the sense of pleasure from knowing. It is a desire to know that becomes obsessive and eclipses other human responsibilities. Given the fascination of astrobiology and its potential to arouse the imagination, there is always this implicit danger, namely that it will become detached from more practical issues and responsibilities on Earth.

The virtue that needs to be cultivated instead is *studiositas*, a love of learning and knowledge that has its proper place. Yet another implicit danger in considering the future of life on other planets is the mode of imprudence expressed as thoughtlessness, acting too quickly without sufficient time for consideration. Like the subsidiary virtues of prudence, such as well-advisedness, negligence is a subsidiary outcome of imprudence. Those who fail to act at all by irresoluteness are also lacking perfected prudence, as are those who cannot act in unexpected circumstances, who lack what is known as *solertia* (acumen) or clear sighted objectivity in the face of the unexpected. Given the nature of unknowns in space exploration, the virtue of solertia is vitally needed in order to act in an ethically responsible way. Sham prudence is prudence directed to the wrong end, namely "the flesh" set up as the final goal of the human life. If the real goal of a particular project is to make financial gains and exploit other planets, regardless of the life forms that they bear, this would be a form of sham prudence. Sham prudence also comes through using false or deceitful means, namely cunning, to attain either a good or evil end. The underlying sin in the case of sham prudence is covetousness, the desire to have what is not rightfully one's own. This tendency would exist where there are exaggerated claims for success in order to win public support and funding, as mentioned above. While it is difficult, given the multifaceted nature of prudence, to do justice to it through a simple definition, Pieper has come close through his summary of many aspects of prudence in the following paragraph:

It holds within itself the humility of the silent, that is to say, of unbiased perception, the trueness to being of memory, the art of receiving counsel, alert, composed readiness for the unexpected. Prudence means studied seriousness and, as it were, the filter of deliberation, and at the same time the brave boldness to make final decisions. It means purity, straightforwardness, candor and simplicity of character, it means standing superior to the utilitarian complexities of mere "tactics" [21] (p. 36).

Given this understanding of prudence, such a task would be almost impossible given the propensity to err (sin) in all its varied facets without Aquinas's belief that perfection in prudence is only possible through the grace of God. The form of prudence to which the Christian aspires is prudence in the light of the three theological virtues of faith, hope, and charity. Prudence as learned and prudence perfected by grace both move from cognition through to decision, volition, and action. This is where the virtue of prudence as *learned* links with the theological virtues of faith, hope, and love. Commentators on prudence in Aquinas seem to have ignored the possibility of prudence acting under the influence of God's grace, it has proved to be an embarrassment to modern readers, preferring to align themselves with secular concepts of prudence. However, if Christian ethics is to move beyond a simple endorsement of secular models, then some sense of the possibility at least of prudence given by grace is not only necessary, but vital in thinking through difficult ethical questions. Such a rich understanding of prudence when combined with love of God in charity leads to a higher plane of prudence, one that is detached from the compulsions of worldly desires. This does not come from human judgment about the world, such disdain would stem from arrogance and pride. Rather, it is the overwhelming love of God that relativizes all human efforts, while at the same time holding a clear perspective on human obligation and responsibility. I prefer to call this form of prudence wisdom, since it can be more readily identified as a theological category [22].

Unlike approaches within Christian ethics that refer to particular traditions or principles that seem unrelated to practical contexts, prudence demands full encounter with experience, including the experience of science, taking time to perceive what is true in the natural world. While Aquinas restricted his idea of taking counsel to other human subjects, in an environmental context it is essential to try as far as possible to perceive from the perspective of all creatures, all of whom are loved by God and under God's providence. Hence, any discussion of the future of life needs to include not just what is good for humanity, but also what might be good for other creatures that share our planetary home. Prudence is also helpful in discriminating between different life forms in terms of their relative intrinsic value. The role of microorganisms in stabilization of the planet might warn against too hasty a dismissal of microbial life as being of lesser value than higher life forms, though evolutionary complexity is indicative of a greater sentience and consciousness that needs to be taken into account.

Consequentialist approaches to the ethics of nature have sought to frame decisions in terms of costs and benefits, or risks. While prudence would include some perception of risks where they are known, the ability to have accurate foresight depends on how far such decisions promote the overall goal of prudence towards goodness. In other words, it is the character of the agents that is as important as the particular consequences of individual decisions made. Hence, the good of humanity is included along with the goods of other creatures. While those who are not Christian will be able to identify with the goal of goodness, a Christian virtue ethic springing from prudence will seek to move to a particular understanding of goodness, one that coheres with the overall goodness for creation, as well as goodness for humanity.

A discussion of the significance of prudence for ethical decision making about the future of life would not be complete without brief mention of the cardinal virtue of justice. Justice is often split off from a consideration of virtue ethics, as it is more commonly associated with rule-based ethics. I suggest, alternatively, from the side of virtue ethics, that when considered as a virtue to be developed justice gives consideration of rules and principles a proper place in an overall ethical framework.

Justice is concerned broadly with the idea that each is given her/his due. Unlike many other virtues, justice specifically governs relationships with others, and also unlike other virtues it is possible to act justly without necessarily having a proper attitude towards that action. A particular rule or pattern for prudence prescribes what is a just deed according to reason, and if this is written down it becomes law. Consideration of the legal framework to protect life on other planets needs to be considered in any international treaty set up in order to regulate space exploration and colonization. The *jus gentium* is constituted by natural reason and shared by all people, equivalent to what we might today term international law.

6.6 Conclusions

I have argued that scientific questions at the beginning of life on Earth still fall short of full explanation as to what might have happened in the earliest history of the Earth. From the perspective of Christian theology any discovery of natural causes for the shift towards life from organic molecular precursors would be welcomed, as it would remove any vestiges of deism or theories of Intelligent Design that are far removed from a theology of nature. I have suggested that attempting to derive God from gaps in our understanding of the nature of functional complexity either at the start of life or throughout evolutionary change is mistaken. Theology is important, however, in as much as from the perspective of faith it can give meaning to the directionality of evolution, challenging the notion of purposelessness that is implicit in much evolutionary philosophy. There are also theological questions about the meaning of humanity and humanity's future raised by our search for other planetary homes. I have argued that from the perspective of Christian theology all of creation needs to be viewed as good, given as a gift of God, though not necessarily for human instrumental use, but good in its own right. Where there is conflict between ethical demands between species, or between different life forms, then the virtue of prudence or practical wisdom is especially relevant. Prudence includes deliberation, judgment, and action, hence it includes elements of scientific forms of reasoning, but seeks to go beyond this by seeking a more holistic approach to complex issues. A Christian virtue approach to the ethics of future life is, it seems to me, a more satisfying way of reasoning compared with principled approaches that can seem detached from secular discussion, or consequentialist approaches that view decision making simply in terms of risks and benefits. Above all, as a human community, we should not give up on the search for wisdom, for wisdom allows an integration of religious and scientific ways of thinking about the Universe in a way that can inform future visions. Such visions need to be derived from a settled sense of meaning and identity that comes from knowing one's place in the cosmos, rather than a desperate search for cosmic immortality that resists human finitude.

References

[1] R. Popa. *Between Necessity and Probability: Searching for the Definition and Origin of Life* (Berlin: Springer-Verlag, 2004).

[2] J. Maynard Smith and E. Szathmáry. *The Origin of Life: From the Birth of Life to the Origins of Language* (Oxford: Oxford University Press, 1999), p. 6.

[3] H. Rolston (ed.). *Biology, Ethics and the Origins of Life* (Boston, MA and London: Jones and Bartlett, 1995), p. 16.

[4] C. De Duve. *Blueprint for a Cell: The Nature and Origin of Life* (Burlington, NC: Patterson, 1991).

[5] L. E. Orgel. The origin of life – a review of facts and speculations. *Trends in Biochemical Sciences*, **23** (1998), 491.

[6] C. Wills and J. Bada. *The Spark of Life: Darwin and the Primeval Soup* (Cambridge, MA: Perseus Publishing, 2000), pp. 39–50.

[7] J. W. Szostak, D. P. Bartel, and P. L. Luisi. Synthesizing life. *Nature*, **409** (2001), 387–390.

[8] T. R. Cech. The origin of life and the value of life. In *Biology, Ethics and the Origins of Life*, ed. H. Rolston (Boston, MA: Wadsworth Publishing, 1995), p. 32.

[9] B. John and G. Cantor. *Reconstructing Nature: The Engagement of Science and Religion* (Edinburgh: T & T Clark, 1998), pp. 141–175.

[10] S. Conway Morris. *Life's Solution: Inevitable Humans in a Lonely Universe* (Cambridge: Cambridge University Press, 2003), pp. 44–68.

[11] A. Peacocke. *Paths From Science Towards God: The End of All Our Exploring* (Oxford: One World, 2001).

[12] K. Ward. *God, Chance and Necessity* (Oxford: One World, 1996), p. 78.

[13] C. Morris. *Life's Solution: Inevitable Humans in a Lonely Universe* (Cambridge: Cambridge University Press, 2003).

[14] W. Pannenberg. *Towards a Theology of Nature: Essays on Science and Faith* (Westminster: John Knox Press, 1993).

[15] M. Peplow. Scientists propose conservation parks on Mars. *Nature* (News on line, November 26, 2004).

[16] M. Midgley. *Science and Poetry* (London: Routledge, 2001).

[17] R. Muers. Pushing the limit: theology and responsibility to future generations. *Studies in Christian Ethics*, **16**(2) (2004), 36–51.

[18] C. Deane-Drummond. *The Ethics of Nature* (Oxford: Blackwells, 2004), pp. 1–53.

[19] J. Pieper. *The Silence of St Thomas*, trans. D. O'Connor (London: Faber and Faber, 1957), p. 65.

[20] Aquinas. *Summa Theologiae, Prudence, vol. 36*, 2a2ae, trans. T. Gilby (London: Blackfriars, 1973).

[21] J. Pieper. *Prudence*, trans. R. Winston and C. Winston (London: Faber and Faber, 1959).

[22] C. Deane-Drummond. *Creation Through Wisdom* (Edinburgh: T & T Clark, 2000), p. 36.

Part II

Extent of life

7

A biologist's guide to the solar system

Lynn J. Rothschild

7.1 Introduction

Astrobiology has life at its core: Where does life come from? Where is it going? Are we alone? While it includes the search for extraterrestrial life – the very bit that has so captured the public's attention – it uses life on Earth as its reference point. Of course this probably has less to do with philosophy, and more to do with practicalities. After all, there is only one place that we know with certainty contains life, and most likely an indigenous biota at that. So, planet Earth remains the reference point. Thus, a search for life elsewhere, even in our own solar system, must include an understanding of the known range of life on Earth. And, even before that, an understanding of what we mean by "life."

Understanding the range of *current* life on Earth, and mapping it to *current* environments in the solar system, is only a start as it lacks the element of time. Life on Earth may have been substantially different when it arose around about 4 billion years ago because the environmental range on Earth was dramatically different. Similarly, the climatic conditions forecast for a billion or so years into the future are bleak for much of life as we know it, including ourselves. Without intervention, the Sun as we know it will not even exist. The Sun will continue to become more luminous as it burns, raising the temperature on Earth to the point it is intolerable for us, and ultimately all life. As the Moon migrates away from the Earth, the Earth will have an increasingly chaotic obliquity, leading to an unstable climate on evolutionary timescales.

This chapter will begin by defining the minimum conditions for life to exist as a way to understanding why environmental extremes are obstacles to life. From there it will survey different environmental factors, briefly describing why they

Exploring the Origin, Extent, and Future of Life: Philosophical, Ethical, and Theological Perspectives, ed. Constance M. Bertka. Published by Cambridge University Press. © Cambridge University Press 2009.

are barriers to life, and how some life has overcome them. In the process, a rough multidimensional envelope for life on Earth will have been sketched.

With this knowledge, we will explore our solar system for potential habitats for life. As there are currently no definitive signs of life elsewhere in the solar system, or, in fact anywhere besides the Earth or transported Earth life (for example in human spacecraft), the exploration will really be to find environments where we know Earth life *could* exist. This allows us to say that since there is nothing to absolutely preclude Earth life from living in location x, there is nothing about life *per se* that would not allow it to live in that location. This does not mean that life does live there, or even that there are not other forms of life – even Earth life that is yet to be discovered – that could live there. And, with the announcement of a cold water geyser on Titan's moon Enceladus in the spring of 2006 [1], it would be foolish to even assume that we are adequately knowledgeable about the habitable environments present in even our own solar system. Thus, this survey must always be taken as a *minimum* range for potential life.

One final distinction. The conditions for the origin of life may be different than those for the maintenance of life. The current working assumption for most scientists is that spontaneous generation – the formation of living from non-living materials – happened at least once, but it is unclear that it could happen again "in the wild" on present-day Earth.

7.2 Extent of known life – extremophiles

In order to understand the possible range for life elsewhere, we must first appreciate the range of life on Earth. This is done by examining various environmental factors such as temperature, that provide limits to life, and reconciling biodiversity with the theoretical limits on life. This approach can be traced back to the British scientist and author, Richard Proctor, and is exemplified in his book *Other Worlds Than Ours* [2].

The theoretical limits on life stem from the fact that life as we know it, and arguably even as we do not know it, is based on organic carbon organized as polymers using liquid water as a solvent ([3, 4], but also see both for a discussion of other solvents for life). Organic carbon, especially long-chain polymers, must be stable. Beyond stability, the biopolymers need to be in a conformation that is functional. But the structure of organic polymers is sensitive to such environmental conditions as temperature, pH, and so on. Thus, the freezing and boiling temperature of aqueous solutions, the ability of organic compounds to react in the aqueous solutions, and the presence or absence of a source of energy should place limits on the environmental range of life. These issues will be discussed for each environment for those who wish a deeper understanding of why it is extreme for life. For example, while we have an intuitive understanding that extremes in temperature are danger-ous or lethal for life, the biochemical basis is less widely understood, and thus is available here.

The study of life in extreme environments, "extremophiles" [5], has to a large extent been driven by the desire to preserve food, to explore the limits of our planet, and, of course, by the search for life elsewhere. Excellent reviews of life in extreme environments exist elsewhere. Comprehensive reviews include Rothschild and Mancinelli [6], Ashcroft [7] and Rothschild [9]. Extremes can be physical (e.g. temperature, radiation, pressure, gravity) or chemical (e.g. pH, salinity, des-iccation, gas concentration). Although not normally considered, one could argue that biological extremes, such as viral load or extreme population density, exist. Table 7.1 summarizes the major physical and chemical extremes for life.

7.3 Temperature

Temperature is a fundamental environmental parameter as it is critical to maintain water – or any other solvent that life might use – in a liquid state. Temperature also has a direct effect on the stability and function of biopolymers.

High temperatures are dangerous because at some point DNA, RNA and pro-teins denature, that is, lose structure (Figure 7.1).

The melting temperature of DNA depends on the length of the sequence (shorter sequences melt faster), the percentage of A-T pairs (A-T pairs melt faster than G-C pairs because they have two rather than three bonds), alternating regions of purine/pyrimidine nucleotides (melt faster), the topology of the DNA (e.g. whether it is a closed circle that is relaxed or supercoiled, or a linear piece, or is heavily nicked), and the amount of salt and which ions are present in the solution. Similarly, each protein and RNA has characteristic denaturation tem-peratures. Depending on the compound and the temperature, the changes may be irreversible or reversible. The white and yolk of an egg turn opaque upon cooking (denaturation), a process that is irreversible, whereas the enzyme lysozyme can renature [10].

How does thermal denaturation work? Envision a protein as a long string of beads with twenty or so different colors. The sequence of beads is its primary structure. If the beads are twisted so that the green ones that are nearby are next to each other as much as possible, or short segments are twisted into helices, this would be analogous to secondary structure. The α helix, β sheet, and turns are common examples of secondary structure. Tertiary structure involves further fold-ing. Quaternary structure entails the linking of different, likewise folded, chains of

Table 7.1 *Limits for life on Earth. This table is current as of publication,*
but new records are reported periodically, so these numbers
are not to be taken as absolute limits

Environmental parameter	Type	Definition	Environments	Known limits
Temperature	Hyper-thermophile	Growth >80 °C	Geyser	113 °C (*Pyrolobus fumarii*)
	Thermophile	Growth 60–80 °C	Hot spring	
	Mesophile	15–60 °C		*Homo sapiens*
	Psychrophile	<15 °C	Ice, snow	*Psychrobacter,* some insects
pH	Alkaliphile	pH >9	Soda lakes	pH 12.5
	Acidophile	Low pH loving	Acid mine drainoff, hot springs	pH 0 (*Cyanidium caldarium, Ferroplasma* sp.)
Salinity	Halophile	Salt loving (2–5 M NaCl)	Salt lakes, salt mines	*Halobacteriacea, Dunaliella salina*
Desiccation	Xerophiles	Anhydrobiotic	Desert	*Artemia salina;* nematodes, microbes, fungi, lichens
Pressure	Barophile	Weight loving	Deep ocean	Unknown 130 MPa (microbes)
	Piezophile	Pressure loving		
Radiation		Tolerates high levels of radiation	Reactors, high solar exposure, e.g. at altitude	*Deinococcus radiodurans*
Oxygen	Aerobe	Tolerates oxygen	Most of Earth today	
Gravity	Hypergravity	>1 g		*None known*
	Hypogravity	<1 g		*None known*
Vacuum		Tolerates vacuum (space devoid of matter)		*Tardigrades, insects, microbes, seeds*
Chemical extremes	Gases, metals	Can tolerate high concentrations of metal (metalotolerant)	Mine drainage	Pure CO_2 (*Cyanidium caldarium*) Cu, As, Cd, Zn (*Ferroplasma acidarmanus; Ralstonia* sp.) CH34 (Zn, Co, Cd, Hg, Pb)

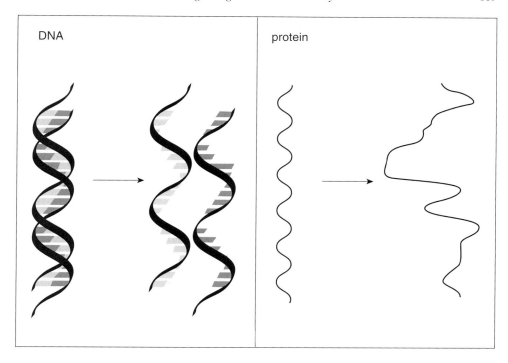

Figure 7.1 Denaturation of biopolymers. Shown are generalized cartoon drawings of the DNA double helix, and a protein in the form of an α helix, going from a native to a denatured state.

beads. As temperature increases, the first bonds affected are the long range ones, the quaternary and tertiary bonds. As heating continues, bonds such as those that stabilize α helix structure will begin to break, allowing water to interact with the hydrogen bonds and with the amide nitrogen and carbonyl oxygens of the peptide bonds. Hydrophobic groups are exposed to the solvent, increasing the amount of water bound to the protein. As the protein will try to minimize its free energy by exposing polar groups and burying hydrophobic groups, it will fold, but the resulting structure is likely to be vastly different from the original protein. As structure is critical for protein function, a different structure may weaken or abolish function.

Similarly, DNA must be double-stranded to function, with moments of single-stranded conformation, such as during replication, carefully controlled. Because DNA normally denatures at high temperatures, hyperthermophiles must evolve ways to avoid this. Closed-circle DNA is more resistant to strand separation, and *in vitro* closed-circle DNA does not denature completely until 107 °C [11]. However, cytosine deaminates at high temperatures and there is spontaneous hydrolysis of the *N*-glycosyl bond. These promote chain breakage. Once the DNA helical structure destabilizes, strand separation is initiated. Strand separation then leads to a four-fold higher rate of depurination relative to double-stranded DNA. Further strand breakage results. Studies on the stability of covalently closed circular DNA suggest that this thermodegradation is a greater hazard than thermodenaturation, but that the intracellular concentration of K^+ or Mg^{++} can help stabilize DNA primary structure [11].

Reverse gyrase, the only protein that is both specific and common to all hyperthermophiles, reduces the rate of double-stranded DNA breakage eight-fold at 90 °C [12]. This activity does not require ATP hydrolysis and is independent of the positive supercoiling activity of the enzyme. Reverse gyrase has a minor nonspecific effect on the rate of depurination, and a major specific effect on the rate of double-strand breakage.

RNA relies heavily on higher-order structure for function [13]. For RNAs that rely on secondary structure for function, a positive correlation exists between increasing (G + C)% and growth temperature. In contrast, mRNA needs to *minimize* RNA-RNA interactions. There is a weak correlation between elevated growth temperature in prokaryotes and the purine-loading index in mRNA such that at higher growth temperatures there is a tendency to substitute adenine for cytosine [14]. Post-translational modification of bases has been shown to stabilize tRNAs from *Pyrococcus furiosus*; cultures grown at 70, 85, and 100 °C had increasing amounts of modified bases [15].

Lipids become more fluid as temperature increases. Changes in lipid composition occur with increasing temperature, such as a switch from 50 to 75 °C [16, 17]. A change in the composition of lipids, for example the addition of more saturated lipids, can stiffen membranes at elevated temperatures [18]. Conversely, the addition of unsaturated lipids at low temperatures helps to keep membranes fluid. Thermophilic and extremely acidophilic archaea have membrane-spanning tetraether lipids (review in [17]). These tetraether membranes have a limited permeability for protons even at the high temperatures of growth and this property makes it possible for thermophilic archaea to maintain a viable proton motive force under the extreme conditions. Ether lipids cannot be degraded easily and are highly stable, which is also a requirement for life under extreme conditions.

With all of the problems of high temperature, it is incredible that there are organisms that live at high temperature [19], and even archaea that can survive well over the boiling temperature of water (Figure 7.2).

While reports exist of microbes surviving over 120 °C, the highest confirmed temperature for life is 113 °C for the crenarchaeotan *Pyrolobus fumarii* [21]. This organism was isolated from a hydrothermally heated black smoker wall in the Mid Atlantic Ridge and grows between pH 4.0–6.5, with an optimum at pH 5.5, and 1–4% salt with an optimum at 1.7%. It can grow at a pressure of 25,000 kPa (250 bar). Exponentially growing cultures survived autoclaving at 121 °C for one hour. Other

Figure 7.2 Octopus Spring, a well-studied habitat of hyperthermophiles in Yellowstone National Park. In the foreground is the edge of the source pool, which is near the boiling point of water at that elevation (~95 °C). The run-off channel, ~82 °C, contains large, fast-growing clumps of pinkish bacterial communities [20]. Photo taken by the author, August 2, 2006.

hyperthermophiles have been found in low salinity up to 3% salt, highly acidic (pH 0.5) to mildly alkaline (pH 8.5) conditions [22]. Hyperthermophiles can withstand drying in the lab and anoxic conditions [21], as the solubility of gases such as oxygen decreases with increasing temperature.

Low temperature presents other problems. Enzyme activity and membrane fluidity is reduced. More critically, freezing occurs at low temperature. As freezing begins, purer water is frozen first, leaving increasingly concentrated solutions, resulting in dehydration. Ultimately ice crystals can form causing physical damage to membranes. Yet, representatives from a wide variety of taxa from fish to bacteria can live in Antarctic waters below 0 °C with the help of antifreeze compounds. Bacteria from South Pole surface snow have been shown to be metabolically active down to –17 °C, as have Antarctic lichens [23].

Even at these low temperatures, a small but apparently sufficient amount of liquid water exists [23]. For example, the accumulation of glycerol in frogs allows a natural tolerance to freezing [24], whereas the antifreeze proteins in fish provide the same function [25, 26].

7.4 pH

The pH of the external and internal environment of a cell is extremely important. The pH of a solution refers to the activity of hydrogen ions (H^+) in solution, although in practice it is measured as the concentration of H^+, specifically:

$$pH \sim -\log_{10} |[H^+]| \tag{7.1}$$

Note that the scale is logarithmic; thus, lemon juice at pH 2 has ten times the concentration of H^+ as orange juice or soda at pH 3. While the pH scale is thought of as from 0 to 14, a pH below 0 or above 14 is possible. For example, acid mine run-off can have a pH of –3.6, corresponding to a molar concentration of H^+ of 3981. The cytoplasm of most cells is near neutral, which is considered pH 7.

Just as high temperature can cause loss of structure and activity in biomolecules, so too can extremes in pH. For example, the shape of proteins is altered by pH, which can hinder or destroy their activity through unfolding, cleavage of salt bridges, and changes in solubility. As the pH of the cytoplasm decreases to near the isoelectric point of proteins, proteins tend to precipitate. At even lower pH, proteins will become positively charged and will repel each other. If the charge density is high enough, intramolecular repulsions may be high enough to cause unfolding of the proteins. If the unfolding is extensive enough to expose hydrophobic groups, the denaturation is irreversible. The Latin American dish ceviche uses this principle: lime or lemon juice is used to "cook" raw seafood by denaturing the proteins.

Some proteins contain acid labile groups, and even relatively mild acid treatment may cause an irreversible loss of function. This generally results from the breaking of specific covalent bonds, and thus is not technically denaturation, but still results in loss of function and is therefore deleterious for the organism. The charges of the individual amino acids in proteins will vary depending on the pH of the environment and the specific acid dissociation constant of the amino acid. Changes in charges with pH affect the activity, structural stability, and solubility of the enzyme. Further, since enzymes contain a large number of acid and basic groups, mainly located on their surface, so, again, pH will affect the total net charge of the enzymes and the distribution of charge on their exterior surfaces, in addition to the reactivity of the catalytically active groups. Extremes of pH will also cause degradation of proteins. In alkaline solution (pH > 8), there may be partial destruction of cystine residues due to base catalyzed β-elimination reactions whereas, in acid solutions (pH < 4), hydrolysis of the labile peptide bonds, sometimes found next to aspartic acid residues, may occur.

Extremes in pH cause denaturation of nucleic acids. At pH 4 the beta-glycosidic bonds to the purine bases are hydrolyzed and purine bases (N7 of guanine, N3 of adenine) are protonated. Protonated purines are good leaving groups hence the hydrolysis. Once this happens, the depurinated sugar can easily isomerize into the open-chain form and in this form the depurinated (or apurinic) DNA is susceptible to cleavage by hydroxyl ions. Between pH 11.65 and 11.8, DNA denatures [27]. RNA is very unstable in alkali solutions due to hydrolysis of the phosphodiester backbone. The 2'OH group in ribonucleotides renders RNA molecules susceptible to strand cleavage in alkali solutions.

The final problem with pH has to do with organisms that use a hydrogen pump to generate ATP. A sufficient concentration of H^+ is critical, and at high pH there is a low concentration of H^+.

Acidophiles and alkaliphiles adapt to living at low and high pH by regulating their internal pH via a proton pump. For example, acidophiles pump protons out of the cytoplasm. As a result, adaptations to extremes in pH require minimal internal evolutionary change. This appears to be a relatively simple adaptation as multiple organisms from all three domains have been found at or near pH 0, and up to at least pH 10.5, and probably pH 12.5. The red alga, *Cyanidium caldarium*, can live down to nearly pH 0 [28]. Yet, soluble proteins from *Cyanidium* are not particularly acid or heat stable compared with those from mesophilic algae such as *Anabaena variabilis* and *Anacystis nidulans* [29].

The archaea *Picrophilus oshimae* and *Picrophilus torridus* both live in dry soils below pH 0.5 [30], and *Ferroplasma acidarmanus*, isolated from acid mine drainage, can grow at pH 0 [31]. A diversity of organisms, from bacteria to cyanobacteria to rotifers, can live at pH 10.5 (e.g. [32]; Figure 7.3). There are microbial communities at pH 12.9 in the soda lakes of Maqarin, Jordan [33]. It should be noted that soda lakes, in general, tend to be highly productive, with high concentrations of cyanobacteria because of the high concentration of dissolved CO_2 [34, 35].

7.5 Salinity and ionic strength

The salt sodium chloride (NaCl), makes up about 0.28% of the human body by weight, according to the Salt Institute. Since cells are surrounded by semi-permeable membranes, a high external concentration of salt should result in the loss of water from the cell leading to dehydration. Conversely, if the salt concentration of the environment is lower than that of the cell, the cell should swell as water rushes in. This has been shown to be true even for human cheek cells *in vivo* where, after eating salted foods, cells had, on average, shrunk, whereas after drinking tap water, many cheek cells swelled and subsequently lysed (Rothschild-Mancinelli and Rothschild, unpublished results; Figure 7.4).

Nucleic acids are negatively charged; thus, some salt increases structural stability. In contrast, proteins consist of a chain of amino acids with different charges. Dilute salt solutions increase the solubility of proteins over pure water because the salts associate with opposite charge groups in the protein, increasing protein

Figure 7.3 Lake Bogoria, a high pH lake in Kenya. The photo is of the north rim of the lake (N 0°19.889, E 36°4.574, altitude 1008 m), where the pH was 11.4 and the temperature 30 °C. Patches of the filamentous cyanobacterium *Arthrospirea fusiformis* are visible on the bottom of the lake near the shore. These are grazed by the lesser flamingo, *Phoeniconaias minor* [36]. Photo taken by the author, January 6, 2007.

Untreated cheek cells   Frito (high salt)   Water (low salt)

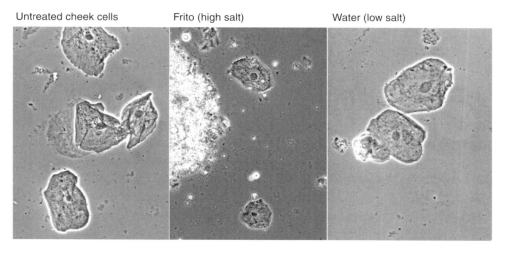

Figure 7.4 Change in volume of human cheek cells as a function of salinity. Photos were taken with under phase and show human cheek cells (left), cells after the subject had eaten five salted Frito chips, and cells after the subject had drunk tap water. Photos taken by the author and B. Rothschild-Mancinelli.

Figure 7.5 High salt environment in salt concentration pond, Cargill Salt Company, located on the southern part of the San Francisco Bay. Evaporation ponds appear red owing to the presence of large concentrations of halophiles, and white from dried salt, which is apparent in the photo. The near shore is the peninsula, the far shore the east bay. Photo taken by the author, July 14, 2007.

hydration. At high salt concentrations, the ions compete with the protein for water, and thus protein precipitates out of solution, a phenomenon called "salting out."

Most plants cannot tolerate high salt, although exceptions include salt marsh grass *Spartina alterniflora*, which can produce biomass up to 0.6 M NaCl [37]. There are a multitude of microbes, especially halophiles, which accumulate compatible solutes or osmolytes to compensate for a high-saline environment. The compatible solutes include low molecular weight organic compounds, and include sugars, amino acids and their derivatives, polyols and their derivatives, betaines, and ectoines. Glycine betaine, for example, is a very common compatible solute in mesophilic bacteria and archaea. With these adaptations, organisms are able to thrive in the high salt of salt evaporation ponds at 5 M salt, roughly ten times the concentration of salt in the ocean (Figure 7.5). Most halophiles are archaea and bacteria, although the green alga, *Dunaliella salina*, can also live up to saturated NaCl by accumulating glycerol and β-carotene. These compounds are harvested commercially.

Ionic strength is defined as half of the total sum of the concentration (c_i) of every ionic species (i) in the solution times the square of its charge (z_i). Thus, the ionic strength (I) = $0.5\Sigma(c_i z_i^2)$. For example, the ionic strength of a 0.1 M solution of $CaCl_2$ is $0.5 \times (0.1 \times 2^2 + 0.2 \times 1^2) = 0.3$ M.

At very low ionic strengths, nucleic acids denature. Ionic strength also affects enzyme activity, especially when the activity of the enzyme (catalysis) depends on the movement of charged molecules relative to each other. In practical terms, ionic strength is usually discussed in conjunction with salinity.

7.6 Desiccation

All known life is based on aqueous solution chemistry; thus, a lack of water is stressful to lethal. For example, water makes up 95–99% of the total molecules in invertebrates, and humans cannot survive even 14% water loss [38]. Thus, all life on Earth requires liquid water, with the possible exception of some lichens, which are thought to be able to survive on water vapor. Water stress is a particular concern to all terrestrial species.

Dessication can damage lipids, proteins, and nucleic acids. Even exposure to a liquid–air interface can cause protein denaturation. Hydrophobic groups tend to associate in the air while the hydrophilic ones remain in the water. The presence of shear helps to unfold the protein and to introduce more air into the solution.

In spite of the critical need for water, some organisms can lose almost all of their body water and still survive in a state of anhydrobiosis. Further, a variety of organisms use desiccation-resistant spores to survive dry periods, and for dispersal, for example by wind. Desiccation tolerance has been reported for a variety of organisms including bacteria, yeast and other fungi, plants, insects, and crustacea (reviewed in [39, 40, 41]). Resurrection plants such as *Craterostigma plantagineum* are unusual in that the plant itself can survive desiccation and can even revive from months in an air-dried state [8]. Some can even survive loss of chlorophyll and thylakoid membranes.

Increasingly, culture collections are using lyophilization to preserve microbes. This requires various adaptations including behavioral, morphological, physiological, and physical. Important among these is the ability to withstand high salinity Thus anhydrobiotic organisms accumulate compatible solutes such as the sugar trehalose in invertebrates and plants [38].

7.7 Pressure

We are quite sensitive to changes in pressure. For example, pilots and divers must be aware that rapid changes in pressure upon ascent can result in gases, generally nitrogen, coming out of solution in the blood forming gas bubbles and, if not treated, will result in death. A similar situation occurs in microbes with gas-filled vacuoles. If decompression occurs too rapidly, the vacuole expands and bursts, causing death.

There is life under high hydrostatic pressure in the deep trenches of the ocean, including microbes, invertebrates, and fish. Some of these are "barophiles," "weight lover," or more recently "piezophiles," that is, organisms that require high pressure [42]. However, most microbes from the deep ocean are able to grow at atmospheric pressure if decompression occurs gradually. Conversely, Sharma *et al.* [43] showed that the surface bacteria *Escherichia coli* and *Shewanella oneidensis* are metabolically active up to 1600 MPa, substantially higher than ambient pressure.

Under high hydrostatic pressure, there is also an increase in the viscosity of the cell membrane, similar to that of decreasing temperature. As in the case of low temperature, most organisms respond by increasing the percentage of monounsaturated lipids in the phosopholipid membrane [42].

7.8 Radiation

Radiation, both ultraviolet (UV) and ionizing radiation, are particularly hazardous to life. They can damage every biopolymer, causing destruction of nucleic acids, proteins and lipids. The peak absorption for proteins is ~280 nm, and that of nucleic acids ~260 nm, so ultraviolet radiation near these wavelengths, or shorter, is especially hazardous. Several types of damage to DNA can be directly induced by radiation, including single- and double-stranded breaks, cross-linkage of DNA with proteins, and modification of the purine (adenine and guanine) and pyrimidine (thymine and cytosine) bases. This makes life in a high-UV environment, such as the surface of the early Earth or Mars, or in space, extremely hazardous. Because of the importance of keeping biopolymers intact and functional, organisms can avoid exposure, for example by living under sand or producing UV-attenuating pigments [44]. However, because radiation damage cannot always be avoided, and there are other sources of DNA damage that are the same as that caused by radiation, there are multiple generalized mechanisms for DNA repair in all organisms, especially nucleotide excision repair [44].

Still, there are some organisms that stand out in their ability to handle radiation damage. In general, organisms that go through periods of desiccation or anhydrobiosis are not able to repair damage while desiccated. When they rehydrate, they must repair the accumulated damage quickly. Beyond this, there are a few organisms that are known for their outstanding radiation tolerance. Among the animals, tardigrades (water bears) are able to tolerate high levels of gamma radiation, but highly irradiated animals did not lay eggs that hatched [46, 47]. Among the bacteria, several members of the genus *Deinococcus*, particularly *D. radiodurans*, are known as the "gold medalists" for radiation resistance (e.g. [48]). The accumulation of Mn(II) helps confer gamma irradiation resistance in *D. radiodurans* [49]. Daly and collaborators [50] showed that a remarkably small number of genes are responsible for extreme radiation resistance, and these are not of novel DNA repair systems but of cell-cleaning systems.

7.9 Oxygen

We are absolutely dependent on a steady supply of oxygen, which we breathe in from our ~21% oxygen environment. Thus, it is hard to appreciate that, like radiation damage, oxidative damage can result in the destruction of all major biomolecules resulting in the sort of nucleic acid damage outlined above, as well as oxidation of proteins and lipid peroxidation. Oxygen is reduced to water in a series of steps, the terminal compound being water. Oxygen and water are not hazardous, but the intermediate so-called "reactive oxygen species" are, particularly the hydroxyl radical ($\cdot$OH).

The reactions, from oxygen to the superoxide anion (O_2^-), hydrogen peroxide (H_2O_2), the hydroxyl radical ($\cdot$OH), and to H_2O, are:

$$O_2 \leftrightarrow O_2^- \leftrightarrow H_2O_2 \leftrightarrow \cdot OH \leftrightarrow H_2O \qquad (7.2)$$

The reduction of O_2 to H_2O occurs during aerobic respiration, and the reverse during oxygenic photosynthesis, which creates reactive oxygen species. Without the superior generation of energy from aerobic over anaerobic respiration, it is unlikely that animals would have arisen, due to their high metabolic demands. Oxidative damage is the price we pay. Current thinking suggests that many, if not all, human disease involves oxidative damage [51].

Interestingly, radiation can produce reactive oxygen species, and therefore oxidative damage can be an indirect effect of radiation exposure.

When oxidative damage occurs to DNA, some of the DNA damage repair mechanisms used to fix radiation damage can be used. In addition, antioxidants exist, such as the enzymes superoxide dismutase, catalases, and peroxidases, as well as low molecular weight antioxidants, such as ascorbate and glutathione [52].

Thus, we could consider an aerobe, especially an organism that is also undergoing oxygenic photosynthesis, to be an extremophile. But are there organisms that live in even higher than 21% oxygen? First, microbial mats and corals with algal members can produce oxygen concentrations within the mat of several hundred times the saturation level (e.g. [53, 54, 55]). Antarctic lakes also have been shown to be supersaturated with oxygen [56, 57]. The cells can respond by producing high levels of the enzymes superoxide dismutase, catalase, and ascorbate peroxidase [53].

Second, during the Phanerozoic, atmospheric oxygen levels are thought to have been higher than at present during two separate times, the most dramatic of these during the late Paleozoic when oxygen levels are thought to have gone over 30% [58].

These higher levels of oxygen affected evolution in several important ways [59, 60]. Gigantism evolved in several taxa of insects during this time, resulting in, for example, dragonflies with a wingspan up to 71 cm [59]. This gigantism may have been partially the result of the use of additional oxygen and partially from the increased lift that a denser atmosphere would have provided [59]. The fossil record also shows a correlation between increases in oxygen concentration and increased size in egg-laying vertebrates and mammals [60]. Experimental studies show increases in body size in trout and alligator, *inter alia*, in response to hyperoxia [60].

In sum, although high oxygen is extreme in that it can increase oxidative damage, many organisms are capable of dealing with well above present levels of oxygen. The need for oxygen for large and energy-intense organisms such as insects, must far outweigh the negative effects of coping with oxidative damage.

7.10 How to look for life – biosignatures

Like any detective work, searching for life elsewhere relies on two main approaches: direct physical evidence of an organism, or indirect evidence. These vary depending on whether the organism is alive (extant) or dead (extinct).

The dream of many is to find evidence of extant life, meaning finding a visible body. However, many microbes are small with a very limited morphology. Most archaea and bacteria are tiny spheres, rods, or spirals, and can be difficult to distinguish from abiotic features. So, if that body has behavior, or reproduces, so much the better. It is possible that an indirect detection of tiny microbial life would be more likely than direct detection. Included here would be the sort of metabolic experiments exemplified by biological experiments on NASA's Viking Landers. Of these, the carbon assimilation experiment (also called the pyrolytic release experiment) looked for the incorporation of inorganic carbon into organic compounds, the gas exchange experiment measured production and/or uptake of CO_2, N_2, CH_4, H_2, and O_2, and labeled release experiments looked for the transformation of organic molecules into gases [61, 62, 63].

Similarly, the detection of extinct life could be based on either direct or indirect evidence. Direct would include finding fossilized components or otherwise preserved artifacts such as inorganic shells or even subcellular components such as magnetosomes. On Earth, trace fossils indicating the migration of organisms are used to infer the former presence of life. Metabolic products, such as certain sterols, may be preserved such as those on Earth. Finally, isotopic ratios, most famously the ratio of ^{12}C to ^{13}C, in organic materials are used to indicate biological (autotrophic) activity.

Figure 7.6 Potential habitable locations in our solar system. Individual planetary images courtesy of NASA and ESA. The planets are shown in relative size to each other, and in relative distance from the Sun.

7.11 Possibilities for life – past or present

More than 2300 years ago the ancient Greeks debated the possibility of life existing beyond Earth.[1] In our own solar system the number of bodies thought to harbor life has gone from nearly all of the planets in the nineteenth century to just a few planets and moons today (Figure 7.6). These are listed, along with some of their more relevant physical and chemical properties in Table 7.2.

Not included in the list are Mercury and our Moon because of their extreme temperatures, lack of atmosphere, and their slim chance of containing liquids. Also unlikely to harbor life are the four Jovian planets – Jupiter, Saturn, Uranus, and Neptune.[2] All are gas giants with extremely cold atmospheres, but all have interiors that must be quite hot thus allowing liquid water at some altitude. Still, the prospect of life is small. Jupiter and Saturn have strong vertical winds that would carry any complex organic molecules to depths where they would be destroyed by heat. Uranus and Neptune have atmospheres even colder than those of Jupiter and Saturn, and also have vertical winds similar, albeit slower, to those of Jupiter and Saturn. But Uranus and Neptune have outer cores of water, methane, and ammonia, and these may even be in liquid form. While the high pressures, unusual combination of liquids, and lack of a candidate energy source suggest that life

---

[1] For an excellent review of the potential for life in our solar system, see J. O. Bennett and S. Shostak. *Life in the Universe*, 2nd edn. (San Francisco, CA and London: Addison-Wesley, 2007).

[2] Summarized in J. O. Bennett and S. Shostak. *Life in the Universe*, 2nd edn. (San Francisco, CA and London: Addison-Wesley, 2007).

Table 7.2 *Potential abodes for life in the solar system*

Body	Average distance from Sun (km)	Average distance from Sun (AU)	Equatorial radius (km)	Water	Organic other than CH_4	Atmosphere
Venus	108,208,930	0.723	6051			CO_2, N_2
Earth	149,597,890	1	6378	+	+	N_2, O_2
Mars	227,936,640	1.6	3397	+	?	CO_2, N_2, Ar
Europa	778,412,020	5.2	1565	+	?	O_2
Titan	1,426,725,400	9.5	2575	?	+	Thick! N_2, (CH_4)
Enceladus	1,426,725,400	9.5	256	+	?	O_2

is unlikely, it cannot be ruled out and will need to wait for future unmanned exploration.

7.12 Venus

Venus is, with respect to astrobiology, Earth's "twin" that has "gone bad." It is the planet closest to Earth, only 30% closer to the Sun, and is similar in size, mass, density, and volume. If its atmosphere were of the same composition as that of Earth, it would have a global average temperature of 35 °C [64].

Proctor [2] and others have discussed the possibility of a Venusian biosphere. However, hopes for life on the surface of Venus were dashed by the discoveries first of NASA's Mariner 2 spacecraft, which flew past Venus in 1962, confirming a blisteringly hot surface, and later of the Russian Venera landers in the 1980s. Additional information has been gained from NASA's Magellan spacecraft, the Russian Venera orbiters, and the European Space Agency's Venus Express mission, which entered Venusian orbit on April 11, 2006.

The surface temperature is the hottest in the solar system, 477 °C, so there is no liquid water. This is the result of a runaway greenhouse effect resulting from a CO_2-rich atmosphere, which trapped incident sunlight and thus raised the surface temperature. While on Earth nearly all of the CO_2 outgassed by the Earth is now locked up in carbonate rock or dissolved in the ocean, CO_2 outgassed by Venus remains in its atmosphere, which is thus 96% CO_2. Further, the atmospheric pressure is >90 times that of Earth at sea level, and contains clouds of sulfuric acid. But during the time prior to the runaway greenhouse effect, Venus should have been Earth's slightly warmer twin in all planetary features, and thus could well have also supported either an indigenous Venusian biota, or organisms transported from Earth or Mars [65]. Crater counts show that the surface of Venus is less than 1 billion years old, and thus it is unlikely that an extinct Venusian biota will ever be discovered.

While the surface is clearly too hot to support life today, there are those who, even in the face of current data, have shifted attention from the surface to the clouds of Venus. First Sagan [66], and more recently Cockell [67], and Grinspoon, Schulze-Makuch, and others [68] have suggested that, in the lower and middle cloud layers of Venus, the temperatures are lower and water vapor is available, and thus these clouds could provide a refuge for life. Perhaps acidophilic sulfate-reducing chemoautotrophs suspended as aerosols could survive there even today. The UV radiation flux would be high, but Schulze-Makuch *et al.* [68] have suggested a way that the flux could be attenuated, allowing life to survive.

7.13 Earth

It is trivial to point out that Earth is infested with life, but for that reason it is instrumental as a base for understanding life elsewhere. It is clearly instructive to determine the limits of life, as discussed above. However, knowing the limits to life for individual extreme environments may give a false impression of the actual envelope for life. Certainly some organisms tolerate multiple extremes. These "polyextremophiles" include organisms that live in hot, acidic environments, such as the red alga, *Cyanidium caldarium*, and, even more impressively, *Sulfalobus acidocaldarius*, an archaean that flourishes at pH 3 and 80 °C. However, there are many combinations where the niche may not exist or cannot be filled. For example, high-salinity, low-pH environments may not exist on Earth, but may be theoretically habitable. A better understanding of possible combinations of polyextremophilic conditions may be the most important next step in the study of extremophiles.

As mentioned above, direct and indirect methods are applied to the discovery of extant and extinct life on Earth, and some could be used to look for life elsewhere. But even in the search for extinct life on Earth, there may be uncertainties in interpretation. The debate over the last few years regarding the authenticity of Precambrian fossils originally described by Schopf (e.g. [69]) and later challenged by Brasier and colleagues [70] and countered by Schopf and colleagues [71] illustrates the difficulty of distinguishing between inorganic artifact and actual fossils in even terrestrial samples.

7.14 Mars

As with Venus, hopes for neighboring civilizations have centered on our other neighbor, Mars. This interest reached a feverish pitch in the minds of scientists and the public at large during the latter half of the nineteenth century by the discovery of "canali" on Mars by Giovanni Schiaparelli [72]. Beginning with NASA's Mariner IV mission, the first close-up peak at the Martian surface, the hopes for a neighboring intelligence or even extensive vegetation were dashed. Mars was dry. Between the cold temperatures and low atmospheric pressures, it was thought that it was physically impossible for the surface of Mars to even support liquid water. But by the end of the 1960s, Mariner 9 had sent back photos of gigantic volcanoes, a grand canyon stretching 4800 km across, and, most important for astrobiology, the relics of ancient riverbeds. And, just to make sure, the NASA Viking Landers which arrived on Mars in 1976 carried cameras. Yet, no macroscopic life has been detected.

But the jury is still out on the possibility of extinct life.

Unlike Venus, the cratering record of Mars shows an ancient surface which preserves ample geomorphology suggesting an earlier epoch of flowing water. Flowing water suggests warmer temperatures. Further, although the Viking Lander GC-MS (gas chromatograph-mass spectrometer) did not detect the presence of organic compounds on the surface, surely the conditions existed on early Mars for the production of organics. Further, as on Earth, there is an exogenous input of organics through meteorites, comets and so on. So, with liquid water, organics, and ample solar energy and if life is inevitable, or even possible given these physical and chemical conditions, surely life may have originated on Mars. As Mars lost its atmosphere and cooled, the life could have gone extinct, may still seek refuge in such places as evaporites or under the surface [43, 73] or even left descendants on the Earth.

In 1996 McKay and colleagues [74] stunned the world with their interpretation of a suite of morphological and chemical features of Alan Hills Meteorite ALH84001. Taken together, these features were interpreted as demonstrating the presence of fossils of Martian microbes. Although this interpretation has not been generally accepted, it remains one of the most exciting stories in the history of looking for life on Mars as it suggested new approaches to the search.

And the jury is still out on the possibility of extant microbial life.

We know there is frozen water in the polar ice caps, a small amount of gaseous water in the atmosphere, and possibly liquid water in the subsurface. NASA's Mars Odyssey gamma-ray spectrometer has allowed the identification of two regions near the poles that are enriched in hydrogen, and are also likely to contain frozen water as they correlate with areas of maximum ice stability [75]. The Mars Global Surveyor returned photos of Mars from 1997, but fell silent in November 2006. But it has excited the world with a stunning posthumous discovery: that there may be liquid water periodically on the surface of Mars even today. Pairs of photos, showing the same gully in December 2001 and April 2005 [76], suggest that water is flowing.

Environments where life is found on Earth and are present on Mars include the polar ice caps, in subsurface communities, in hydrothermal systems, or in endoliths or evaporites [73]. All circumvent the high flux of UV radiation on the surface. Primarily for these reasons, and the likelihood of eventual human settlement, Mars remains a prime target for missions, if not to specifically look for life, to better understand the environment of the planet. By understanding the environment, we can better judge its habitability.

7.15 Europa

If liquid water is the *sine qua non* of life, then Europa, the second large satellite of Jupiter, is one of the most exciting objects in our solar system. Europa has long been thought to have water ice on its surface [77], a model that gained support from the geological and geophysical observations from the Galileo spacecraft (reviewed in [78]). Recent estimates suggest a deep liquid water ocean, 100 km in depth, beneath an icy shell of ~ 15 km in thickness [79], with a volume of the ocean about twice that of Earth's [80, 81]. The temperature at bottom could be that at which water attains maximum density, $4\,°C$.

The second prerequisite for life, the presence of organic carbon, is also thought to be present in substantial quantities on Europa, although this has not been confirmed either [78]. The organic inventory is unlikely to have come from exogenous input (e.g. by comets, meteorites, and interplanetary dust particles) because without an atmosphere to decelerate these bodies, any organics would likely be destroyed.

Photosynthesis is difficult, but as life on Earth arose prior to photosynthesis, we know that does not preclude the origin of life. Note that the ability to originate life is not the same as sustenance. Could life arise on Europa?

Unlike the planets discussed above, Europa has a very short history as a body of astrobiological interest because it was thought to be too far from the Sun to have sufficient heat to maintain liquid water (reviewed in [82]). However, tidal flexing between Europa and Jupiter could provide a mechanism to generate enough energy to keep a liquid water ocean beneath the surface. It should be added that both Ganymede, the largest moon in the solar system, and even Callisto, the furthest of the Galilean moons from Jupiter, may contain liquid water. However, even if some salty, liquid water is present on either, it would be below a very thick layer of ice, and thus even less likely to harbor life.

In 1975, Consolmagno proposed that life might be present in a liquid water ocean on Europa. Since then, a series of models for a potential Europan biota have been described. Water inclusions near the surface have been postulated to be capable of containing photosynthetic organisms that could harvest biogenic energy from the weak sunlight (5% as intense as on Earth) at Europa [83, 84, 85].

Chyba and Phillips [86] thought the probability of a photosynthetic organism on Europa unlikely and instead see methanogenesis at hydrothermal vent sites in Europa's putative ocean more likely. Unfortunately, methanogenesis-based communities on Earth are not able to support as high a biomass as those based on photosynthesis. McCollom [87] estimates a potential annual biomass production from methanogenesis on Europa from vents at $< 10^8 - 10^9$ g per year, which he compares to the terrestrial primary production based on photosynthesis of $< 10^{17}$ g per year.

Radiation could produce both oxidants and simple organics at the surface. Survival of these organics depends on the competing mechanisms of "gardening," which incorporates the organics beneath the surface, and "sputtering" by incoming radiation, which ejects surface particles. Chyba and Phillips [86] conclude that organics could make their way into a subsurface ocean. Megafauna, such as squid, on Earth needs substantial amounts of oxygen for aerobic metabolism. Such levels of dissolved oxygen in Europa's ocean are unlikely, thus, by analogy with the Earth, suggesting that a megafauna would not exist in Europa's ocean. They suggest that further missions are needed, but the number one priority must be to find organics.

A severe paucity of oxidants cautions against expectations that diverse, thriving life will be found in a Europan ocean. Gaidos *et al.* [88] assumed that there was no interchange between ocean and surface because of an ice crust tens to hundreds of kilometers thick. This would make life difficult energetically since even hydrothermal vent communities on Earth use mostly oxidized species (SO_4^{2-}, O_2, and CO_2) transported from the surface. This problem could be circumvented with methanogenesis and elemental sulfur reduction, but these are low energy yielding processes. The only plausible oxidants may be oxidized metals such as ferric iron. However, Johnson *et al.* [89] used data and modeling to suggest that oxidants might be able to penetrate the ice crust into the regolith.

Kargel *et al.* [90] suggested that biomass could be high enough to support a dense, well-developed ecosystem if their calculations on oxidant delivery to the ocean are correct. The ocean is likely to be stratified: at a minimum, the top of the ocean must be in equilibrium with ice, and the solute abundances are most likely to be extremely high and saturated, so the seafloor could be covered with precipitated salts. If these assumptions are true, the temperature of the ocean at the base of the ice would be between -57 and $-7\,°C$, well within the range of life on Earth. Temperature could be warmer at depth, particularly if there are hydrothermal vents. Pressure depends on the thickness of the icy shell, and could be up to about 200 MPa. Most other parameters, such as pH, are poorly constrained. Extensive discussions compare possible niches on Europa with limits for terrestrial life, and find there are no "show-stoppers" – almost every physiological stress can be dealt with as long as there is liquid water. There are extremophiles that are good analogues for a potential Europan organism [91], including the bacterial genus *Halomonas* [90], which can grow in a wide range of salt concentrations (0.01% to saturated brine), grow at temperatures between –5 and 45 °C, and tolerate high pH and millimolar levels of heavy metals. It has been isolated from deep-sea, high-pressure environments (22 MPa), deep-sea sediments (34 MPa), and sea ice brines. Further, *Halomonas* can grow on low concentrations of organic compounds (0.002% w/v).

However, even if life can survive in these conditions, it is unknown whether life could originate on Europa.

7.16 Titan

The second-largest moon in the solar system, Saturn's moon Titan is a frigid – 180 °C. Since 1944 there has been spectroscopic evidence identifying methane in Titan's atmosphere. The flyby of Titan by the two Voyager spacecraft in 1980 and 1981 showed that Titan was surrounded by an opaque, reddish "smog" made of the photochemical products of methane. In January 2005 the European Space Agency's Huygens probe descended to the surface of Titan. Complex organics were found in the atmosphere [92]. The surface showed traces of once flowing liquid, similar to the geomorphological evidence used to suggest liquid water on ancient Mars [93]. The infrared reflectance spectrum measured for the surface is unlike any other in the solar system, and is consistent with an organic material such as tholins, while a blue slope in the near-infrared suggests another, unknown constituent [93]. Absorption from water ice is seen.

Although there is intriguing organic chemistry on Titan, and even the possibility of a subsurface ocean of liquid water and ammonia, life on Titan is unlikely. McKay and Smith [94] have explored the possibility of a methane-based life on Titan, and showed the energetics to be favorable. However, they acknowledge that there are two major problems. The first is the low reaction rate in liquid methane, but this could conceivably be overcome with catalysts. More troublesome is the lower solubility of organics in methane relative to water.

7.17 Enceladus

Tiny Enceladus, the sixth largest of the Saturnian moons, is suddenly on the radar screen. Earlier Earth-based and Voyager observations suggested an icy – and complex – surface. Cassini's cameras showed images of folded mountain ridges, cracked white ice plains and dark green organic material. The south polar area is still being resurfaced by cryovolcanism and fresh snowfall. An underground heat source lies beneath the south polar "tiger stripes" of organic material [95]. From these warm vents water vapor, ice, and dust spew forth. And with liquid water and organics, Enceladus becomes a prime candidate to search for evidence of life or even prebiotic chemistry [1].

7.18 How could life evolve on different planets?

This survey of the solar system has focused on habitats suitable for the survival of life. But the conditions for the origin of life may be quite different than that of its maintenance. It is uncertain whether life could arise on present-day Earth. If the conditions for the origin of life differ from those suitable for its continuance, it means that a host body either needs to change as the Earth did, or life may need to come from elsewhere. For example, an indigenous biota on Mars could have seeded the early Earth. This sort of transfer of life, panspermia, is currently being tested experimentally through studying the survival of organisms in space on satellites (e.g. [40, 96, 97, 98, 99]) to simulate transport on a meteorite or comet, and through modeling of the impacts that would be required for lift-off and for delivery (e.g. [100]). These experiments lend scientific credibility to the idea that such transfer is at least plausible.

7.19 The future

The lessons from the last decade or two are that, on one side, the range for life on Earth is far vaster than we realized, although all niches may not be filled – or indeed be "fillable." On the other, we know less about our own solar system than we thought. With the energetic exploration of Earth by biologists, and the planets and moons by the planetary community aided by a vigorous mission program by the various space agencies, we may in the coming decades discover that we are not the only living creatures in even our own neighborhood.

Acknowledgements

I am indebted to Connie Bertka for the invitation to contribute to this volume, as well as her extraordinary patience in seeing this work through. NASA's Astrobiology Program has supported this work.

References

[1] C. C. Porco, P. Helfenstein, P. C. Thomas, *et al.* Cassini observes the active South Pole of Enceladus. *Science*, **311** (2006), 1393–1401.
[2] R. Proctor. *Other Worlds Than Ours* (Longmans, 1870), pp. 134.

[3] S. A. Benner, A. Ricardo, and M. A. Carrigan. Is there a common chemical model for life in the universe? *Current Opinions in Chemical Biology*, **8** (2004), 672–689.

[4] Committee on the Limits of Organic Life in Planetary Systems, Committee on the Origins and Evolution of Life. *The Limits of Organic Life in Planetary Systems* (Washington, DC: The National Academies Press, 2007).

[5] R. MacElroy. Some comments on the evolution of extremophiles. *Biosystems*, **6** (1974), 74–75.

[6] L. J. Rothschild and R. L. Mancinelli. Life in extreme environments. *Nature*, **409** (2001), 1092–1101.

[7] F. Ashcroft. *Life at the Extremes: The Science of Survival* (London: Flamingo Press, 2001), p. 326.

[8] D. Bartels. Desiccation tolerance studied in the resurrection plant *Craterostigma plantagineum*. *Integrative and Comparative Biology*, **45** (2005), 696–701.

[9] L. J. Rothschild. Extremophiles: defining the envelope for the search for life in the universe. In *Planetary Systems and the Origins of Life*, eds. R. E. Pudritz, P. Higgs and J. Stone (Cambridge: Cambridge University Press, 2007), pp. 123–146.

[10] S. Arai and M. Hirai. Reversibility and hierarchy of thermal transition of hen egg-white lysozyme studied by small-angle x-ray scattering. *Biophysical Journal*, **76** (1999), 2192–2197.

[11] E. Marguet and P. Forterre. DNA stability at temperatures typical for hyperthermophiles. *Nucleic Acids Research*, **22** (1994), 1681–1686.

[12] M. Kampmann and D. Stock. Reverse gyrase has heat-protective DNA chaperone activity independent of supercoiling. *Nucleic Acids Research*, **32** (2004), 3537–3545.

[13] J. H. A. Nagel, A. P. Gultyaev, K. J. Öistämö, K. Gerdes, and C. W. A. Pleij. A pH-jump approach for investigating secondary structure refolding kinetics in RNA. *Nucleic Acids Research*, **30** (2002), e63.

[14] R. J. Lambros, J. R. Mortimer, and D. R. Forsdyke. Optimum growth temperature and the base composition of open reading frames in prokaryotes. *Extremophiles*, **7** (2003) 443–450.

[15] J. A. Kowalak, J. J. Dalluge, J. A. McCloskey, and K. O. Stetter. The role of posttranscriptional modification in stabilization of transfer RNA from hyperthermophiles. *Biochemistry*, **33** (1994), 7869–7876.

[16] P. H. Ray, D. C. White, and T. D. Brock. Effect of growth temperature on the lipid composition of *Thermus aquaticus*. *Journal of Bacteriology*, **108** (1971), 227–235.

[17] S. V. Albers, J. L. van de Vossenberg, A. J. Driessen, and W. N. Konings. Adaptations of the archaeal cell membrane to heat stress. *Frontiers in Bioscience*, **5** (2000), D813–D820.

[18] R. Singleton Jr. and R. E. Amelunxen. Proteins from thermophilic microorganisms. *Bacteriological Reviews*, **37** (1973), 320–342.

[19] T. D. Brock. Life at high temperatures. *Science*, **158** (1967), 1012–1019.

[20] L. Reysenbach, G. S. Wickham, and N. R. Pace. Phylogenetic analysis of the hyperthermophilic pink filament community in Octopus Spring, Yellowstone National Park. *Applied and Environmental Microbiology*, **60** (1994), 2113–2119.

[21] E. Blöchl, R. Rachel, S. Burggraf, D. Hafenbradl, H. W. Jannasch, and K. O. Stetter. *Pyrolobus fumarii*, gen. and sp. nov., represents a novel group of archaea, extending the upper temperature limit for life to 113 °C. *Extremophiles*, **1** (1997), 14–21.

[22] H. Huber and K. O. Stetter. Hyperthermophiles and their possible potential in biotechnology. *Journal of Biotechnology*, **64** (1998), 39–52.

[23] E. J. Carpenter, S. Lin, and D. G. Capone. Bacterial activity in South Pole snow. *Applied and Environmental Microbiology*, **66** (2000), 4514–4517.

[24] W. D. Schmid. Survival of frogs in low temperature. *Science*, **215** (1982), 697–698.

[25] P. L. Davies and C. L. Hew. Biochemistry of fish antifreeze proteins. *The Federation of American Societies for Experimental Biology Journal*, **4** (1990), 2460–2468.

[26] A. Clarke. Evolution at low temperatures. In *Evolution on Planet Earth: The Impact of the Physical Environment*, eds. L. Rothschild and A. Lister (London: Academic Press, 2003), pp. 187–208.

[27] M. Ageno, E. Dore, and C. Frontali. The alkaline denaturation of DNA. *Biophysical Journal*, **9** (1969), 1281–1311.

[28] J. Seckbach. The Cyanidiophyceae: hot spring acidophilic algae. In *Enigmatic Microorganisms and Life in Extreme Environments*, ed. J. Seckbach (Dordrecht: Kluwer Academic Publishers, 1999), pp. 427–435.

[29] I. Enami. Mechanisms of the acido- and thermophily of *Cyanidium caldarium* Geitler V. Acid and heat stabilities of soluble proteins. *Plant and Cell Physiology*, **19** (1978), 869–876.

[30] C. Schleper, G. Puehler, I. Holz, *et al. Picrophilus* gen. nov., fam. nov.: a novel aerobic, heterotrophic, thermoacidophilic genus and family comprising archaea capable of growth around pH 0. *Journal of Bacteriology*, **177** (1995), 7050–7059.

[31] K. J. Edwards, P. L. Bond, T. M. Gihring, and J. F. Banfield. An archaeal iron-oxdizing extreme acidophile important in acid mine drainage. *Science*, **287** (2000), 1796–1799.

[32] R. F. Martins, W. Davids, W. A. Al-Sond, F. Levander, P. Radström, and R. Hatti-Kaul. Starch-hydrolyzing bacteria from Ethiopian soda lakes. *Extremophiles*, **5** (2001), 135–144.

[33] K. Pedersen, E. Nilsson, J. Arlinger, L. Hallbeck, and A. O'Neill. Distribution, diversity and activity of microorganisms in the hyper-alkaline spring waters of Maqarin in Jordan. *Extremophiles*, **8** (2004), 151–164.

[34] W. D. Grant, W. E. Mwatha, and B. E. Jones. Alkaliphiles, ecology, diversity and applications. *FEMS Microbiology Reviews*, **75** (1990), 255–270.

[35] H. C. Rees, W. D. Grant, B. E. Jones, and S. Heaphy. Diversity of Kenyan soda lake alkaliphiles assessed by molecular methods. *Extremophiles*, **8** (2004), 63–71.

[36] E. A. Vasquez, E. P. Glenn, G. R. Guntenspergen, J. J. Brown, and S. G. Nelson. Salt tolerance and osmotic adjustment of *Spartina alterniflora* (Poaceae) and the invasive M haplotype of *Phragmites australis* (Poaceae) along a salinity gradient. *American Journal of Botany*, **93** (2006), 1784–1790.

[37] D. M. Harper, R. B. Childress, M. M. Harper, *et al.* Aquatic biodiversity and saline lakes: Lake Bogoria National Reserve, Kenya. *Hydrobiologia*, **500** (2003), 259–276.

[38] M. Watanabe. Anhydrobiosis in invertebrates. *Applied Entomology and Zoology*, **41** (2006), 15–31.

[39] L. M. Crowe and J. H. Crowe. Anhydrobiosis: a strategy for survival. *Advances in Space Research*, **12**(4) (1992), 239–247.

[40] J. H. Crowe, F. A. Hoekstra, and L. M. Crowe. Anhydrobiosis. *Annual Review of Physiology*, **54** (1992), 579–599.

[41] R. L. Mancinelli, M. R. White, and L. J. Rothschild. Biopan-survival I: exposure of the osmophiles *Synechococcus sp.* (Nageli) and *Haloarcula sp.* to the space environment. *Advances in Space Research*, **22**(3) (1998), 327–334.

[42] A. A. Yayanos. Microbiology to 10,500 meters in the deep sea. *Annual Review of Microbiology*, **49** (1995), 777–805.

[43] A. Sharma, J. H. Scott, G. D. Cody, *et al.* Microbial activity at gigapascal pressures. *Science*, **295** (2002), 1514–1516.

[44] L. J. Rothschild and L. J. Giver. Photosynthesis below the surface in a cryptic microbial mat. *International Journal of Astrobiology*, **1** (2003), 295–304.

[45] C. Petit and A. Sancar. Nucleotide excision repair: from *E. coli* to man. *Biochimie*, **81** (1999), 15–25.

[46] K. I. Jönsson, M. Harms-Ringdahl, and J. Torudd. Radiation tolerance in the eutardigrade *Richtersius coronifer*. *International Journal of Radiation Biology*, **81** (2005), 649–656.

[47] D. D. Horikawa, T. Sakashita, C. Katagiri, *et al.* Radiation tolerance in the tardigrade *Milnesium tardigradum*. *International Journal of Radiation Biology*, **82** (2006), 843–848.

[48] J. R. Battista. Against all odds: the survival strategies of *Deinococcus radiodurans*. *Annual Review of Microbiology*, **51** (1997), 203–224.

[49] M. J. Daly, E. K. Gaidamakova, V. Y. Matrosova, *et al.* Accumulation of Mn(II) in *Deinococcus radiodurans* facilitates gamma-radiation resistance. *Science*, **306** (2004), 1025–1028.

[50] K. S. Makarova, *et al. Deinococcus geothermalis*: the pool of radiation resistance genes shrinks. *PLoS ONE,* Issue 9 (2007), e955.

[51] R. A. Jacob and B. J. Burri. Oxidative damage and defense. *American Journal of Clinical Nutrition*, **63** (1996), 985S–990S.

[52] O. Blokhina, E. Virolainen, and K. V. Fagerstedt. Antioxidants, oxidative damage and oxygen deprivation stress: a review. *Annals of Botany*, **91** (2003), 179–194.

[53] N. Shashar, Y. Cohe, and Y. Loya. Extreme diel fluctuations of oxygen in diffusive boundary layers surrounding stony corals. *Biological Bulletin*, **185** (1993), 455–461.

[54] C. M. Burke. Benthic microbial production of oxygen supersaturates the bottom water of a stratified hypersaline lake. *Microbial Ecology*, **19** (1995), 163–171.

[55] M. Kühl, C. Lassen, and N. P. Revsbech. A simple light meter for measurements of PAR (400 to 700 nm) with fiber-optic microprobes: application for P vs $E0$(PAR) measurements in a microbial mat. *Aquatic Microbial Ecology*, **13** (1997), 197–207.

[56] R. A. Wharton Jr., C. P. McKay, G. M. Simmons, and B. C. Parker. Oxygen budget of a perennially ice-covered Antarctic lake. *Limnology and Oceanography*, **31** (1986), 437–443.

[57] H. Craig, R. A. Wharton Jr., and C. P. McKay. Oxygen supersaturation in ice-covered Antarctic lakes: biological versus physical contributions. *Science*, **255** (1992), 318–321.

[58] R. A. Berner, D. J. Beerlind, R. Dudley, J. M. Robinson, and R. A. J. Wildman. Phanerozoic atmospheric oxygen. *Annual Review of Earth and Planetary Science*, **31** (2003), 105–134.

[59] J. B. Graham, N. M. Aguilar, R. Dudley, and C. Gans. Implications of the late Palaeozoic oxygen pulse for physiology and evolution. *Nature*, **375** (1995), 117–120.

[60] R. A. Berner, J. M. VandenBrooks, and P. Ward. Oxygen and evolution. *Science*, **316** (2007), 557–558.

[61] N. H. Horowitz, G. L. Hobby, and J. S. Hubbard. Viking on Mars: the carbon assimilation experiments. *Journal of Geophysical Research*, **82** (1977), 4659–4661.

[62] H. P. Klein. The Viking biological experiment on Mars. *Icarus*, **34** (1978), 666–674.

[63] H. P. Klein. The Viking mission and the search for life on Mars. *Reviews of Geophysics and Space Physics*, **17** (1979), 1655–1662.

[64] J. O. Bennett and S. Shostak. *Life in the Universe*, 2nd edn. (San Francisco, CA and London: Addison-Wesley, 2007).

[65] B. Gladman, L. Dones, H. F. Levison, and J. A. Burns. Impact seeding and reseeding in the inner solar system. *Astrobiology*, **5** (2005), 483–496.

[66] C. Sagan. The planet Venus. *Science*, **133** (1961), 849–858.

[67] C. S. Cockell. Life on Venus. *Planetary and Space Science*, **47** (1999), 1487–1501.

[68] D. Schulze-Makuch, D. H. Grinspoon, O. Abbas, L. N. Irwin, and M. A. Bullock. A sulfur-based survival strategy for putative phototrophic life in the Venusian atmosphere. *Astrobiology*, **4** (2004), 11–17.

[69] J. W. Schopf. Microfossils of the Early Archean Apex chart: new evidence of the antiquity of life. *Science*, **260** (1993), 640–646.

[70] M. D. Brasier, O. R. Green, A. P. Jephcoat, *et al.* Questioning the evidence for Earth's oldest fossils. *Nature*, **416** (2002), 76–81.

[71] J. W. Schopf, A. B. Kudryavtsev, D. G. Agresti, T. J. Wdowiak, and A. D. Czaja. Laser-Raman imagery of Earth's earliest fossils. *Nature*, **416** (2002), 73–76.

[72] K. Zahnle. Decline and fall of the Martian empire. *Nature*, **412** (2001), 209–213.

[73] L. J. Rothschild. Earth analogs for Martian life. Microbes in evaporites, a new model system for life on Mars. *Icarus*, **88** (1990), 246–260.

[74] D. S. McKay, E. K. Gibson Jr., K. L. Thomas-Keprta, *et al.* Search for past life on Mars: possible relic biogenic activity in Martian meteorite ALH84001. *Science*, **273** (1996), 924–930.

[75] W. V. Boynton, W. C. Feldman, S. W. Squyres, *et al.* Distribution of hydrogen in the near surface of Mars: evidence for subsurface deposits. *Science,* **297** (2002), 81–85.

[76] M. C. Malin, K. S. Edgett, L. V. Posiolova, S. M. McColley, and E. Z. Noe Dobrea. Rate and contemporary gully activity on Mars. *Science*, **314** (2006), 1573–1577.

[77] P. M. Cassen, R. T. Reynolds, and S. J. Peale. Is there liquid water on Europa? *Geophysical Research Letters*, **6** (1979), 731–734.

[78] C. F. Chyba and C. B. Phillips. Europa as an abode of life. *Origins of Life and Evolution of Biospheres*, **32** (2002), 47–68.

[79] H. J. Melosh, A. G. Ekholm, A. P. Showman, and R. D. Lorenz. The temperature of Europa's subsurface water ocean. *Icarus,* **168** (2004), 498–502.

[80] R. T. Pappalardo, *et al.* Does Europa have a subsurface ocean? Evaluation of the geological evidence. *Journal Geophysical Research*, **104** (1999), 24015–24055.

[81] D. J. Stevenson. Europa's ocean: the case strengthens. *Science*, **289** (2000), 1305–1307.

[82] R. Greenberg. *Europa – the Ocean Moon: Search for an Alien Biosphere* (Berlin and Chichester, UK: Springer-Praxis, 2005).

[83] R. T. Reynolds, S. W. Squyres, D. S. Colburn, and C. P. McKay. On the habitability of Europa. *Icarus*, **56** (1983), 246–254.

[84] J. I. Lunine and R. D. Lorenz. Light and heat in cracks on Europa: implications for prebiotic synthesis. *Lunar and Planetary Science*, **28** (1997), 855–856.

[85] R. Greenberg, P. Geissler, B. Tufts, and G. Hoppa. Habitability of Europa's crust: the role of tidal-tectonic processes. *Journal of Geophysical Research*, **105**(E7) (2000), 17551–17562.

[86] C. F. Chyba and C. B. Phillips. Possible ecosystems and the search for life on Europa. *Proceedings of the National Academy of Sciences of the United States of America*, **98** (2001), 801–804.

[87] T. M. McCollom. Methanogenesis as a potential source of chemical energy for primary biomass production by autotrophic organisms in hydrothermal systems on Europa. *Journal of Geophysical Research*, **104** (1990), 30,729–30,742.

[88] E. J. Gaidos, K. H. Nealson, and J. L. Kirschvink. Life in ice-covered oceans. *Science*, **284** (1999), 1631–1633.

[89] R. E. Johnson, T. I. Quickenden, P. D. Cooper, A. J. Mckinley, and C. G. Freeman. The Production of oxidants in Europa's surface. *Astrobiology*, **3** (2003), 823–850.

[90] J. S. Kargel, J. Z. Kaye, J. W. Head, *et al.* Europa's crust and ocean: origin, composition, and the prospects for life. *Icarus*, **148** (2000), 226–265.

[91] G. M. Marion, C. H. Fritsen, H. Eicken, and M. C. Payne. The search for life on Europa: limiting environmental factors, potential habitats, and Earth analogs. *Astrobiology*, **3** (2003), 785–811.

[92] G. Israël, *et al.* Complex organic matter in Titan's atmospheric aerosols from in situ pyrolysis and analysis. *Nature*, **438** (2005), 796–799.

[93] M. G. Tomasko, *et al.* Rain, winds and haze during the Huygens probe's descent to Titan's surface. *Nature*, **438** (2005), 765–778.

[94] C. P. McKay and H. D. Smith. Possibilities for methanogenic life in liquid methane on the surface of Titan. *Icarus*, **178** (2005), 274–276.

[95] J. R. Spencer, J. C. Pearl, M. Segura, *et al.* Cassini encounters Enceladus: background and the discovery of a South Polar hot spot. *Science*, **311** (2006), 1401–1405.

[96] G. Horneck, H. Bücker, G. Reitz, *et al.* Microorganisms in the space environment. *Science*, **225** (1984), 226–228.

[97] G. Horneck. Responses of *Bacillus subtilis* spores to space environment: results from experiments in space. *Origins of Life and Evolution of Biospheres*, **23** (1993), 37–52.

[98] G. Horneck. European activities in exobiology in Earth orbit: results and perspectives. *Advances in Space Research*, **23** (1999), 381–386.

[99] W. Schulte, R. Demets, P. Baglioni, P. Rettberg, R. von Heise-Rotenburg, and J. Toporski. BIOPAN and ESPOSE: space exposure platforms for exo/astrobiological research in Earth orbit with relevance for Mars exploration. *Geophysical Research Abstracts*, **8** (2006), 06643.

[100] G. Horneck, D. Stöffler, and U. Eshweiller. Bacterial spores survive simulated meteorite impact. *Icarus*, **149** (2001), 285–290.

8

The quest for habitable worlds and life beyond the solar system

Carl B. Pilcher[1] and Jack J. Lissauer

There are infinite worlds both like and unlike this world of ours … We must believe that in all worlds there are living creatures and plants and other things we see in this world.

Epicurus (c. 300 BCE)

There are more things in heaven and earth, Horatio, than are dreamt of in your philosophy.

Shakespeare, *Hamlet*, Act I, Scene 5

We shall not cease from exploration
And the end of all our exploring
Will be to arrive where we started
And know the place for the first time.

T. S. Eliot, *Four Quartets*

One of the most basic questions that has been pondered by Natural Philosophers for (at least) the past few millennia concerns humanity's place in the universe: are we alone? This question has been approached from a wide variety of viewpoints, and similar reasoning has led to widely diverse answers. Aristotle believed that earth, the densest of the four elements, fell towards the center of the universe, so no other worlds could possibly exist. In contrast, Epicurus and other early atomists surmised that the ubiquity of physical laws implied that innumerable Earth-like planets must exist in the heavens.

Many aspects of the question of human uniqueness remain ill-constrained, but others have yielded to scientific investigation. Copernicus, Kepler, Galileo, and Newton convincingly demonstrated that the Earth is not the center of the universe,

---

[1] The first draft of this article was written when CBP was the Senior Scientist for Astrobiology in the Universe Division, Science Mission Directorate, NASA Headquarters, Washington, DC.

Exploring the Origin, Extent, and Future of Life: Philosophical, Ethical, and Theological Perspectives, ed. Constance M. Bertka. Published by Cambridge University Press. © Cambridge University Press 2009.

and that other worlds qualitatively similar to Earth orbit the Sun. Telescopic observations, and more recently interplanetary spacecraft, have told us a great deal about
these neighboring worlds. In the past dozen years, more than 300 planets have
been discovered in orbit about stars other than our Sun. These planets are much
more massive than our Earth, and most occupy quite different orbits, making them
unsuitable for life as we know it. However, technologies are currently being developed to locate and characterize Earth-like planets around other stars. Future studies
will reveal what, if anything, is special about our planet.

8.1 Finding worlds around other stars

Hundreds of planets have been discovered around other stars since 1995, and far
larger numbers are likely to be found in the upcoming decades. These objects are
referred to as extrasolar planets, extra-solar planets, or more informally as exoplanets. Several methods for detecting extrasolar planets are being used or studied for possible future use. As distant planets are extremely faint, most methods
are indirect, in the sense that the planet is detected through its influence on the
light that we detect from the star that it orbits. The various methods are sensitive to different classes of planets, and provide us with complementary information about the planets they do find, so most or all of them are likely to provide
valuable contributions to our understanding of the diversity of planetary system
characteristics.

    As of late 2008, more than 300 planets are known to be orbiting other stars
[1, 2, 3]. Figure 8.1 shows schematically those known in late 2005. All of the currently known planets are more than five times as massive as Earth, and most are
giants, made primarily from hydrogen and helium, lacking solid surfaces, and more
akin to Jupiter and Saturn than to Earth and its nearest neighbors, Venus and Mars.
Almost all known extrasolar planets orbit closer to their stars than does Jupiter
(the closest giant planet to our Sun), and many are nearer their star than Mercury,
the innermost planet of our solar system, is to the Sun. And most exoplanets that
aren't located *very* close to their stars travel on more elliptical paths than do any
planets in the solar system. All of the extrasolar planets thus far discovered induce
variations in stellar reflex motion much larger or occurring on a much shorter timescale than would a planetary system like our own, and the surveys accomplished to
date are strongly biased towards detecting high-mass and short-period planets [1].
Planets have so far been found around only 1 in 15 surveyed stars, so it is possible
that most Sun-like stars host planetary systems similar to our own. However, our
solar system represents a biased sample of a different kind, because it contains a
planet with conditions suitable for life to evolve to the point of being able to ask
questions about other planetary systems [4].

The 178 known nearby exoplanets

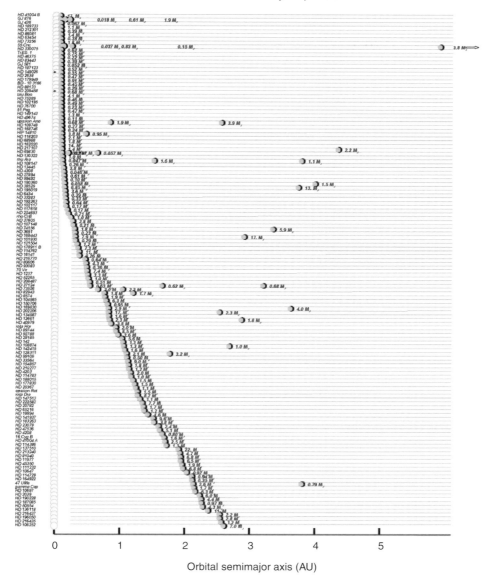

Orbital semimajor axis (AU)

Figure 8.1 Schematic of extrasolar planetary systems known as of late 2005 [5].
Figure posted at http://exoplanets.org/massradiiframe.html.

8.1.1 Radial velocity detections

Although there are several ways to detect extrasolar planets, almost all of the discoveries to date have been made from ultra-precise measurements of the radial
velocities of the planets' parent stars. An orbiting planet causes a small periodic

motion of its star – as Newton put it, for every action, there is an equal and opposite reaction, and the small motion of the massive star balances the much larger motion of the planet in the opposite direction. In the radial velocity technique, the wavelengths of the absorption lines in the stellar spectrum as viewed at the telescope are compared to those of the reference spectrum (typically iodine gas), providing a measurement of the Doppler shift of the star toward or away from the observer. The leading observer groups can now achieve better than 1 meter/second (three billionths of the speed of the light that is being observed!) precision in their measurements of stellar velocities. This corresponds to the ability to detect both Jupiter and Saturn in our solar system, although measurements would have to be made for a planetary orbital period (12 years for Jupiter, 30 years for Saturn) in order to have a firm detection. Figure 8.2 shows the radial velocity curve for a particularly well-studied extrasolar planet discussed further below.

High-mass planets orbiting close to their parent stars produce large variations in their star's motion on short timescales. It is thus not by chance that most known extrasolar planets range from a few tenths of to about ten times Jupiter's mass, and many are remarkably close to their stars, often closer than Mercury is to the Sun. In comparison, Earth's mass is 0.3% that of Jupiter. Figure 8.3 shows that the mass distribution of the known extrasolar planets increases strongly

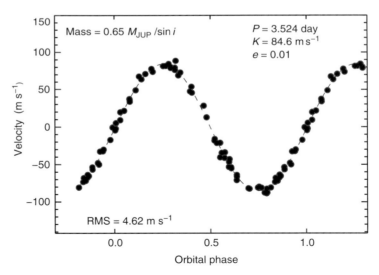

Figure 8.2 Radial velocity curve for HD 209458 [6, 7]. The measurements were made between 1999 and 2007 at the Keck telescope. The measurements have been merged to show a single planetary orbital period of 3.524 days. Measurements obtained when the planet was in transit, blocking part of the light from the star, have been removed for clarity. Figure courtesy Geoff Marcy.

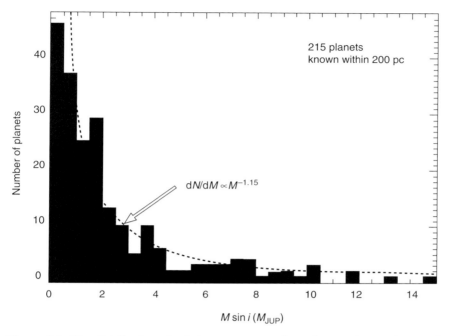

Figure 8.3 The distribution of minimum masses of exoplanets observed around stars in our region of the galaxy [5]. As more massive planets are easier to detect, the actual distribution of objects must be even more strongly concentrated towards low-mass planets. Figure courtesy Geoff Marcy.

toward lower mass, with no sign that we have sampled down to a mass at which the number of planets starts decreasing. These exoplanets are clearly not Earth-like, but their mass distribution suggests that many Earth-size planets may exist as well.

8.1.2 Transits

The other extrasolar planet detection technique achieving major successes today is the measurement of planetary transits. A transit, which is an incomplete eclipse, occurs when an extrasolar planet passes in front of its parent star as seen from our vantage point. If the orbital planes of extrasolar planets are oriented randomly with respect to our line of sight, then the probability that a planet will transit ranges from about 10% for planets orbiting within one-twentieth of an astronomical unit (AU, Earth's mean distance from the Sun) from their stars to about 0.5% for a planet at 1 AU. Moreover, as transiting planets partially obscure their stars once per orbit, those that are close to their star transit with higher cadence. These factors

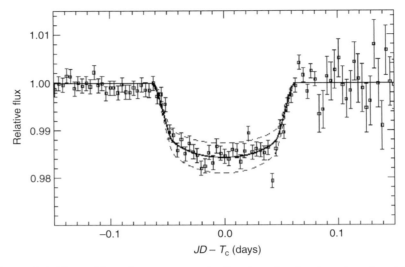

Figure 8.4 Ground-based measurement of a transit of extrasolar planet HD 209458 b [8].

make detection via transit even more biased in favor of close-in planets than is the radial velocity technique. During a transit, the planet blocks out a portion of its star's light, causing the star to appear to dim. A Jupiter-size planet blocks about 1% of the light from a star like the Sun. This dimming can easily be measured with a small ground-based telescope equipped with a modern charge-coupled-device (CCD) detector.

The first extrasolar planet observed to transit its star [8, 9], HD 209458 b,[2] was initially discovered by means of radial velocity measurements (see Figure 8.2). A ground-based measurement of the transit made with a 4-inch telescope is shown in Figure 8.4 [8].

The depth of the transit is about 1.6% and the precision of each data point (a 5-minute average of two separate measurements) is about 0.15%. When this planet is in transit, some of its star's light passes through its atmosphere on the way to Earth. A spectrograph on the Hubble Space Telescope (HST) has been used to detect the effects of the atmosphere on the light from HD 209458, producing the

---

[2] Extrasolar planets have not yet been assigned official names. They are generally referred to using a convention that is an extension of the system used for multiple star systems. Many different algorithms have been used to name individual stars; most are based upon catalog identifier and number, a few on discoverer; for the brightest stars, classical names or constellation name with a Greek letter prefix are often used. However, it is standard to designate the primary star within a bound multiple star system with an "A" following its name, the secondary with a "B," etc. Extrasolar planets are designated analogously, using lower case letters beginning with "b" and assigned in the order in which the planets are detected. Thus, HD 209458's planet is known as HD 209458 b.

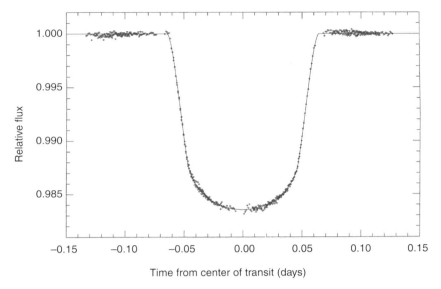

Figure 8.5 Hubble Space Telescope composite measurement of the transit of extrasolar planet HD 209458 b. The steep slope of the curve at the beginning (end) of the transit is caused by the planet moving to cover more (less) of the star's disk. The continuing gradual reduction in brightness towards mid-transit is caused by the center of the star appearing brighter than the star's limb [10].

first measurements of the composition of an extrasolar planet's atmosphere, which was found to contain hydrogen and sodium.

HST also has the capability for making very precise measurements of the stellar dimming caused by transits of HD 209458 b. Compare Figure 8.5 [10] with the ground-based transit measurement shown in Figure 8.4. The HST data, a composite of measurements taken during four separate transits, has a precision of about 0.01% (1×10^{-4}). This is close to the precision needed to detect the much smaller dimming caused by an *Earth-size* planet transiting its parent star! Although HST is capable of this high precision, its narrow field-of-view is not well suited to searching large numbers of stars for transits of Earth-size planets. This is the task for which the Kepler mission has been designed.

Kepler, a NASA exoplanet mission launched on March 6, 2009, is a relatively small (0.95-meter aperture), wide-field telescope with a large array of 42 charge-coupled-device (CCD) detectors similar to those in digital cameras. Kepler can be thought of as an ultra-precise, shutterless, 95-megapixel digital camera which stares at a roughly square region of sky more than ten times the diameter of the full Moon on a side. During its 3.5 to 6 year mission it will monitor the brightness of 100,000 stars, searching for the tiny (roughly 0.01%) dimming caused by Earth-size transiting planets. If Earth-size planets around other stars are common, Kepler

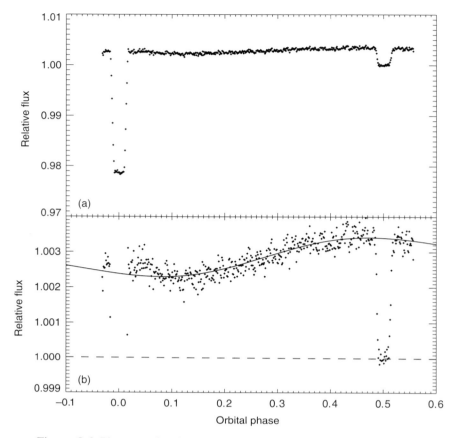

Figure 8.6 Photometric observation of the combined 8 micron radiation from the star HD 189733 and its transiting planet, HD 189733 b. The orbital phase is measured relative to the midpoint of the transit. The observed flux is normalized to that of the star alone, with the range in panel (a) being large enough to show the full depth of the transit and panel (b) showing a magnified view that emphasizes the smaller variations from the secondary eclipse (centered at orbital phase 0.5) and the increase in radiation emitted from the planet as the hemisphere facing the star comes into view. Note that the transit light curve is nearly flat-bottomed, because the disk of the star is nearly uniform in brightness when observed at 8 microns [11].

is expected to detect tens to hundreds. Its particular focus is Earth-size planets in the habitable zones of their parent stars, i.e. within the range of distance from the star in which liquid water is expected to be stable on the planet's surface. If a planet has surface liquid water, it could in principle support water-based life, which could in turn alter the surface and atmosphere of the planet in observable ways. Kepler is an important first step toward finding out how common these planets are. Kepler can also detect planets as small as Mars in the habitable zones of some stars, and is expected to detect hundreds of gas giant planets the size of Jupiter or Saturn. If Kepler launches as scheduled and performs as designed, around 2013 we will know a great deal about the abundance of Earth-size planets orbiting at distances from their stars appropriate to have surface liquid water and hence the potential to support life.

The Spitzer Space Telescope, which detects infrared (heat) radiation, has observed both transits and the secondary eclipse that occurs when a planet passes behind the star. This is illustrated in Figure 8.6 for HD 189733 b. These data provide information on the temperature of the planet as a function of longitude. In some transiting planets, atmospheric winds seem to redistribute heat rapidly, while in other cases there is a large day–night contrast. In these latter cases, the atmosphere cools so rapidly that winds cannot redistribute energy efficiently around the planet and air parcels cool substantially after they rotate from underneath the substellar point.

Variations in the depth of the transit of HD 209458 b with wavelength observed with the Hubble Space Telescope revealed the presence of sodium in the upper atmosphere of this close-in giant planet [12]. Sodium was also discovered more recently in the atmosphere of HD 189733 b [13]. The very large depth of the transit of HD 209458 b at the wavelength of a major diagnostic feature in ultraviolet light suggests that hydrogen associated with the planet extends over an area larger than the size of the planet's gravitational reach; this would imply that hydrogen is escaping the planet at a considerable rate, albeit not so rapidly as to have removed a substantial fraction of the planet's mass over its lifetime [14]. Spitzer infrared spectra of both HD 209458 and HD 189733 during and surrounding their planets' secondary eclipses indicate the presence of water vapor in the atmosphere of HD 189733 b and possibly HD 209458 b as well [15].

8.1.3 *Astrometry*

A third technique, astrometry – the precise measurement of positions of stars – is expected to begin producing extrasolar planet detections soon. Astrometry is complementary to the radial velocity technique that has produced most of the exoplanet detections to date. Radial velocity measures the movement of a star toward and away from the observer, i.e. along the observer's line of sight to the star. Astrometry measures the star's movement perpendicular to the line of sight. Whereas radial velocity is most sensitive to massive planets close to their stars (which cause rapid, but small stellar motions), astrometry is most sensitive to massive planets far from their stars (which cause large, albeit slow, stellar displacements).

The most precise method for measuring stellar positions is called interferometry. In an interferometer, the light from separate telescopes is combined in a way that makes them behave optically as if they were part of one large telescope. An interferometer can be used to make ultra-precise measurements of the angular distance between stars that are closely spaced on the sky. If one of these is a stable reference star and the other is a star with one or more orbiting planets, the periodic change in the stars' angular separation can reveal the presence of the planet(s).

Figure 8.7 shows the apparent motion of the Sun as it would be seen from a distance of 10 parsecs (33 light-years) on a line of sight perpendicular to the plane of the solar system. The amplitude of the Sun's motion shown in Figure 8.7 is very small: approximately one milliarcsecond (10^{-3} arcseconds), which is thirty thousand times less than the apparent diameter of the full Moon. Most of this motion is due to the presence of the four giant planets, with Jupiter making the largest contribution, followed by Saturn. The presence of the Earth causes additional motion less than the thickness of the line showing the Sun's movement. Detecting the displacement of a Sun-like star at 10 parsecs by an Earth-like planet requires a precision

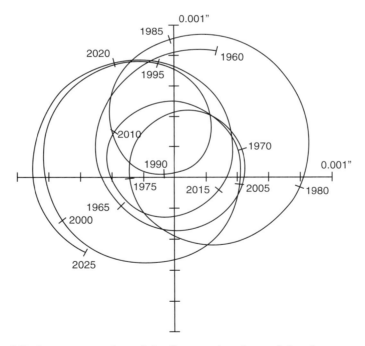

Figure 8.7 Apparent motion of the Sun on the plane of the sky as seen from 10 parsecs (33 light-years) on a line of sight perpendicular to the plane of the solar system. An observer twice as far away would see the same pattern but with half of the amplitude.

equivalent to measuring the diameter of a bacterium from a distance of 200 km! Fortunately, interferometry can in principle achieve this level of astrometric precision. This technique has been demonstrated with lower precision at ground-based observatories [16]; a flight mission like the Space Interferometer Mission [17] can achieve the precision required to detect Earth-like planets around nearby stars.

8.1.4 Microlensing

According to Einstein's general theory of relativity, the path of light from a distant star that passes by a massive object (lens) between the source and the observer is bent. The bending angle is typically very small, hence the effect is known as microlensing. The lens amplifies the light from the source by a substantial factor when the light passes very close to the line of sight. Microlensing is currently being used to investigate the distribution of faint stellar and substellar mass bodies within our galaxy. The brightness of the source can increase several fold for a period of weeks during a microlensing event, and the pattern of brightening can be used to estimate properties of the lens. If the lensing star has planetary companions, then these less massive bodies can produce characteristic blips on the observed light provided the line of sight passes extremely close to the planet. This technique is most sensitive to planets orbiting a few AU from their stars. As of 2008, six planets had been detected by microlensing: four are giant planets, including a system containing Jupiter and Saturn analogues [18], and two are of intermediate mass, about ten times as massive as Earth.

8.2 Planetary habitability and life detection

8.2.1 Habitability and the diversity of life on Earth

None of the planets detected to date are thought to be habitable. They all appear to be either gas giants like Jupiter and Saturn or intermediate-mass planets like Uranus and Neptune, none of which have the conditions needed for habitability. Those conditions are generally inferred by extrapolating from what is known about life on Earth. For example, all life on Earth uses liquid water as a solvent. This has led to the concept of a planetary habitable zone, i.e. the range of distance from a star in which liquid water is expected to be stable on a planet's surface, as discussed above in the context of the Kepler mission. Too far and the water freezes (although subsurface liquid water might support life), too close and it boils and is ultimately lost to space, as is thought to have happened on Venus. Although water may not be the only solvent capable of supporting life (see discussion of "Weird life" in Section 8.2.4), it is the only solvent that we know with certainty *can* support life, and hence focusing on water-based life is a reasonable place to start our search for life beyond the solar system. The focus on finding *surface* liquid water on extrasolar planets reflects the fact that we can only hope to observe the surfaces and atmospheres of these worlds, so finding subsurface water is probably not a practical goal.

Other conditions for habitability include a useable source of energy and suitable environmental conditions, e.g. temperature, pressure, radiation level, acidity/alkalinity (pH), etc. The Earth's biosphere derives the bulk of its energy from sunlight. Thus any planet with a surface exposed to the light of its star in principle has an adequate source of energy. However, studies of the diversity of life on Earth have shown that many organisms use sunlight only indirectly or perhaps even not at all. These studies have also shown that some life can thrive under environmental conditions that are extreme by human standards. This understanding of the diversity of life on Earth has greatly broadened our perspective on the possibility of life on other planets.

A modern perspective on the diversity of life on Earth is illustrated by the Universal Tree of Life (Figure 8.8, adapted from Pace [19]). This Tree is based on molecular studies, particularly of the gene coding for a component of the small subunit (SSU) of ribosomal RNA (rRNA). The vast majority of the genetic diversity in this Tree is reflective of microbial life. Most macroscopic life is confined to three small "twigs" at the extreme of the Eukaryotic domain representing the animal (Homo), plant (Zea), and fungal (Coprinus) kingdoms.

Life that does not use sunlight directly derives energy from chemical reactions (the transfer of electrons) between reduced and oxidized compounds. Animals specialize in using sugars like glucose as the reduced electron donor (food) and oxygen as the final electron acceptor. But microbes, particularly bacteria and archaea, can use a wide variety of reduced organic and inorganic compounds such as hydrogen, methanol, acetate, and ferrous iron as electron donors, and a wide range of oxidized compounds such as carbon dioxide, sulfur, nitrate, and ferric iron as electron acceptors. This broad microbial metabolic diversity is accompanied by a broad physiological diversity, as has been shown in the study of extremophiles – organisms that live under extremes of temperature, pressure, pH, salinity, desiccation, or radiation [20].

The asterisks in Figure 8.8 indicate one type of extremophile: thermophiles and hyperthermophiles that have optimal growth temperatures between 55 °C and ~105 °C. These microbes derive their energy from a variety of chemical sources (see Table 8.1 of [21] for the chemical reactions associated with these diverse microbial metabolisms).

The result of this remarkable microbial diversity is that microbial life is found in many environments, including deep subsurface rock, submarine hydrothermal vents, and the Antarctic dry valleys, where life was not imagined to exist just a few decades ago. Since environments similar to some of these may exist on other planets in our solar system and on planets around other stars, it appears there may be many places beyond Earth capable of sustaining life, at least of the microbial variety we know on Earth.

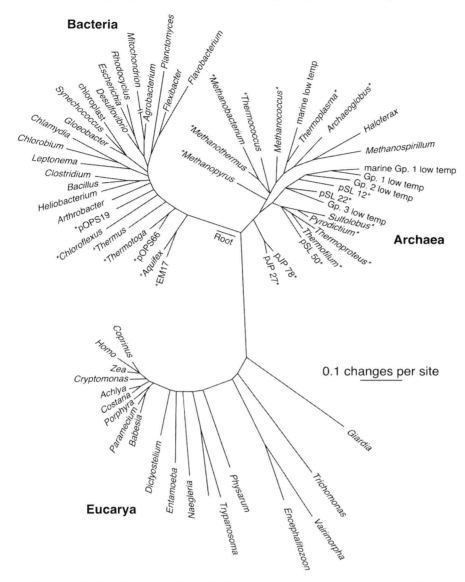

Figure 8.8 The Universal Tree of Life, adapted from Pace [19]. Linear distances along branches correspond to degrees of genetic difference, as shown.

8.2.2 Characterizing extrasolar terrestrial planets

The next step after detecting Earth-size (terrestrial) planets around other stars, particularly those in the habitable zone, is determining their atmospheric and surface compositions and any other characteristics that might help us understand how Earth-like they really are and, particularly, whether any might harbor life. Since we

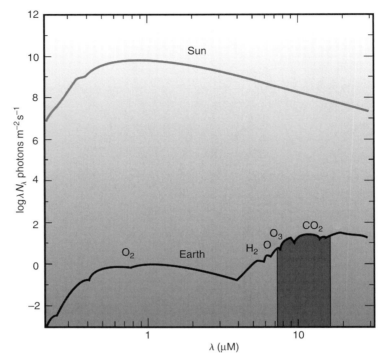

Figure 8.9 The brightness of the Sun and Earth as a function of wavelength (λ), seen from a distance of 10 parsecs (33 light-years). The Sun is 10 million times brighter than Earth at mid-infrared wavelengths (λ~10 μm), and even brighter by an additional factor of 1000 (a brightness ratio of 10 *billion*) at visible wavelengths (λ<1 μm). Both spectral regions contain absorptions of gases that may serve as biosignatures [22].

can't send probes over interstellar distances, our tools are necessarily astronomical. Although some properties of transiting extrasolar terrestrial planets may be measurable using the techniques currently being applied to transiting gas giants using the Hubble and Spitzer Space Telescopes, in general we will have to separate the light of the planet from that of its parent star. This task is daunting because of the enormous brightness difference involved. Figure 8.9 illustrates this for the Sun and Earth, whose brightness ratio is about *ten billion* in the visible part of the spectrum and around ten million in the infrared. In addition, there is background radiation expected due to dust in the planetary system under study that makes the problem akin to trying to observe a firefly next to a searchlight on a foggy night!

There are a number of technological solutions to this challenge being studied. A mission concept known as the Terrestrial Planet Finder (TPF) emerged in the late 1990s [22] and was modified in the early 2000s [23] to include two missions: a visible-light coronograph and an infrared interferometer to observe

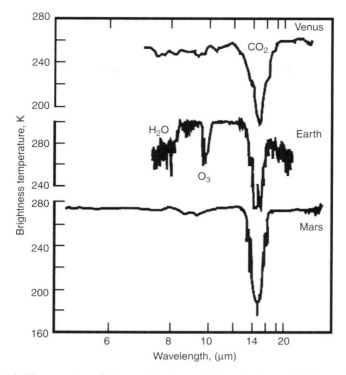

Figure 8.10 The spectra of Venus, Earth, and Mars in the mid-infrared portion of the spectrum.

complementary portions of the planetary spectrum. Because these missions are extremely difficult and expensive, a number of mission concepts with more limited objectives have emerged.[3] The current concept is for a program of extrasolar planet characterization missions that would lead to the originally envisioned TPF capability. That capability would provide low-resolution spectra of Earth-like extrasolar planets adequate to measure some of their main atmospheric and possibly surface constituents.

8.2.3 Biosignatures

What might we expect to detect with this capability? A good starting point is to consider Venus, Earth, and Mars, three "Earth-like" planets in this solar system. Figure 8.10 shows the spectra of all three planets in the mid-infrared, one of two spectral regions that would be measured with TPF capability. Radiation emitted by an Earth-like planet at these wavelengths comes entirely from its atmosphere. Both

[3] For example, see presentations made at the Navigator Program Forum – 2007 on *Small- and Mid-Scale Exoplanet Space Missions*, available online at http://planetquest.jpl.nasa.gov/NavigatorForum/agenda.cfm.

Venus and Mars show a prominent spectral feature centered near a wavelength of 15 μm due to carbon dioxide (CO_2), the main atmospheric component on both planets. Earth also shows this feature, as well as features at shorter wavelengths due to water vapor (H_2O) and ozone (O_3). Water vapor is of course present because of the large amount of surface liquid water on Earth. Ozone is produced by sunlight acting on molecular oxygen (O_2). Molecular oxygen is present on Earth because of life; specifically, because photosynthetic organisms such as plants and cyanobacteria release oxygen as a waste product as they derive useable energy from sunlight. Ozone is therefore an important biosignature, i.e. a component of an extrasolar planetary atmosphere that would indicate the possible presence of life, since it is difficult to produce large amounts of molecular oxygen, ozone's precursor, in the absence of life. However, an ozone spectral feature may be present when there is only a small amount of molecular oxygen, so care must be taken when making a quantitative interpretation.

Figure 8.11 shows a simulation of the spectrum of an extrasolar planet like modern Earth as it might be observed with a TPF-like mission designed to work in the mid-infrared. The spectral features of water, carbon dioxide, and ozone are all readily discernible. TPF would thus be able to detect two key indicators of life (biosignatures) on an Earth-like planet: atmospheric water, indicating the presence of large amounts of surface liquid water, and ozone, indicating a potentially large source of molecular oxygen.

But molecular oxygen was not always abundant in Earth's atmosphere. There is strong evidence that atmospheric molecular oxygen increased sharply in abundance

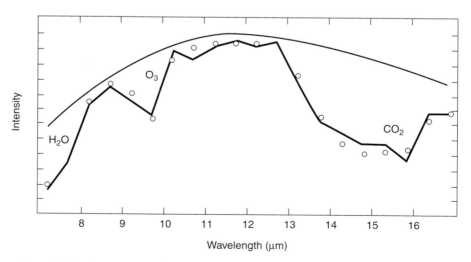

Figure 8.11 A spectrum of an Earth-like planet 10 parsecs (33 light-years) away as might be obtained with TPF [24].

toward modern levels about 2.2 billion years ago [25]. Prior to that time there was very little atmospheric oxygen, even though life was present on Earth as long ago as 3.5–3.8 billion years [26]. It is thought that before the rise of oxygen, methane was much more abundant in Earth's atmosphere than it is today [27]. This higher abundance would have reflected both the absence of a chemical sink for methane, i.e. reaction with atmospheric oxygen, and a large biological methane source. Biologically produced methane might have served as a greenhouse gas to keep the young Earth warm despite a solar luminosity estimated to have been 30% less than the present at the time Earth was formed (the "faint young Sun paradox").

An early biological source of methane is consistent with the Universal Tree of Life (Figure 8.8), which indicates that microbes producing methane, principally from hydrogen and carbon dioxide, were among the earliest forms of life on Earth. It is thus plausible that large abundances of methane might be a biosignature on a young Earth-like world around another star [25]. But methane may also be produced on such a world abiotically, so interpreting a methane spectral feature in an extrasolar planet spectrum will be challenging.

Another compound produced by Earth's earliest microbial life is hydrogen sulfide (H_2S), the gas that has given rotten eggs a bad name. Hydrogen sulfide is also readily produced abiotically, so it is not a good biosignature. But compounds that combine H_2S and methane, some of which are called mercaptans or thiols, may be useful indicators of life [21].

It may even be possible, with TPF-like capability, to see indicators of life on a planetary surface at visible and near-infrared wavelengths. Figure 8.12 shows a spectrum of a deciduous leaf [28]. The so-called "red edge," the sharp increase in reflectivity at $0.7\,\mu m$, is the result of leaf structure. This feature is large enough to be detectable for Earth even in a hemispherically averaged spectrum, as has been shown through observations of "Earthshine," sunlight reflected by Earth and observed after it has been reflected a second time from the "dark" portion of the Moon [29].

8.2.4 Weird life?

Might there be life so different from life on Earth that the reasoning outlined above of what constitutes a biosignature does not apply? For example, what about life that doesn't use liquid water as a solvent? A recent National Research Council report [30] considered non-aqueous solvents and other possible characteristics of "weird life."[4] The report's authors concluded that non-aqueous solvents such as formamide (formed from cosmically abundant hydrogen cyanide and water), liquid

[4] See also P. D. Ward and S. A. Benner. Alien biochemistries. In *Planets and Life: The Emerging Science of Astrobiology*, eds. W. T. Sullivan III and J. A. Baross (Cambridge: Cambridge University Press, 2007), pp. 537–544.

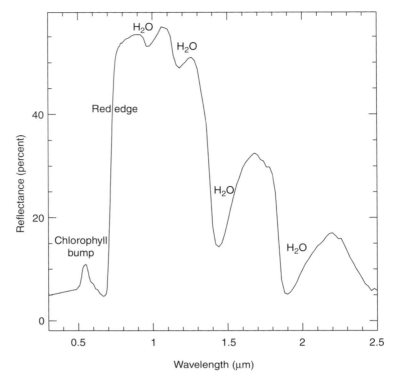

Figure 8.12 Spectrum of a deciduous leaf [28].

water–ammonia mixtures (thought to be present on Titan), and low-temperature solvents such as liquid nitrogen and liquid hydrogen warrant further research.

What about life that is not carbon based? The NRC report's authors noted that silicon-based polymers are stable in liquid nitrogen, but that silicon cannot replace carbon in water-based life because of the reactivity of silicon–hydrogen compounds with water. One should therefore remain open to the possibility of extraterrestrial life based on completely different chemistry than that of life on Earth, but at the same time not regard a focus on water- and carbon-based life as overly constrained and geocentric.[5]

The chemistry of life on Earth is remarkably specific. All Earth life is based on a two-biopolymer system (DNA and proteins) using the same genetic code, the same small group of left-handed amino acids and right-handed sugars, and the same energy-carrying molecule. This commonality is attributed to all Earth life having a common ancestor that possessed these characteristics. Why the common ancestor had these characteristics and not others is a matter of considerable conjecture.

---

[5] For an interesting fictional account of life not based on molecular chemistry, see F. Hoyle. *The Black Cloud* (London: Heinemann, 1957).

Some of these characteristics may be essential for life, but it seems likely that at least some are dependent either on chance or on the particular circumstances under which life developed on Earth. So it seems plausible that extraterrestrial life could be both carbon- and water-based, yet very different from life on Earth. Such life would share functional chemical characteristics with Earth life and would be expected to produce some of the same spectral signatures, for example by producing similar gaseous waste products.

8.2.5 *"Complex" life*

Does an inference that microbial life may be common lead to the inference that animal life – or its macroscopic, multicellular extraterrestrial equivalent – may be common as well? In *Rare Earth*, Ward and Brownlee [31] argue against that supposition. They make the case that "The existence of larger and more complex life occurred only late in Earth history, it occurred only in restricted environments, and the evolution and survival of this more fragile variant of terrestrial life seem to require a highly fortuitous set of circumstances that could not be expected to exist commonly on other planets". These fortuitous circumstances include the presence of a large moon – the product of a chance collision – which stabilizes the tilt of Earth's rotational axis and hence its climate; and the presence of the requisite amounts of planetary heating and surface liquid water, as well as the right mantle and crustal composition, to sustain plate tectonics as a mechanism for climate stabilization (CO_2 regulation) and the creation of continental habitat.

Although some of these arguments are enticing, the Shakespearean admonition at the beginning of this chapter may be appropriate. For example, the Rare Earth argument dismisses the possibility of habitable planets around M-stars, the most common stars in our galaxy. M-stars are much less massive than the Sun and therefore much fainter as well. Their lower luminosity means that a planet must be closer to be in the star's habitable zone. So close, in fact, that the planet will likely be "tidally locked," i.e. it will keep one face pointed constantly toward the star much as the Moon keeps one face pointed constantly toward Earth. The Rare Earth argument envisions extreme cold on the dark side of the planet causing the atmosphere to freeze, eliminating any possibility of habitability.

But the atmosphere of a planet in an M-star's habitable zone may transfer heat effectively enough from the permanent day side to the permanent night side to prevent the night-side atmosphere from freezing [32]. This possibility was discussed at a workshop on M-star planet habitability sponsored by the NASA Astrobiology Institute and reported by Tarter *et al.* [33].[6] In our own solar system, Venus rotates

---

[6] See also J.J. Lissauer. Planets formed in habitable zones of M dwarf stars probably are deficient in volatiles. *Astrophysical Journal*, **660** (May 10, 2007), L149–L152.

very slowly (243-day period), yet its day and night sides are of similar temperature because of the atmosphere's effectiveness in transferring heat from one side of the planet to the other. Similarly, Earth's atmosphere transfers heat from the equatorial regions, which receive most of Earth's sunlight, to the polar regions, which would otherwise be far colder than they are. A habitable planet around an M-star might thus have a climate that varied with distance from the terminator, the circular boundary between night and day, much as Earth's climate varies with distance from the equator.

8.2.6 *The Search for Extraterrestrial Intelligence (SETI)*

No discussion of life in the universe would be complete without consideration of the possibility of detecting intelligent life beyond the solar system. (This discussion draws on material from the SETI 2020 Roadmap [34] and a more recent exposition by Tarter [35].) SETI addresses the most profound dimension of the age-old question, "Are we alone?" With respect to SETI, the question might be more precisely phrased as, "Is there anyone to talk to?" For though we want to know if there is intelligent life out there, we can only search for life that develops the ability and desire to communicate over interstellar distances.

Although there might be "messages in a bottle" drifting in interstellar space – something akin to those we sent aboard the Pioneer 11 and Voyager spacecraft – it seems more likely that interstellar communication will involve signals rather than artifacts. SETI has accordingly focused on searching for electromagnetic signals, particularly in the microwave region of the spectrum where the background radiation from astronomical sources is low. Searches have begun in the optical spectral region more recently.

The first search, Frank Drake's Project Ozma, targeted two nearby solar-type stars in the early 1960s using an 85-meter diameter radio antenna. Its later counterpart, Project Phoenix, targeted 1000 solar-type stars between 1998 and 2004 with resolution and sensitivity orders of magnitude greater than that possible in SETI's early days. Using the 305-meter Arecibo Observatory and the 76-meter Lovell Telescope at Jodrell Bank, Project Phoenix was 10^{14} times more powerful than Project Ozma in terms of the parameter space (frequency, resolution, polarization, sensitivity) examined.

An even greater advance in targeted search capability is being made in the development of the Allen Telescope Array (ATA) – a SETI system which will ultimately have 10,000 square meters of collecting area – named in 2001 for Paul Allen who provided funding for the project's technology development phase. The Allen Family Foundation has subsequently provided funds for the project's initial construction phase. There are currently 42 six-meter diameter antennas operating, with a planned final array of 350. Beyond the ATA, early planning has begun for the "Square Kilometer Array," which could examine stars in a major fraction of our galaxy.

Highly sensitive targeted microwave searches are complemented by lower sensitivity sky surveys. The main current survey is SERENDIP IV, which uses Arecibo in the northern hemisphere. From 1998 to 2003, it was complemented by Southern SERENDIP, which used the Parkes focal plane array in Australia. (SERENDIP stands for "The Search for Extraterrestrial Radio Emissions from Nearby Developed Intelligent Populations.") By observing from both hemispheres, these surveys could in principle examine the entire sky. However, both surveys are "commensal," meaning they are conducted at the same time as conventional radio astronomy investigations, which dictate the portion of sky observed. Other surveys (BETA, Argus) aim for more systematic sky coverage, but at lower sensitivity than the SERENDIP surveys.

The development of high power lasers and fast, inexpensive optical detectors has led to reconsideration of "optical SETI" or OSETI. Searches for nanosecond optical laser pulses (which are distinguishable from naturally occurring sources) are now being conducted by a number of groups.[7]

8.2.7 The Drake Equation

The equation that Frank Drake formulated in the 1960s and which now bears his name is perhaps the most well known effort to characterize the uncertainties in estimating an as yet unknowable quantity. The equation is

$$N = R^* \times f_p \times n_e \times f_l \times f_i \times f_c \times L \tag{8.1}$$

where N is the number of communicating civilizations in our galaxy, R^* is the average rate of Sun-like star formation (~2 per year), f_p is the fraction of these stars with planets (~1/2), n_e is the average number of rocky planets within the Continuous Habitable Zone ($\lesssim 1$), f_l is the fraction of these planets on which life develops, f_i is the fraction of those on which intelligence arises, f_c is the fraction of intelligent species that develop the ability and desire to communicate (e.g. they construct radio telescopes or laser systems), and L is the average lifetime (in years) of a communicating civilization.

When the equation was formulated, the only quantity known with any reliability was R^*, the rate of Sun-like star formation in our galaxy. Today some additional terms can be estimated as indicated above. In addition, some scientists active in origin-of-life research argue that $f_l \sim 1$.

---

[7] See also http://www.setileague.org/general/optical.htm.

Steven Dick, currently the NASA Historian and author of *The Biological Universe: The Twentieth-Century Extraterrestrial Life Debate and the Limits of Science* [36], has made an interesting observation about Drake's equation [37]. He notes that the first three terms on the right-hand side of the equation may be thought of as astronomical, reflecting the processes of cosmic evolution. The next two reflect the processes of biological evolution beginning with the origin of life. The final two terms are cultural, reflecting the development of a civilization and its cultural evolution. Dick argues that cultural evolution must be seen as an integral part of cosmic evolution and the Drake equation. He goes further to consider a "postbiological" universe in which intelligence is predominantly technological (artificial) and its improvement and perpetuation are the central driving forces of cultural evolution, with fundamental implications for SETI strategies.

The link between cosmic and biological evolution (the first two sets of terms in the Drake equation) is being explored today as well by cosmologists, who are attempting to account for the apparent "fine tuning" of the physical constants in our universe that appears to be necessary for the existence of life.[8] Rees [38] provides a short and accessible summary of how multiverse cosmologies, in which our universe is one of many, perhaps an infinite number, provide a conceptual framework for understanding a universe tuned in a way that allows us to exist and observe the tuning.

8.3 A concluding thought

In a galaxy of more than 100 billion stars, in which a sizable fraction (perhaps most) may have planets, and a universe of 100 billion such galaxies, it seems unlikely that life has arisen only on one pale blue dot. However, the challenges of detecting life beyond Earth, and particularly beyond the system of planets circling our Sun, are extreme at our current level of technological development. Nonetheless, efforts are underway to detect and characterize habitable worlds around nearby stars and examine them for evidence of life and even technology. We are able to use these tools of science to address that age-old question, "Are we alone?" Whatever the answer, it is likely to be profound. As Buckminster Fuller observed, "Sometimes I think we're alone. Sometimes I think we're not. In either case, the thought is staggering."

Acknowledgements

J. J. L. was supported by the NASA Astrobiology Institute under the NASA Ames Investigation "Linking our Origins to our Future."

---

[8] For example, see J. D. Barrow and F. J. Tipler. *The Anthropic Cosmological Principle* (Oxford: Oxford University Press, 1986), p. 706.

References

[1] S. Udry, D. Fischer, and D. Queloz. A decade of radial-velocity discoveries in the exoplanet domain. In *Protostars and Planets V*, eds. B. Reipurth, D. Jewitt, and K. Keil (Tucson, AZ: University of Arizona Press, 2007), pp. 685–699.

[2] http://exoplanets.org/

[3] http://exoplanet.eu/

[4] G. W. Wetherill. Possible consequence of absence of "Jupiters" in planetary systems. *Astrophysics and Space Sciences*, **212** (1994), 23–32.

[5] R. P. Butler, J. T. Wright, G. W. Marcy, *et al*. Catalog of nearby exoplanets. *Astrophysical Journal*, **646** (2006), 505–522.

[6] G. W. Marcy, R. P. Butler, D. A. Fischer, and S. S. Vogt. Properties of extrasolar planets. In *Scientific Frontiers in Research on Extrasolar Planets*, ASP Conference Series, Vol. 294, eds. D. Deming and S. Seager (2003), pp. 1–16.

[7] G. W. Marcy, R. P. Butler, D. A. Fischer, and S. S. Vogt. A Doppler planet survey of 1330 FGKM stars. In *Extrasolar Planets, Today and Tomorrow*, ASP Conference Series, Vol. 321, eds. J.-P. Beaulieu, A. Lecavelier, and C. Terquem (2004), pp. 3–14.

[8] D. Charbonneau, T. M. Brown, D. W. Latham, and M. Mayor. Detection of planetary transits across a sun-like star. *Astrophysical Journal*, **529** (2000), L45–L48.

[9] G. W. Henry, G. W. Marcy, R. P. Butler, and S. S. Vogt. A transiting "51 peg-like" planet. *Astrophysical Journal*, **529** (2000), L41–L44.

[10] T. M. Brown, D. Charbonneau, R. L. Gilliland, R. W. Noyes, and A. Burrows. Hubble space telescope time-series photometry of the transiting planet of HD 209458. *Astrophysical Journal*, **552** (2001), 699–709.

[11] H. A. Knutson, D. Charbonneau, L. E. Allen, *et al*. A map of the day–night contrast of the extrasolar planet HD 189733b. *Nature*, **447** (2007), 183–186.

[12] D. Charbonneau, T. M. Brown, R. W. Noyes, and R. L. Gilliland. Detection of an extrasolar planet atmosphere. *Astrophysical Journal*, **568** (2002), 377–384.

[13] S. Redfield, M. Endl, W. D. Cochran, and L. Koesterke. Sodium absorption from the exoplanetary atmosphere of HD 189733 b detected in the optical transmission spectrum. *Astrophysical Journal Letters*, **673** (2008), L87–L90.

[14] A. Vidal-Madjar, A. Lecavelier Des Etangs, J.-M. Désert, *et al*. An extended upper atmosphere around the extrasolar planet HD209458b. *Nature*, **422** (2003), 143–146.

[15] J. J. Fortney and M. S. Marley. Analysis of Spitzer spectra of irradiated planets: evidence for water vapor? *Astrophysical Journal Letters*, **666** (2007), L45–L48.

[16] M. M. Colavita, J. K. Wallace, B. E. Hines, *et al*. The Palomar testbed interferometer. *Astrophysical Journal*, **510** (1999), 505–521.

[17] http://planetquest.jpl.nasa.gov/SIM/sim_index.cfm

[18] B. S. Gaudi, D. P. Bennett, A. Udalski, *et al*. Discovery of a Jupiter/Saturn analog with gravitational microlensing. *Science*, **319** (2008), 927–930.

[19] N. Pace. A molecular view of microbial diversity and the biosphere. *Science*, **276** (1997), 734–740.

[20] L. J. Rothschild and R. L. Mancinelli. Life in extreme environments. *Nature*, **409** (2001), 1092–1101.

[21] C. B. Pilcher. Biosignatures of early earths. *Astrobiology*, **3** (2003), 471–486.

[22] C. A. Beichman, N. J. Woolf, and C. A. Lindensmith (eds.). *The Terrestrial Planet Finder (TPF): A NASA Origins Program to Search for Habitable Planets* (JPL Publication 99–003, 1999); available online at http://planetquest.jpl.nasa.gov/TPF/tpf_book/index.cfm.

[23] P. R. Lawson and W. A. Traub (eds.). *Earth-Like Exoplanets: The Science of NASA's Navigator Program* (JPL Publication 06–5 Rev A, 2006); available online at http://planetquest.jpl.nasa.gov/TPF/tpf_what_is.cfm.

[24] J. R. P. Angel and N. J. Woolf. An imaging nulling interferometer to study extrasolar planets. *Astrophysical Journal*, **475** (1997), 373–379.

[25] J. Kasting. The rise of atmospheric oxygen. *Science*, **293** (2001), 819–820.

[26] J. W. Schopf (ed.). *Earth's Earliest Biosphere: its Origin and Evolution* (Princeton, NJ: Princeton University Press, 1983), pp. 543.

[27] A. A. Pavlov, L. L. Brown, and J. F. Kasting. UV shielding of NH_3 and O_2 by organic hazes in the Archean atmosphere. *Journal of Geophysical Research*, **106**(E10) (2001), 23267–23287.

[28] S. Seager, E. L. Turner, and E. B. Ford. Vegetation's red edge: a possible spectroscopic biosignature of extraterrestrial plants. *Astrobiology*, **5** (2005), 372–390.

[29] N. J. Woolf, P. S. Smith, W. A. Traub, and K. W. Jucks. The spectrum of earth-shine: a pale blue dot observed from the ground. *Astrophysical Journal*, **574** (2002), 430–433.

[30] National Research Council. *The Limits of Organic Life in Planetary Systems* (Washington, DC: The National Academies Press, 2007).

[31] P. D. Ward and D. Brownlee. *Rare Earth: Why Complex Life is Uncommon in the Universe* (New York: Copernicus, Springer-Verlag, 2000).

[32] M. M. Joshi, R. M. Haberle, and R. T. Reynolds. Simulations of the atmospheres of synchronously rotating terrestrial planets orbiting M dwarfs: conditions for atmospheric collapse and the implications for habitability. *Icarus*, **129**(2) (1997), 450–465.

[33] J. L. Tarter, P. R. Backus, R. L. Mancinelli, *et al.* A reappraisal of the habitability of planets around M dwarf stars. *Astrobiology*, **7** (2007), 30–65.

[34] R. D. Ekers, D. K. Cullers, J. Billingham, and L. K. Scheffer (eds.). *SETI 2020: A Roadmap for the Search for Extraterrestrial Intelligence* (Mountain View, CA: SETI Press, 2002).

[35] J. C. Tarter. Searching for extraterrestrial intelligence. In *Planets and Life: The Emerging Science of Astrobiology*, eds. W. T. Sullivan III and J. A. Baross (Cambridge: Cambridge University Press, 2007), pp. 513–536.

[36] S. J. Dick. *The Biological Universe: The Twentieth-century Extraterrestrial Life Debate and the Limits of Science* (Cambridge: Cambridge University Press, 1999).

[37] S. J. Dick. Cultural evolution, the postbiological universe and SETI. *International Journal of Astrobiology*, **2**(1) (2003), 65–74.

[38] M. Rees. Cosmological challenges: are we alone, and where? In *The Next Fifty Years: Science in the First Half of the Twenty-first Century* (New York: Vantage Books, 2002), pp. 18–28.

9

A historical perspective on the extent and search for life

Steven J. Dick

The premise of my chapter is that history can be not only enlightening in itself but also significant in a variety of contemporary contexts. One clear example of this is in connection with the theological, ethical, and philosophical implications of the search for life, because these issues have been addressed again and again throughout history as the possibility of life beyond Earth has been raised.[1] It is true that most of the historical debate has centered on extraterrestrial intelligence rather than microbial life. But at least part of the interest in the search for microbial life is that it is an indication of the prevalence of intelligence in the universe. This explains in part the uproar over the claims for nanofossils in the Mars rock in 1996 – much more was at stake than primitive life on Mars itself. This connection between microbes and intelligence is made especially in the popular mind, despite the fact that many evolutionists would argue that the gap between microbes and intelligence is greater than that between life and non-life. Thus, Peter Ward and Donald Brownlee have recently, and famously, argued that the universe may well be full of microbes, but not intelligence, making ours a "Rare Earth" [1]. If this is true, the study of the implications of microbial life beyond Earth is all the more important.

Having said that, I will present the major issues in the broader tradition known for much of its history as the "plurality of worlds" debate, which transformed in

---

[1] For the history of the extraterrestrial life debate, including its religious implications, on which this essay is based, see S.J. Dick. *Plurality of Worlds: The Origins of the Extraterrestrial Life Debate from Democritus to Kant* (Cambridge: Cambridge University Press, 1982); M.J. Crowe. *The Extraterrestrial Life Debate 1750–1900: The Idea of a Plurality of Worlds from Kant to Lowell* (Cambridge: Cambridge University Press, 1986); S.J. Dick. *The Biological Universe: The Twentieth Century Extraterrestrial Life Debate and the Limits of Science* (Cambridge: Cambridge University Press, 1996); S.J. Dick. *Life on Other Worlds* (Cambridge: Cambridge University Press, 1998); S.J. Dick and J.E. Strick. *The Living Universe: NASA and the Development of Astrobiology* (New Brunswick, NJ: Rutgers University Press, 2004); K. Guthke. *The Last Frontier: Imagining Other Worlds from the Copernican Revolution to Modern Science Fiction* (Ithaca, NY: Cornell University Press, 1990).

Exploring the Origin, Extent, and Future of Life: Philosophical, Ethical, and Theological Perspectives, ed. Constance M. Bertka. Published by Cambridge University Press. © Cambridge University Press 2009.

the twentieth century into exobiology, bioastronomy and astrobiology. I will argue (1) that this tradition of discussion originated with, and was sustained by, cosmological worldviews through the middle of the eighteenth century; (2) was dominated for the following century by philosophical and religious explorations; (3) received its scientific foundations in modern terms late in the nineteenth century; and (4) began its formation as a discipline only in the Space Age. As I proceed, I will point out the major theological arguments, many of which are still heard today. And I will end by concluding that astrobiology, or what I call the "biological universe," is itself a kind of cosmological worldview, which has broad societal implications no less than other cosmological worldviews.

9.1 The cosmological connection

My first point is that the history of the extraterrestrial life (ETL) debate is not frivolous, but intellectually significant. That may seem an innocuous statement, but when I proposed some 30 years ago to write the history of the ETL debate in a *history of science* department, I was told that there were two problems – it had no intellectually significant history, and it wasn't science! This is undoubtedly why so little had been done on the subject's history. Virtually the only substantial historical treatment of the theme as of the early 1970s was a chapter in Arthur O. Lovejoy's classic 1936 book *The Great Chain of Being* [2], in which he argued that the philosophical principle of plenitude was the driving force for the belief in other worlds, rather than concepts related to science. That principle of plenitude states that the essence of Nature, or of God, demands that anything that can be done will be done, so that as many worlds as can be made will be made. That theme indeed played a role, but my subsequent research showed how the belief in other worlds was actually tied to major cosmological traditions.[2]

The idea of other worlds was already present in one of the earliest cosmological worldviews, ancient atomism. Constructed in the fourth and fifth centuries BC by Leucippus, Democritus, and Epicurus, this system held that, in the words of Epicurus:

There are infinite worlds both like and unlike this world of ours. For the atoms being infinite in number, as was already proved, are borne on far out into space. For those atoms which are of such a nature that a world could have been created by them ... have not been used up either on one world or a limited number of worlds ... So that there nowhere exists an obstacle to the infinite number of worlds [3] (p. 10).

---

[2] The arguments in this section are elaborated, with full references, in S.J. Dick. *Plurality of Worlds: The Origins of the Extraterrestrial Life Debate from Democritus to Kant* (Cambridge: Cambridge University Press, 1982).

The "world" of Epicurus was the Greek kosmos, meaning an ordered system, as opposed to chaos. The entire visible universe composed one *kosmos*; the atomists here proposed the remarkable idea that an infinite number of such worlds exist completely beyond the human senses – but not beyond human reasoning. Since the infinite number of atoms was not exhausted in our finite world, as our world was created by the chance collision of atoms in an entirely natural process, so must other worlds be created in like manner. Some may recognize here the modern concept of a multiverse – other universes that cannot be seen, and cannot be proven, except as one option inferred from the anthropic principle. (The anthropic principle says the universe is fine-tuned for life. There are two possible reasons: either a designer did it, or there is an ensemble of universes and we happen to live in one that is hospitable for life.)

This atomist doctrine would eventually be spread throughout Europe by the Roman poet Lucretius (ca. 99–55 BC), whose *De rerum natura* (*On the Nature of Things*) supported the belief of Epicurus, and enunciated a principle of plenitude as applied to other worlds: "when abundant matter is ready, when space is to hand, and no thing and no cause hinders, things must assuredly be done and completed." It is notable here, however, that what can be done is defined by the physical principles that one accepts. It is almost inconceivable that the concept of infinite worlds would have arisen in the atomist system in the absence of the physical principles of the atomist cosmology.

Atomist physical principles, however, were not destined to win the day, or even the millennium; it was almost two thousand years before they would be revived in the sixteenth and seventeenth centuries with the birth of modern science. In the meantime a far more elaborate cosmology was constructed by Aristotle (383–322 BC), who gave new meaning to the word "kosmos" with his highly structured system. Aristotle's cosmology placed the Earth at the center of a nested hierarchy of celestial spheres, from the spheres of the Moon and planets to the sphere of the fixed stars. In this system the Earth was ever-changing and corruptible, as could be seen by experience, while the region of the celestial crystalline spheres was eternally unchanging. And the Earth was more than a physical center, it was also the center of motion. According to one of the basic tenets of Aristotle's cosmology – the doctrine of natural motion and place – everything in the cosmos moved with respect to that single center: the element earth moved naturally toward the Earth, and the element fire moved naturally away, while air and water assumed intermediate natural places. Aristotle's belief in the impossibility of more than a single kosmos was directly tied to this basic tenet. In his cosmological treatise *De caelo* (*On the Heavens*) he reasoned that if there were more than one world the elements of earth and fire would have more than one natural place toward which to move, a physical and logical contradiction.

   The issue of a plurality of worlds was thus reduced to a confrontation with the most basic assumptions of Aristotle's system. Either he must reject his doctrine of natural motion and place, on which he had built his entire physics, along with his belief in four elements, on which his theory of matter rested, or conclude that the world was unique. The choice was not difficult; indeed, he must have taken comfort in reaching a conclusion so diametrically opposed to the atomists, whose system differed from his in so many other ways.

   Now this might seem like an esoteric set of arguments, but it was exactly this set of arguments that was to be discussed for the next two millennia. It was Aristotle's geocentric system that was transmitted to the Latin West, where it was repeatedly commented upon in the context of the Christian system, and where it famously formed the reference frame for Dante's *Divine Comedy* in the early 1300s. Dante required only one world in his work. But for Christianity the impossibility of a plurality of worlds directly confronted its omnipotent God: suppose God wished to create another world; how could he do so given the principles of Aristotle? Either Aristotle was wrong, and by his own admission wrong in some very basic principles, or God's power was severely limited. For the first century of Aristotle's introduction into the West, his conclusion of a single world was largely accepted, and the concept of God's omnipotence redefined. Thomas Aquinas, among others, argued that God's perfection and omnipotence could also be found in the unity of the world.

   But by the late thirteenth and fourteenth centuries, university scholars claimed that the plurality of worlds was not theologically impossible because God can act beyond the Aristotelian laws of Nature. God could create another world and reorder its elements, so that they would move according to Aristotelian law but with respect to their own world. William of Ockham – of Ockham's razor fame – went still further by altering Aristotle's doctrine of natural place to state that the elements in each world would return to the natural place within their own world, and this without any intervention from God. By 1377 the Paris master and bishop of Lisieux, Nicole Oresme, had completely reformulated the doctrine of natural place to state that as long as heavy bodies were situated in the middle of light bodies, no violence would be done to the doctrine of natural place. In one stroke, he thus abandoned the Earth–outer sphere relation so central to Aristotle. Other worlds were possible, and without any supernatural intervention. But while possible, almost all the medieval Scholastics stressed, God in reality had not created more than one world. Neither the modern doctrine of the plurality of worlds, nor the universe of modern science, emerged from the Middle Ages, despite significant advances over Aristotle.

   In the fifteenth and sixteenth centuries, other troubling questions that Oresme and his contemporaries never addressed came to the fore. The German theologian, philosopher and cardinal, Nicholas of Cusa, asserted that inhabitants exist in every region of the heavens, and that they differ in nature according to their location. Cusa's contemporary, William of Vorilong, reaffirmed God's power to create other worlds, but then asked a bold question: "whether men exist on that world, and whether they have sinned as Adam sinned." He answered that such creatures "would not exist in sin and did not spring from Adam … As to the question whether Christ by dying on this earth could redeem the inhabitants of another world, I answer that he is able to do this even if the worlds are infinite, but it would not be fitting for Him to go into another world that he must die again" [3] (pp. 40–43). A century later, in the mid-sixteenth century, Philip Melancthon, the systematic expositor of Martin Luther's theology, forcefully expressed the initial Protestant position. Any sound mind, he insisted, would totally reject the possibility of other worlds, based both on Aristotle and on Scriptural doctrine. Scripture made no mention of the creation of man on any other planet other than Earth, he argued. "It must not be imagined that there are many worlds, because it must not be imagined that Christ died and was resurrected more often, nor must it be thought that in any other world without the knowledge of the Son of God, that men would be restored to eternal life" [3] (p. 89).

Here, then was the crux of the matter. While the existence of other worlds might be consonant with plenitude and Divine omnipotence, it had grave implications for certain Church doctrines such as Redemption, Incarnation and the implied one-to-one relationship between man and his Creator. Moreover, extraterrestrial inhabitants were nowhere to be found in the pages of Scripture. A Pandora's box of puzzling questions and implications was cause for grave concern. These concerns would be expressed again and again in the plurality of worlds debate, and have no agreed answer even today.

Like the move to an infinite universe, the transition to the rudiments of a biological universe was not made through successive rebuttals to Aristotle's doctrine of a single world. Rather, it stemmed from the complete overthrow of Aristotle's geocentric universe and its replacement with the Copernican system of the world. It was this heliocentric system that gave birth to the new tradition of the plurality of worlds, where "world" (*mundus*) was redefined to be an Earth-like planet, which now took on the kinematic or motion-related functions of the single Earth in the old geocentric system. Just as the kinematic implications of the decentralization of the Earth led to the birth of a new physics, so the "planetary physics" implications of that move led to the birth of the concept of the biological universe. All discussions of life on other worlds since then, whether consciously or not, recall that fainter echo that Copernicus set in motion: if the Earth is a planet, then the planets may be Earths. And if the planets are Earths, they may harbor intelligence.

Copernicus himself did not pursue the implications of his system, but the ideas of the Italian philosopher Giordano Bruno showed just how far such implications might go, though he was largely influenced by more metaphysical arguments. An avowed Copernican, Bruno nevertheless gave himself full credit for his universe filled with inhabited worlds. In his *De l'infinito universe e mondi* (*On the Infinite Universe and Worlds*, 1584), Bruno pointed to the metaphysical concept of unity as the source of his belief. The unity of the universe shattered the old Aristotelian spheres, as well as any celestial-terrestrial dichotomy, and it led to innumerable worlds via his conviction that both the greatness of Divine power and the perfection of Nature lay in the existence of infinite worlds. Greatly influenced by the atomists, their principle of plenitude also entered Bruno's argument: Nature could not help but produce infinite worlds. For his belief in other worlds, as well as other supposed heresies, the Roman Catholic Church burned Bruno at the stake in 1600. While it is unclear how much of a role the idea of infinite worlds played in his condemnation, the idea was prominent in his writings.

Though not tied directly to his Copernicanism, Bruno's was a view that would undoubtedly have profoundly disturbed Copernicus. And yet, it was the view toward which his system led, in a march that had not been completed by the twentieth century. The invention of the telescope began the long trend of attempts at empirical verification of the Copernican implication that the planets were Earth-like. The response of Aristotelians to such ideas was outraged astonishment, and even among Copernicans predictable caution was the watchword. Galileo himself at first denied such implications, and even in his defense of the Copernican system (*Dialogue on the Two Chief World Systems*, 1632), sought to downplay the possibility of life on the Moon, admitting only that if there was lunar life, it would be "extremely diverse and far beyond all our imaginings." The Copernican tide, however, could not be stemmed. Six years later, in the less repressive atmosphere of Anglican England, Bishop John Wilkins penned his *Discovery of a World in the Moone* (1638), where Galilean caution was thrown to the wind. Copernicanism was not synonymous with inhabited planets, but it did give theoretical underpinning to the possibility of habitable planets. The proof or disproof of this Copernican implication that the planets are Earth-like and habitable is a theme that runs from the sixteenth century through the Viking landers to the present day.

All important cosmological worldviews of the seventeenth century and later incorporated the Copernican system as a basic truth. Such was the case with the first complete physical system proposed since Aristotle, that of the French philosopher Rene Descartes. His *Principia Philosophiae* (*Philosophical Principles*, 1644), also greatly influenced by a revived atomism, offered a mechanical philosophy in which particles in motion formed the basis for a rational cosmology. For the plurality of worlds tradition it did even more, for it was through the Cartesian cosmology that the quest for a biological universe was first carried to other solar systems, and in a fashion so graphic that it remained an ingrained concept to the present day. That Cartesian cosmology was as follows: rejecting the void space of his atomist predecessors (and his Newtonian successors), Descartes proposed that the universe was a plenum, filled with atoms in every nook and cranny. A consequence of this was that, once set in motion by God, the particles of the plenum formed into vortices, systems analogous to our solar system, centered around every star. Though Descartes himself, again for religious reasons, was careful not to mention that these vortices could contain inhabited planets, his application of Cartesian laws to the entire universe, and the graphic vortex cosmology was plain for all to see.

Descartes' followers were not slow to realize the implications, and some of them elaborated on the nature of the whirling vortices. But none was more bold than his countryman Bernard le Bovier de Fontenelle. His *Entretiens sur la pluralité des mondes* (*Conversations on the Plurality of Worlds*, 1686), a treatise that exploits both the Copernican and the Cartesian theories to shed light on the question of life on other worlds, was explicit in its reason for postulating the possibility of life beyond the solar system:

If the fix'd Stars are so many Suns, and our Sun the centre of a Vortex that turns round him, why may not every fix'd Star be the centre of a Vortex that turns round the fix'd Star? Our Sun enlightens the Planets; why may not every fix'd Star have Planets to which they give light? [3] (p. 126)

The *Entretiens* was spread throughout Europe in an extraordinary number of editions, and with it the idea of a plurality of solar systems became ingrained in the European consciousness. The *Kosmotheoros* of the Dutch astronomer Christiaan Huygens did the same. Strongly motivated by his experience as an observational astronomer, and explicitly building on the foundation of the Copernican theory, Huygens also proposed a plurality of solar systems, based primarily on the analogy that the fixed stars were suns (another of Descartes' tenets), and only loosely tied to Cartesian vortices. Although Cartesian vortices would soon be swept away by the Newtonian system, the general idea of multiple planetary systems would not.

It is ironic that, of all these cosmological worldviews, the scientific principles of the Newtonian worldview – the system that we have inherited in modified form – entailed extraterrestrial life least of all. The ancient atomists held that an infinite number of atoms in an infinite universe must necessarily form an infinite number of worlds, given the example of our finite world; the Copernican principle of the non-centrality of the Earth opened the way to the implication of other Earths; and the Cartesian plenum led directly to vortices that common sense dictated were similar to our own solar system. Although a mechanical system like that of Descartes, Newton's atoms and void, with each body subject to universal gravitation according to fixed laws, did not necessarily imply other solar systems.

There was no mechanical necessity for the formation of systems as there had been in Descartes' system; indeed, under Newtonian principles the whole question has proved to be one of the greatest complexity to the present day.

One could, of course, argue that since our solar system exists, and the laws of gravitation are universal, other solar systems should exist. This, however, was only the grossest analogy, almost equivalent to assuming what one wished to prove. Newton himself declined to expound any rational cosmogony that might shed light on the question; in fact in his letters to the theologian Richard Bentley he indicated that he did not consider it possible to settle this question based on the principles of his system. Instead, he insisted that the formation of all ordered systems was contingent upon God's will, contenting himself with the observation in the second edition of his *Principia* (1713) that "if the fixed stars are the centers of other like systems, these, being formed by the like wise counsel, must be all subject to the dominion of the One ... " [3] (p. 147).

Despite the arguments of Newton and others to retain the Deity in a system subject to natural law, it became increasingly clear that his laws actually lessened the need for a Deity. In this atmosphere it is not surprising that the concept of a plurality of worlds was put to good use – as a proof of God's existence and glory drawn from his works in Nature, part of the enterprise known as natural theology. If God could no longer be given a role in maintaining his universe, then the concept of a plurality of inhabited planets could be made to reflect the glory, power, and wisdom of the Creator, in addition to natural theology arguments from design for the existence of God. In one Newtonian treatise after another, the theological view of an inhabited universe was joined to the physical principles of Newton's system. Again and again, a universe full of inhabited solar systems was applauded as one "far more magnificent, worthy of, and becoming the infinite Creator, than any of the other narrower schemes" [3] (p. 154). Once this decision had been made, overwhelming all Scriptural objections, other arguments could be adduced in its favor. One of the most frequent was the argument of teleology, of purpose in the universe, clearly set forth already by Bentley when he wrote "All Bodies were formed for the Sake of Intelligent Minds: As the Earth was principally designed for the Being and Service and Contemplation of Men; why may not all other Planets be created for the like uses, each for their own Inhabitants who have Life and Understanding" [3] (p. 149).

This satisfying vision of the universe, operated by Newtonian laws and reflecting the power of the Deity by spreading intelligence through the universe, was transmitted to the modern world. The proof of other solar systems by observation, and the proof of their likely formation by Newtonian principles, remained a desired goal in the centuries to follow. But the basic predisposition toward a universe of inhabited solar systems was set, almost within the lifetime of Newton himself. Indeed the Newtonian system was carried to its ultimate extreme in this respect already in the 1750s and 1760s when the philosopher-cosmologists Thomas Wright, Immanuel Kant, and Johann Lambert argued for hierarchy of ordered systems, stretching from our own solar system to the system of the Milky Way Galaxy itself and even beyond. Their universe, "animated with worlds without number and without end," brings us back almost full circle to the ancient Greek atomist view of infinite kosmoi, systems of stars separated from one another, each filled with a multitude of inhabited worlds. But now the laws of motion of such systems had been fathomed, somehow extracted from Nature by the very mind that was now projected throughout the universe.

Far from being frivolous, the extraterrestrial life debate emerged over two millennia from five of the world's great cosmologies.

9.2 Philosophical explorations

Following the triumph of the Newtonian system in the mid-eighteenth century, the extraterrestrial life debate was carried on not so much on a cosmic scale as on a scale of worldviews a level or more below the cosmic. Though sometimes discussed in elaborations of Newtonian science such as the Laplacian nebular hypothesis of the origin of the solar system, more often it fell in the domain of philosophical explorations, both secular and religious. If cosmic worldviews gave birth to the idea of extraterrestrial life, it was theology, philosophy and literature, in their traditional role of examining the human condition, that were led to explore the ramifications of that idea born of that cosmic context. I will focus here on just three of these explorations associated with Christianity, anthropocentrism and chance and necessity respectively.

9.2.1 Christianity

Much of the late eighteenth and nineteenth century plurality of worlds debate – at least in the West – may be understood as a struggle with that widespread theological worldview known as Christianity. If in the Newtonian system the plurality of worlds concept was reconciled with theism via natural theology, this was not equivalent to a reconciliation with Christianity; as Michael Crowe succinctly states in *The Extraterrestrial Life Debate 1750–1900*, "structures of insects or solar systems may evidence God's existence, but they are mute as to a Messiah" [5] (page unknown). Three choices were logically open to Christians who pondered the question of other worlds: they could reject other worlds, reject Christianity, or attempt to reconcile the two. Historically, all three of these possibilities came to pass in the eighteenth and nineteenth centuries.[3]

Though the Scriptural and doctrinal problems raised by the issue had been widely discussed throughout the seventeenth century, only to be overwhelmed by natural theology, no one more forcefully expressed the continuing difficulties of the plurality of worlds doctrine for Christianity than Thomas Paine, that agitator for freedom on two continents. In his influential *Age of Reason* (1793), which saw numerous American editions by the end of the century, Paine bluntly stated that "to believe that God created a plurality of worlds at least as numerous as what we call stars, renders the Christian system of faith at once little and ridiculous and scatters it in the mind like feathers in the air. The two beliefs cannot be held together in the same mind; and he who thinks that he believes in both has thought but little of either" [4, 5 (p. 163)]. With millions of worlds under his care, Paine argued, could we really believe that the Messiah came to save human beings on this small world? Or did the redeemer hop from one world to the next, when the number of worlds was so great that he would be forced to suffer "an endless succession of death, with scarcely a momentary interval of life"? Despite the force of this argument, few would reject Christianity as did Paine. But few would reject plurality of worlds either, a testimony to its entrenchment by the end of the eighteenth century. This left but one alternative: the two systems and all they implied would have to coexist.

That other worlds could be incorporated into Christianity, despite Paine, was demonstrated by the Scottish theologian Thomas Chalmers. His *Astronomical Discourses* (1817) [5] (pp. 182–190) incorporated plurality of worlds into evangelical religion, and his countryman Thomas Dick made it a staple of Christianity in a number of works during the first half of the nineteenth century. But Paine's objections would not disappear. By mid-century the compatibility of plurality of worlds with Christianity was once again called into serious question, this time by William Whewell, philosopher, scientist and Master of Trinity College, Cambridge. His treatise *Of a Plurality of Worlds: an Essay* [5] (pp. 265–299), which appeared anonymously in 1853, was the most learned, radical, and influential anti-pluralist treatise of the century. Unlike Paine and others, Whewell argued that it was pluralism, not Christianity, that should be rejected. To the argument that all the vast space must have some purpose, he countered that geology reveals human existence on Earth to be but a short "atom of time" compared to the age of the Earth; therefore why could not intelligence be confined to the "atom of space" that was

---

[3] For details and full citations see M.J. Crowe. *The Extraterrestrial Life Debate 1750–1900: The Idea of a Plurality of Worlds from Kant to Lowell* (Cambridge: Cambridge University Press, 1986).

the Earth? And although the universe was indeed vast (about 3000 light-years by his estimate), Whewell argued that the possible locales for inhabitants had been vastly overrated. The stars might not be like the Sun, the planets might not be like the Earth; in short, Whewell saw the analogies as greatly exaggerated in the case of other worlds. No longer could the Copernican discovery that the planets were, in effect, other Earths of itself make the case for the plurality of inhabited worlds; greater attention had to be given to the details of the planets' physical conditions. Whewell's treatise generated a tremendous amount of debate, but in the end it did little to weaken support for a plurality of worlds among scientists or religious believers.

Innumerable other discussions of the relation of plurality of worlds to Christianity were penned throughout the nineteenth century. Almost without fail, the point of all of them is some kind of reconciliation with an occasional conclusion that one or the other would have to go. Attempts at reconciliation with the doctrines of Incarnation and Redemption never achieved any consensus, however, with most insisting that Christ's incarnation on Earth was of sufficient import to save any extraterrestrials. This made a planet-hopping Christ unnecessary, but it strained credulity. In addition, plurality of worlds became a central doctrine for at least two nineteenth century religions: the Mormons and Seventh-day Adventists. Yet another religion, the Swedenborgians, had held it as one of their beliefs since the mid-eighteenth century.

Although the idea of other worlds was also incorporated into many secular philosophies, Christianity holds the distinction of being the philosophical worldview that most influenced the plurality of worlds doctrine in the nineteenth century, at least in the Western world. Despite all the discussion, the long-sought resolution between Christianity and pluralism was elusive. Efforts to achieve that resolution would continue in the twentieth century, but in a far less dominant role, as science and religion went their separate ways.

9.2.2 Anthropocentrism

Sometimes separate from the religious issues, the emotional issue of human status was inextricably linked to all discussions of inhabited worlds. Pluralism and anthropocentrism had long been locked in a battle that had not been completely decided by the dawn of the twentieth century. Committed anthropocentrists were likely to be the staunchest foes of pluralism, no matter what the evidence, and pluralists – whether they liked it or not – contributed significantly to the demise of anthropocentrism. For all the appeal to scientific argument, the continuing battle between anthropocentrism and other worlds pervades twentieth century discussions of extraterrestrial life, and carries the Darwinian debate on the status of humanity into the universe at large. Nowhere is this more evident than in the work of Alfred Russel Wallace (1823–1913), cofounder with Darwin of the theory of evolution by natural selection. His influential work *Man's Place in the Universe: A Study of the Results of Scientific Research in Relation to the Unity or Plurality of Worlds* [6] incorporated many of the biological problems that would be elaborated in ever more subtle form throughout the century. Although Wallace's book in some ways marks a signal advance in the debate about other worlds, its failure is marked by the dominance of its anthropocentric worldview over all other arguments. Convinced of the nearly central position of the Sun in the universe, Wallace first sought – and found – the significance of this fact in the uniqueness of life, and then adduced arguments in favor of the view that life was found beyond the Earth neither in our solar system nor in others. Fifty years after Whewell's treatise, Wallace confidently concluded that "Our position in the material universe is special and probably unique, and … it is such as to lend support to the view, held by many great thinkers and writers today, that the supreme end and purpose of this vast universe was the production and development of the living soul in the perishable body of man" [7]. Although professing a scientific approach, Wallace's book serves as a lesson on the limits of science when worldviews dominate empirical evidence. It is a lesson the twenty-first century should take to heart.

During the twentieth century the tug-of-war between anthropocentrism and belief in other inhabited worlds was profoundly affected by radical changes in the astronomical worldview. While it was still possible as that century began for scientists such as Wallace to argue for an anthropocentric universe based on the Earth's privileged physical position in the cosmos, by 1930 advances in astronomy had destroyed this argument. The resultant worldview – an expanding universe of enormous dimensions in which the solar system was at the periphery of one galaxy among millions – tipped the scales strongly toward the presumption of other worlds for the rest of the century. "The assumption of mediocrity" (a loaded term that has the overtone not of the quality of being average, but of defect) became an underlying current of thought favoring other inhabited worlds, superseding the assumption of uniqueness that had opposed it. The hopes for anthropocentrism at the beginning of the century, and its rapid demise thereafter, constitute one of the profound shifts in twentieth century thought. In this sense the proponents of extraterrestrial life therefore champion not only a scientific theory, but an entire philosophy (some would say a religion of its own).

9.2.3 Chance and necessity

Finally, the philosophical question that dominated the problem of the origin of life beyond the Earth was the likelihood of its origin, which bears on the role of chance and necessity in living systems. On this question, after all, hinged the whole enterprise of exobiology, for if life was a chance occurrence with very low odds, it need never happen again, no matter how big the universe and how expansive the time available. One might never come to the question of the nature of life in the universe if it could not exist elsewhere than on Earth. This philosophical question of probabilities loomed throughout the origin-of-life debate, but never brought consensus.

It is one of the hallmarks of the extraterrestrial life debate that exobiologists, perhaps driven by their desire for a universal biology, placed their faith in the necessity, or at least the high probability, of the origin of life under proper conditions. The principle of the high probability of the chance formation of life became one of their major assumptions, and one they sought to justify as widely held. One sees it already in the first modern treatment of extraterrestrial life by British Astronomer Royal Sir Harold Spencer Jones, who wrote in 1940 that although the nature of the steps from complex organic substances to the first living cell were unknown, "Nevertheless, it seems reasonable to suppose that whenever in the Universe the proper conditions arise, life must inevitably come into existence. This is the view that is generally accepted by biologists" [8].[4] As critics pointed out, "reasonable to suppose" is far from proof, and proof would require empirical evidence.

Transcending all problems of observation and methodology is the historical fact that from the ancient Greeks to the nineteenth century (and even by the time of Kant a century before), the physical world had been transformed, in thought only, into the biological universe, one of the great revolutions of Western thought. By the middle of the eighteenth century, cosmological worldviews had brought the idea of life beyond the Earth into the mainstream of European consciousness, where for a time some thinkers reconciled it with the existence of God because its vision of universal life was compatible with an omnipotent God. A century later it had been accepted, reconciled or rejected by numerous authors, who explored more fully its compatibility not with God, but with Christian dogma. And at the threshold of the twentieth century the theory of an evolutionary universe, the development of astrophysics, and refined observational techniques combined to lay the scientific foundations on which the twentieth century extraterrestrial life debate would rest.

9.3 Scientific foundations: evolution and astrophysics

Two fundamental developments shook science in the mid-nineteenth century: first, in 1859 Charles Darwin (prodded by A. R. Wallace's independent discovery)

---

[4] An excellent historical treatment of origin-of-life research, and its chance and necessity thesis, is I. Fry. *The Emergence of Life on Earth: A Historical and Scientific Overview* (New Brunswick: Rutgers University Press, 2000).

published his theory of the origin of species and evolution by natural selection, and secondly, in the early 1860s the new technique of spectroscopy was applied to astronomy. Though these, and less sweeping developments in other fields, did not cause an immediate and radical change in the character of the plurality of worlds debate, they did signal the beginnings of a long-term change that would bring the subject of other worlds increasingly under the purview of science.

In hindsight, much of the progress in the twentieth century debate may be seen as beginning with evolution and astrophysics. Natural selection provided the basis for a discussion of the evolution of life under differing conditions beyond the Earth, while the more general idea of evolution gave impetus to the idea of the physical evolution of the universe, what we now call cosmic evolution. And by examining the chemical fingerprints present in the analysis of starlight, spectroscopy produced a tool to study the nature of the planets and stars in ever-increasing detail. This in turn provided for the first time a means for determining the possibility of life on the basis of physical conditions, and for demonstrating the idea of cosmic evolution.

Spectroscopy would have the more immediate and profound effect on the fate of the biological universe. Though the arguments of analogy and uniformity of nature had long given credence to the belief that the building blocks for matter and life were alike through the universe, now for the first time this crucially significant truth could be observationally proven. Many of the spectroscopic pioneers themselves did not fail to see the connection of their subject to life in the universe, including Sir William Huggins. To detail the subsequent events would be to give the history of planetary astrophysics.

The Darwinian theory of evolution had a more gradual effect, but one eventually even more significant because it encouraged the idea of cosmic evolution.[5] The concept of cosmic evolution – the connected evolution of planets, stars, galaxies *and* life – now provides the grand context within which the enterprise of exobiology is undertaken. The intellectual basis for this guiding principle of cosmic evolution had its roots in the nineteenth century. Historians of science have shown how a combination of Laplace's nebular hypothesis and Darwinian evolution gave rise to the first tentative expressions of parts of this worldview. The philosophy of Herbert Spencer extended it to the evolution of society, and some Spencerians proposed to extend evolutionary principles to the development of life on other planets. In England and America Richard A. Proctor, and in France Camille Flammarion, also made use of Darwinian ideas to promote the question of life on other worlds. The evolutionary view could, however, be carried too far. In the United States Percival

---

[5] The history of the concept of cosmic evolution is given in chapter 1 of S.J. Dick and J.E. Strick. *The Living Universe: NASA and the Development of Astrobiology* (New Brunswick, NJ: Rutgers University Press, 2004), from which parts of this section are adapted.

Lowell concluded that because Mars was an older and more highly evolved planet, so were its superhuman inhabitants. The dangers of the scientific approach when applied to such extremely difficult problems as the observation of planetary surface features, the search for planetary systems, and the origin and evolution of life, constitute one of the *leitmotifs* of the plurality of worlds tradition.

Such a set of general ideas was a long way from a research program. In the first half-century after Darwin's theory was published, cosmic evolution did not find fertile ground among astronomers, who were hard-pressed to find evidence, other than the spectroscopic confirmation of the widely assumed "uniformity of nature." Astronomers recognized and advocated parts of cosmic evolution, as in the study of stellar evolution. But even Percival Lowell limited his evolution of worlds to physical, not biological evolution, Martian canals notwithstanding. In the first half of the twentieth century, astronomers were *still* limited to the uniformity of nature argument confirmed by spectroscopy.

For the most part, biologists were also reluctant cosmic evolutionists. In his book on *Man's Place in the Universe* [6], Wallace held that the universe was static, humanity's position unique, and Earth likely the only inhabited planet. Ten years later Harvard biochemist L. J. Henderson investigated how the environment on Earth became fit for life, and concluded "The properties of matter and the course of cosmic evolution are now seen to be intimately related to the structure of the living being and to its activities; they become, therefore, far more important in biology than has been previously suspected. For the whole evolutionary process, both cosmic and organic, is one, and the biologist may now rightly regard the universe in its very essence as biocentric" [9].[6] Clearly, Henderson grasped the essentials of cosmic evolution, used its terminology, and believed his research into the fitness of the environment supported it. But the idea was largely undeveloped, and would lie dormant for almost a half-century.

Cosmic biological evolution first had the potential to become a research program in the 1950s and 1960s when its cognitive elements had developed enough to become experimental and (to some extent) observational sciences, and when the researchers in these disciplines first realized they held the key to a larger problem that could not be resolved by any one part, but only by all of them working together. Harvard College Observatory Director Harlow Shapley was an early modern proponent of this concept, and already in 1958 spoke of it in now familiar terms [10]. The Earth and its life, he asserted, are "on the outer fringe of one galaxy in a universe of millions of galaxies. Man becomes peripheral among the billions of stars in his own Milky Way; and according to the revelations of paleontology and geochemistry

---

[6] On this aspect of Henderson's work, see I. Fry. On the biological significance of the properties of matter: L. J. Henderson's theory of the fitness of the environment. *Journal of the History of Biology,* **29** (1996), 155–196.

he is also exposed as a recent, and perhaps an ephemeral manifestation in the unrolling of cosmic time." Shapley went on to elaborate his belief in billions of planetary systems, where "life will emerge, persist and evolve." Shapley's belief was unproven then, and remains to be proven today. The transition from belief to proof is tantamount to discovering whether cosmic evolution commonly ends with planets, stars and galaxies, or with life, mind and intelligence. Put another way, does cosmic evolution produce not only a physical universe, but also a "biological universe"?

The idea of cosmic evolution spread rapidly over the next 40 years in the writings of astronomers like Carl Sagan, both as a guiding principle for astronomers and as an image familiar to the general public. NASA enthusiastically embraced, elaborated and spread the concept of cosmic evolution from the Big Bang to intelligence as part of its SETI and exobiology programs in the 1970s and 1980s. And when in 1997 NASA published its Origins program Roadmap, it described the goal of the program as "following the 15 billion year long chain of events from the birth of the universe at the Big Bang, through the formation of chemical elements, galaxies, stars, and planets, through the mixing of chemicals and energy that cradles life on Earth, to the earliest self-replicating organisms – and the profusion of life" [13]. With this proclamation of a new Origins program, cosmic evolution became the organizing principle for most of NASA's space science effort.

By the end of the century cosmic evolution was elaborated in ever greater theoretical and descriptive detail, and was viewed as playing out on an incomparably larger stage than the static universe conceived by Wallace and most astronomers at the beginning of the century. The Hubble Deep Field, a product of NASA's Hubble Space Telescope, symbolized the vast new evolving universe and its biological possibilities. The scientific problem was to determine, in relatively nearby space, whether those possibilities were real or only wishful thinking.

9.4 Formation of a discipline

In the 1950s and 1960s four cognitive elements – planetary science, the search for planetary systems, origin-of-life studies, and SETI – converged to give birth to the field of exobiology. At first quite separate in terms of researchers, techniques and common goals, these fields over four decades gradually became integrated, in large measure because of the scientific and public desire to search for life beyond Earth. During the early Space Age planetary science produced ground-based claims of vegetation on Mars and spacecraft exploration of the planets began. The search for planetary systems remained an embryonic activity, and produced its first modern claim of extrasolar planets in 1963. Origin-of-life studies began their modern era with the Urey–Miller experiments in 1953. SETI received its underpinnings with a seminal article in *Nature* in 1959, and the first radio telescope search for artificial extraterrestrial signals in 1960. All of these studies followed their own trajectories, but gradually became inextricably intertwined as parts of the larger scientific problem of life in the universe [11, 12]. By the mid-1960s an embryonic exobiology program was in existence at NASA, heralding the embryonic discipline of exobiology.

The establishment of exobiology as a discipline can also be gauged by the claims of its practitioners – realizing, of course, that a certain amount of self-interest is at play in proclaiming one's subject a valid discipline if one is seeking Federal funding. No one would claim that a field existed in the first half of the twentieth century. Even by 1955, when Otto Struve pondered the use of the word "astrobiology" to describe the broad study of life beyond the Earth, he explicitly decided against conferring disciplinary status: "The time is probably not yet ripe to recognize such a completely new discipline within the framework of astronomy. The basic facts of the origin of life on Earth are still vague and uncertain; and our knowledge of the physical conditions on Venus and Mars is insufficient to give us a reliable background for answering the question" of life on other worlds [14]. But the imminent appearance of "exobiology" as a discipline in its own right was evident in 1960 when Joshua Lederberg coined the term and set forth an ambitious but practical agenda based on space exploration in his article entitled "Exobiology: experimental approaches to life beyond the Earth" [15]. Over the next 20 years numerous such proclamations of a new discipline were made. By 1979, NASA's SETI chief, John Billingham, wrote that "over the past twenty years, there has emerged a new direction in science, that of the study of life outside the Earth, or exobiology. Stimulated by the advent of space programs, this fledgling science has now evolved to a stage of reasonable maturity and respectability" [16].

After the Viking mission's failure to find life on Mars, by the 1990s many events conspired to revitalize exobiology: the Mars rock, the Mars Global Surveyor observations of the gullies of Mars, the Mars Odyssey detection of water near the surface, the Galileo observations of Europa, circumstellar matter, extrasolar planets, life in extreme environments including hydrothermal vents, and complex interstellar organics [12] (chapter 5). All these elements fed into NASA's new astrobiology program, which emerged from a deep organizational restructuring at NASA in 1995. Astrobiology now is a much more robust science than exobiology was 40 years ago. Astrobiology places life in the context of its planetary history, encompassing the search for planetary systems, the study of biosignatures, and the past, present and future of life. Astrobiology science added new techniques and concepts to exobiology's repertoire, raised multidisciplinary work to a new level, and was motivated by new and tantalizing (but still circumstantial!) evidence for the possibilities of life beyond Earth.

9.5 Societal impact: biological universe as worldview

Astrobiology's image of a biological universe raises deep questions of the destiny of life and societal impact at many levels. As we have seen, the idea of *intelligent* life in the universe has already generated a long history of discussion of potential impact, especially in the theological arena. The societal implications of astrobiology have been taken up in the last decade in a variety of venues, including the Foundation for the Future and Templeton Foundation studies. NASA itself has sponsored a number of societal impact studies, in accordance with the National Aeronautics and Space Act goal of identifying the impact of the space program on society. In conjunction with the launching of the NASA SETI program, in 1991–1992 NASA held a systematic series of workshops on the cultural aspects of SETI. And shortly after the astrobiology endeavor was launched, another group gathered to address broader concerns [17, 18].[7]

One finding that came out of these studies is that historical analogs form a useful basis for discussion, not in the form of the usually disastrous physical cultural contacts on Earth (Pizarro and the Incas, Cortes and the Aztecs, among a long litany of others), but by studying the reception of scientific worldviews [19]. I argue this is particularly true because the biological universe has achieved the status of a worldview, a "biophysical cosmology" that asserts the importance of both the physical and biological components of the universe. Like all cosmologies, it makes a claim about the large-scale nature of the universe; that life is not only a possible implication, but a basic property, of the universe in the sense that the universe seems to be fine-tuned for life. Like all cosmologies, it redefines our place in the universe. And most importantly, like other cosmologies, in the twentieth century the biophysical cosmology became increasingly testable – even if it still embodied philosophical assumptions along with scientific theory and observation. Thus today we are in a situation similar to Galileo's 400 years ago when he marshaled the evidence for and against the heliocentric and geocentric worldviews. Today our worldview hangs in the balance between the physical universe in which the Earth is unique or rare in its possession of life, and the biological universe in which life

---

[7] See also S. Dick. Cultural implications of astrobiology: a preliminary reconnaissance at the turn of the millennium. In *Bioastronomy '99: A New Era in Bioastronomy*, eds. G. Lemarchand and K. Meech (San Francisco, CA: Astronomical Society of the Pacific, 2000), pp. 649–659. The 1991–1992 NASA study was published as *Social Implications of the Detection of an Extraterrestrial Civilization* (SETI Press, prepared 1994, published 1999). The report of the Workshop on Societal Implications of Astrobiology, held November 16–19, 1999 at NASA Ames, is available as a NASA Technical Memorandum at http://astrobiology.arc.nasa.gov/workshops/societal/, revised January 20, 2001.

is a common outcome of cosmic evolution. Astrobiology is the science that will decide which of these worldviews is true.

Acknowledgements

My thanks to Connie Bertka and Ernan McMullin for their comments.

References

[1] P. Ward and D. Brownlee. *Rare Earth: Why Complex Life is Uncommon in the Universe* (New York: Copernicus, 2000).

[2] A. O. Lovejoy. *The Great Chain of Being: A Study of the History of an Idea* (Cambridge, MA: Harvard University Press, 1936), ch. 4, paperback edition, 1971.

[3] S. J. Dick. *Plurality of Worlds: The Origins of the Extraterrestrial Life Debate from Democritus to Kant* (Cambridge: Cambridge University Press, 1982).

[4] *Basic Writings of Thomas Paine* (New York, 1942), pp. 67–78.

[5] M. J. Crowe. *The Extraterrestrial Life Debate 1750–1900: The Idea of a Plurality of Worlds from Kant to Lowell* (Cambridge: Cambridge University Press, 1986).

[6] A. R. Wallace. *Man's Place in the Universe: A Study of the Results of Scientific Research in Relation to the Unity or Plurality of Worlds* (London: Macmillan, 1903).

[7] A. R. Wallace. Man's place in the universe. *The Independent*, **55** (1903), p. 474.

[8] Sir H. Spencer Jones. *Life on Other Worlds* (New York: Macmillan, 1940), p. 57.

[9] L. J. Henderson. *The Fitness of the Environment* (New York: Macmillan, 1913), p. 312.

[10] H. Shapley. *Of Stars and Men* (Boston, MA: Beacon Press, 1958).

[11] On the formation of exobiology as a discipline, see S. J. Dick. *The Biological Universe: The Twentieth Century Extraterrestrial Life Debate and the Limits of Science* (Cambridge: Cambridge University Press, 1996).

[12] S. J. Dick and J. E. Strick. *The Living Universe: NASA and the Development of Astrobiology* (New Brunswick, NJ: Rutgers University Press, 2004).

[13] NASA. *Roadmap for NASA's Office of Space Science Origins Theme* (Washington, DC: NASA, 1997).

[14] O. Struve. Life on other worlds. *Sky and Telescope*, **14** (February, 1955), 137–146.

[15] J. Lederberg. Experimental approaches to life beyond the Earth. In *Science in Space*, eds. L. V. Berkner and H. Odishaw (New York: McGraw-Hill, 1961), pp. 407–425.

[16] J. Billingham. *Life in the Universe* (Cambridge, MA: MIT Press, 1981), p. ix.

[17] S. J. Dick (ed.). *Many Worlds: The New Universe, Extraterrestrial Life and the Theological Implications* (Philadelphia and London: Templeton Foundation Press, 2000).

[18] A. Tough (ed.). *When SETI Succeeds: The Impact of High-Information Contact* (Bellevue, WA: Foundation for the Future, 2000).

[19] S. J. Dick. Consequences of success in SETI: lessons from the history of science. In *Progress in the Search for Extraterrestrial Life*, ed. G. S. Shostak (San Francisco, CA: Astronomical Society of the Pacific, 1995), pp. 521–532.

10

The search for extraterrestrial life: epistemology, ethics, and worldviews

Mark Lupisella

The search for and discovery of extraterrestrial life, especially an independent origin of life, raise interesting philosophical issues (most or all of which can be connected to important practical issues), in at least three interrelated areas: (1) epistemology, (2) value theory (especially ethics), and (3) worldviews. This chapter samples a variety of views in these areas, touching slightly on some policy and theological connections, both of which are covered more extensively elsewhere in this volume.

The first section will explore epistemological areas such as (a) dealing with the limitation of knowing only one kind of biology, (b) challenges of discerning an independent origin of life, and (c) challenges for assessing the biological status of a region or entire planet. The ethical considerations of the second section will explore (a) the role of an independent origin of life vs. interplanetary transport, (b) ethical views ranging from anthropocentric to cosmological, and (c) potential policy implications. The third section will touch briefly on basic worldviews that revolve around (a) randomness and chance (an "accidental" universe), (b) purpose and meaning (a deliberate universe), and (c) a "bootstrapped" universe in which meaning and purpose emerge in the universe through valuing cultural beings (a "cultural cosmos").

10.1 Epistemology

Extraterrestrial life poses unique challenges to the boundaries, application, and confidence in our knowledge. Some interrelated epistemological questions to consider are: (1) How can we deal with the limitations of knowing only one kind of biology?; (2) What will be the criteria for discerning an independent origin

Exploring the Origin, Extent, and Future of Life: Philosophical, Ethical, and Theological Perspectives, ed. Constance M. Bertka. Published by Cambridge University Press. © Cambridge University Press 2009.

of life?; and (3) What will be the criteria for assessing the biological status (for example whether life exists, its distribution, nature, etc.) of a region or perhaps entire planet?

10.1.1 The "one data point" problem

The extent to which extraterrestrial life questions can be confidently addressed rests in large measure on the extent to which we are confident in applying terrestrial biology to extraterrestrial biological possibilities. All life on Earth known to date shares the same basic biochemistry and strongly suggests descent from a common origin. In this sense, we only know one kind of biology – not just in terms of biochemistry, but also in terms of broader ecological and evolutionary principles as well – that may or may not be universal. This raises an interesting question: How can we best apply knowledge from what is effectively a data set of one?

We'd like to be as comprehensive as possible in our search so (a) we don't miss something we might be unfamiliar with and so (b) we avoid causing adverse effects to indigenous life forms (for example due to contamination), especially in the case where it may be an independent origin. Theoretically, an infinite number of lines can be drawn through one data point, implying that high confidence from extrapolation is impossible. But we can nevertheless look under the proverbial lamppost (where the light of knowledge points us) by bounding the search space with appropriate terrestrial biology constraints, such as the need for liquid water, perhaps carbon, etc. These are reasonable measures being taken by today's practitioners that may prove to be successful. However, given the high stakes and the ability for life to surprise us (for example extremophiles, prions, etc.) this may also be unduly limiting.

Clearly, a careful and thorough search for extraterrestrial life is the preferred method for learning more broadly about biology, but how careful and thorough we have to be are interesting challenges in themselves and relate to the other two epistemological questions posed above. In the absence of finding a second data point soon, we might consider trying to broaden the possibility space artificially through perhaps artificial life studies [1] or synthetic life [2] to at least broadly inform how universal terrestrial biology and ecological principles might be, and what the possibility space(s) might look like.

The one-data-point challenge also extends to broader and potentially longer-term "exoecological" questions, for example how to best ensure the integrity of possible extraterrestrial life in light of potential adverse effects from contamination and other ecological impacts, particularly from a human presence. Will cohabitation be possible? If so, how we will know and what level of confidence can and should we obtain? How confident can we be in any conception of life based on terrestrial biology?

10.1.2 Discerning an independent origin of life

Discovering an independent origin of life could be very important for understanding whether or not life is common in the universe. A relatively close independent origin can be interpreted to suggest that origins of life are likely to have arisen wherever the conditions permit. This is, in part, because finding a second example of a phenomenon in a relatively small space implies a dramatically increased likelihood that the phenomenon exists throughout the larger space in which basic conditions are similar (in this case, the galaxy or universe). This "1, 2, infinity" argument is not definitive of course, but it is a reasonable guideline for potential implications.

Discerning an independent origin of life could also have important scientific implications for theoretical biology, which would benefit substantially from understanding alternative biologies on which comparative biology could be based. This could dramatically improve our understanding of our own biology. Unfortunately, in cases where it isn't obvious, it may be very difficult, if not impossible to discern an independent origin of life. Bruce Jakosky notes in a personal email:

My first reaction is that this is an even more difficult problem than you allude to. The last common ancestor of terrestrial life certainly shared the usual characteristics that bind all terrestrial life together – use of RNA and DNA and ATP, and so on. However, if you're willing to go back further, one envisions an RNA world and even a pre-RNA world, with somehow simpler and simpler entities going all the way back to prebiotic chemistry on Earth. I suspect that if you pulled something from the pre-RNA world, it might not share enough of the biochemistry with modern organisms to be recognized as necessarily related to it. If we found something on Mars that had been transplanted early enough, or equivalently if it diverged from terrestrial life during an RNA or pre-RNA world, how would we recognize it as not having come from Earth at a very early time? If it's separated early enough, it may not be possible to distinguish it from an independent origin, especially if subsequent evolution of the Martian side allowed a more robust biochemistry that differed from terrestrial biochemistry to have evolved [3].

This is partly why contamination challenges could turn out to be important. Minimizing variables such as contamination that could complicate the search and discovery of extraterrestrial life and the discerning of an independent origin would be wise steps given what is at stake. Indeed, NASA has already begun to address planetary protection issues associated with a human Mars mission [4].

10.1.3 Biological assessment criteria

It will likely be important to assess the biological status of a location, region, or possibly an entire planet before executing certain activities, particularly sending humans. If so, what level of confidence should we have in that biological assessment, and what is required to establish that confidence? How many and what kinds of missions and experiments would be required to be reasonably confident that life does or does not exist in any given area of interest – including an entire planet, particularly if contamination from a human presence could spread globally?

This is an interesting epistemological challenge because again, we effectively have only one data point – one example of how life inhabits a planet. Is life an "all-or-nothing" phenomenon? How densely should we expect life to occupy an area, region, or entire planet? That is, would we expect life generally to spread and tend strongly toward ecological saturation, perhaps densely occupying niches or consuming a planet, as it has on Earth? Should we expect that sparse inhabitation is unlikely, that life tends to either extremes of extinction or saturation? Or should we assume that life elsewhere could be distributed sparsely, perhaps tenuously, in what may or may not be stable niches? While these questions are not necessarily answerable to the degree we'd like given our knowledge today, there are at least theoretical expectations (perhaps with likelihood probabilities) that can be researched further and adopted as operational working hypotheses, or at least as broad guidelines. This chapter tries to argue that a conservative posture should drive those broad guidelines – for example that life may be sparsely distributed and hard to detect. Assessments of these kinds of questions could drive the confidence level and the criteria for obtaining that confidence in a biological assessment. For example, if life might be sparsely distributed, we would probably require many life detection attempts. But how many? Of what kind? How deep should subsurface life detection missions go?

So, how important is preserving the integrity of extraterrestrial life? How much confidence do we want to have regarding the biological status of any given locale or of the entire planet before possibly jeopardizing indigenous ecosystems with potentially intrusive activities such as a human mission? Many would likely agree a conservative approach is warranted and feasible, but that may not be enough for such an approach to be successfully implemented because circumstances and broader interests often conspire to relax cautious strategies. If we have some sense for this ahead of time, perhaps pre-detection policy guidelines outlining confidence levels and criteria for achieving that confidence for biological assessments could help avoid undesirable outcomes. Ultimately, such guidelines will be driven by how much we value extraterrestrial life and why.

10.2 Ethical considerations for primitive extraterrestrial life

"Primitive" life here means essentially microbial life as we understand it here on Earth, or any life that is relatively simple in its appearance, function, and behavior. While an entire chapter could be devoted to parsing distinctions of this kind, and while compelling arguments can be made for why such distinctions might be irrelevant and misguided (due, for example, to our biased conceptions of what constitutes "simple" life), the interesting practical philosophical question is why and how human beings might value life that is difficult to identify with – such as microbial life. It isn't obvious why extraterrestrial microbes should have any kind of value beyond scientific value. The next sections will explore basic ethical views, some of which turn on whether extraterrestrial life might be of an independent origin.

10.2.1 Interplanetary transport vs. independent origin

There are many philosophical perspectives in value theory, many of which depend on distinguishing between intrinsic value and instrumental value, where intrinsic value suggests something has value in and of itself, independent of anything external to it, and instrumental implies the value of something is derived from having utility or value to something else. Not all of these "metaethical" views (that is, frameworks and methods for developing ethical positions) can be covered in this chapter, but one way to help inform our discussion is to explore what is arguably a unique challenge presented by extraterrestrial life, namely the distinction of whether extraterrestrial life has independently originated, i.e. a "second genesis," (which for some may imply intrinsic value) or whether it is phylogenetically related to life on Earth via interplanetary transport, i.e. panspermia. Some have suggested that this is an important distinction:

> Both the scientific and ethical implications of finding life on Mars are much more vast if that life-form represents a second and independent genesis of life.
>
> (Richard Randolph, Margaret Race, and Chris McKay [5])

> Although the discovery that life had spread, for example, from Earth to Mars or vice versa through a panspermia mechanism, would be extremely interesting scientifically, it would have no significant philosophical or theological implications.
>
> (Paul Davies [6])

> And what were the odds against life getting started? We still do not know whether life's emergence is natural, or whether it involves a chain of accidents so improbable that nothing remotely like it has happened on another planet anywhere else in our galaxy. That is why it would be so crucial to detect life, even in simple and vestigial forms, elsewhere in our solar system – on Mars or under the ice of Europa. If it had emerged twice within our solar system, this would suggest that the entire galaxy would be teeming with life. That momentous conclusion would follow provided that the two origins were indeed independent.
>
> (Martin Rees [7])

Distinguishing "kinds," and particularly "natural kinds," has had a rich history in philosophy. An independent origin of life would, for many, intuitively and deeply represent a compelling distinction of kind, giving rise to a unique attribution of value (perhaps intrinsic value) and respect. Justification for this and related sentiments will not be analyzed exhaustively here, but several views will be highlighted to provide a sense of the variety and richness of thinking that is emerging.

10.2.2 Anthropocentrism

Robert Zubrin, the founder of the Mars Society, acknowledges the unique value of extraterrestrial life, especially scientific value, but nevertheless stresses that we don't hesitate to kill terrestrial microbes under many circumstances, adding to the case for substantial devaluation of Martian microbes relative to human interests [8]. While this is an understandable sentiment, it is also reasonable to consider that extraterrestrial life, especially of independent origin, could be unique, valuable, and worthy of respect in a way that terrestrial microbes are not.

The ecologist Frank Golley has argued that activities in space such as the colonization and terraforming of Mars will be unavoidable since it is consistent with the dominant myths and metaphors of Western civilization [9]. Unfortunately these dominant myths and the exploration that results from them have often had serious adverse effects on indigenous environments and life, including human beings. Indeed the dominant myths of "manifest destiny" have featured prominently in public discussions of humanity's relationship to Mars [10, 11].

10.2.3 Non-human rights and non-interference

If there is life on Mars, I believe we should do nothing with Mars. Mars then belongs to the Martians, even if they are only microbes.

(Carl Sagan [12])

Although the notion of rights is not explicitly articulated in Carl Sagan's sentiment, his perspective can be associated with a rights-based metaethics. While the justification of intrinsic value has been philosophically problematic, rights-based ethical views nevertheless often depend on conceptions of intrinsic value. J. Baird Callicott [13] writes: "The assertion of 'species rights' upon analysis appears to be the modern way to express what philosophers call 'intrinsic value' on behalf of non-human species. Thus the question, 'Do nonhuman species have a right to exist?' transposes to the question, 'Do nonhuman species have intrinsic value?' "

Chris McKay has appealed to an intrinsic value of life principle and hence suggests that Martian microbes, particularly of independent origin, have a right to life – "to continue their existence even if their extinction would benefit the biota of Earth" [14].[1] Deep Ecology views tend to have as a central tenet, biological egalitarianism, according to which all organisms have an equal right to life [15]. If one claims that other animals, and in particular, Martian microbes, have rights,

---

[1] See also C. McKay. Let's put Martian life first. *The Planetary Report*, **21** (July/August 2001), 4–5; and C. McKay and M. Marinova. The physics, biology, and environmental ethics of making Mars habitable. *Astrobiology*, **1** (2001), 89–109.

but that there are no degrees of rights, how are we to assess situations that involve conflicting interests between humans and other life forms? Indeed, for those who think Martian life has rights, a compromise might not be satisfactory. Only a non-interference policy would be acceptable [16] (p. 227). However, degrees of rights or degrees of value (perhaps even degrees of intrinsic value) may provide a more pragmatic framework for considering these issues [17].

10.2.4 Respecting values of rational beings

While the focus of this chapter is not extraterrestrial intelligence, astrobiology nevertheless prompts us to consider values of other potential rational beings, especially as they might apply to non-intelligent or primitive life forms. For example, if we take a view of ethics in which rational beings are the only moral agents, might the possible existence of rational extraterrestrial beings prompt the consideration of broader ethical views that might be important to them – such as a conservation ethic that extends to non-rational living beings (for example extraterrestrial microbes), perhaps as part of a broader environmental/cosmic ethic? This is similar to considering values of our fellow human beings that might go beyond our own values. If others value something for plausible reasons, shouldn't we be prompted to consider respecting those values? Extraterrestrial intelligent beings may consider life in the universe, perhaps independent origins of life in particular, to be extremely valuable, perhaps intrinsically or "cosmically" valuable. Perhaps such values should be considered as we formulate our own views regarding how we should move out into the solar system and beyond.

10.2.5 Cosmological value and ethics

If we consider cosmic and stellar evolution, we can also find an affinity with Martian microbes and fulfill Leopold's concept of common heritage.

(Alan Marshall [16] (p. 234))

In exploring ethical issues regarding the extraterrestrial environment, several writers have suggested the need for a "cosmocentric ethic" because they conclude that existing ethical theories exclude the extraterrestrial environment since they are geocentric and cannot be applied to extraterrestrial environments [18, 19, 20, 21, 22].[2] While many philosophers would disagree about the extent to which ethical

---

[2] Martyn Fogg writes: "the concept of terraforming is inspiring enough to perhaps generate a formal effort toward extending environmental ethics to the cosmic stage" (M. Fogg. *Terraforming: Engineering Planetary Environments* (Warrendale, PA: SAE International, 1995), p. 490).

theories are narrowly constrained to geocentric application, the relatively new context or "lens" of space does nevertheless appear to raise interesting and novel ethical challenges, and provides us with an additional perspective with which to re-examine ethics and value theory in general.

Exploring a broader-based ethic such as a cosmocentric ethic may be helpful in sorting through issues regarding the moral considerability of primitive extraterrestrial life as well as other ethical issues that will confront humanity as we move out into the solar system and beyond [23, 24]. But as with environmental ethics, an important challenge for a cosmocentric ethic is justifying intrinsic value [25].[3] Indeed, part of the usefulness of appealing to the universe as a basis for an ethical view is that a justification of intrinsic value (and perhaps degrees thereof) might be possible since it could be based on what is for many a compelling objective absolute – the universe itself.

Systemic nature is valuable as a productive system, with Earth and its humans on one, even if perhaps the highest in richness or complexity, of its known projects. Nature is of value for its capacity to throw forward all the storied natural history. On that scale, humans on Earth are latecomers, and it seems astronomically arrogant for such late products to say that the system is only of instrumental value, or that not until humans appear to do their valuing does value appear in the universe.

(Holmes Rolston III [26])

Holmes Rolston offers a view that appeals to the "formed integrity" of "projective nature." This view suggests that the universe creates objects of formed integrity (for example objects worthy of a proper name) which have intrinsic value and which should be respected. Robert Haynes points out, however, that such a view appears to conflict with modifying the Earth, even to the benefit of humans [18].

The systemic interdependent connectedness of ecosystems is often cited as a foundation justifying the value of parts of the larger whole, since a subset contributes to the maintenance of the larger whole. Consider Leopold's egalitarian ecosystem ethic: "A thing is right when it tends to preserve the integrity, stability, and beauty of the biotic community. It is wrong if it tends to do otherwise" [27]. Freya Mathews suggests that intrinsic value can be grounded in self-realization, which is a function of interconnectedness. The universe qualifies for selfhood and hence self-realization (again, for which interconnectedness plays a critical role) and humans participate in this cosmic self-realization.[4]

---

[3] J. Baird Callicott writes: "In addition to human beings, does nature (or some of nature's parts) have intrinsic value? That is the central theoretical question in environmental ethics" (J.B. Callicott. Intrinsic value in nature: a metaethical analysis. *The Electronic Journal of Analytic Philosophy*, **3** (Spring 1995)).

[4] Mathews articulates selfhood and self-realization generally and in a cosmic sense in chapter 3, and the associated ethical implications in chapter 4, in *The Ecological Self* (London: Routledge, 1991).

Connectedness, then, which may be a fundamental property of the universe, may hold promise as a foundation for a cosmocentric ethic. The property of connectedness may be realized and instantiated by relations and interactions – supporting the notion of a relational metaphysic [28]. Such a view might favor robust and ever-increasing degrees of relations and interaction, perhaps manifested as the ongoing emergence of new relations and interactions as the universe evolves (for example objects of formed integrity, "emergence properties," etc.), giving rise to diversity and complexity, as well as many other possible measures of relationship and interaction, all of which would contribute to the realization of the nature of the universe (i.e. its self-realization). While these metaphysical musings are certainly speculative, they are at least consistent with, and might explain or even predict a projective universe, and perhaps a universe that manifests a law of increasing organized complexity as well [29], which could then serve as a more direct intellectual grounding for what might be thought of as degrees of cosmological intrinsic value [30].

In making choices consistent with this view, humanity might help preserve and propagate many forms of diversity here on Earth and throughout the universe, keeping in mind broad conceptions of such ideals and that the universe could have meaning, value and purpose that goes well beyond life (indeed, Rolston suggests "it is less short-sighted but still seriously myopic to value the system only for its production of life, although this is of great moment within it" [26] (p. 155)). Freeman Dyson writes: "Diversity is the great gift which life has brought to our planet and may one day bring to the rest of the Universe. The preservation and fostering of diversity is the great goal which I would like to see embodied in our ethical principles and in our political actions" [31].

10.2.6 Values, ethics, and policy: a "prime directive"?

I can think of nothing so positively transforming of human consciousness as the discovery, study, and conservation of life somewhere off the Earth.

(J. Baird Callicott [13])

Callicott invokes a form of "weak anthropocentrism," a view first articulated by Bryan Norton, which suggests that things that transform and ennoble human nature have enough value to require their preservation [32].

A Mars biota will be valuable for scientific reasons, and this to some extent will ensure preservationist policies. But more than that, Martian life is intrinsically valuable.

(Alan Marshall [16] (p. 233))

There is currently no NASA policy, or international protocol, for the proper handling of non-intelligent extraterrestrial life. We believe that such a policy should be developed now, before these discoveries are made. Such a policy would be informed by an ethical analysis concerning our obligations as space explorers.

(Richard Randolph, Margaret Race, and Christopher McKay [5])[5]

NASA and others have begun to address broader issues associated with astrobiology, and this volume is part of that process. NASA held a workshop on the societal aspects of astrobiology in 1999 where ethical issues were explored [34]. In 1992, the National Research Council recommended that the question of life on Mars should first be addressed by robotic missions before sending human missions [35]. As per the concerns raised previously, can such a recommendation be satisfied? If so, how? If it is costly and time-consuming, will scientific value be enough to ensure the preservation of microbial extraterrestrial life? How might our policies reflect a balance of broader non-scientific interests?

If and when life is discovered beyond Earth, non-scientific dimensions may strongly influence decisions about the nature and scope of future missions and activities. It is appropriate to encourage international discussion and consideration of the issues prior to an event of such historical significance.

(Margaret Race and Richard Randolph [36])

Some suggest the need for broad space environmental protection policies [37],[6] which by default would have as a subset, protection of indigenous life. Mark Williamson notes that the Outer Space Treaty "should be amended, supplemented or otherwise reinforced to offer a degree of environmental protection that is currently lacking" [38]. Ivan Almar suggests that planetary scientists should "survey and evaluate all known planetary environments with regards to their scientific value (or even uniqueness), sensitivity to intervention, difficulty or ease of access by space probes, etc." to produce "a limited list of sites and objects open to scientific investigation but closed to exploitation of natural resources." He suggests further that "these international scientific preserves or 'wilderness areas' should be legally protected within the frame of an international environment-protection treaty" [39]. Others suggest it may be possible to explore worlds in such a way as to be able to reverse contamination from human exploration.[7]

---

[5] In M. Lupisella and J. Logsdon. Do we need a cosmocentric ethic? Paper IAA-97-IAA.9.2.09 presented at the International Astronautical Federation Congress. American Institute of Aeronautics and Astronautics, Turin (1997), they suggest the need for similar pre- and post-detection policies regarding primitive extraterrestrial life.

[6] P. Sterns and L. Tennen write: "A balance must be found between the impact of any mission and the scientific results or other benefits which may be obtained thereby. Furthermore, certain activities may be sufficiently detrimental to the environment to require restrictions and prohibitions thereof, regardless of any benefits which otherwise may be realized."

[7] Chris McKay considers such a possibility in a paper in progress and Margaret Race noted that reversibility could act as a guideline for human exploration during this AAAS workshop. I have expressed concern about the feasibility of "biological reversibility" but nevertheless agree it is something worth considering.

Chris McKay suggests that Martian life's right to exist and evolve on its planet of origin outweighs the human settlement of Mars (partly because it is probably not *necessary* for humans to go to Mars) but he does contemplate interference to the extent that the rights of Martian life "confer upon us the obligation to assist it in obtaining global diversity and stability" [14]. This kind of "garden narrative"[8] offers a compelling balance of values that can accommodate human aesthetic and scientific interests while at the same time prioritizing the continued existence and evolution of Martian life. He writes: "There are no absolute criteria for assigning value to life and diversity or for comparing the value of a human outpost against the removal of an indigenous Martian biota.... Nonetheless, we should be intellectually prepared to face this issue and to defer to indigenous Martians – however microscopic – and even revive them if necessary and assist them in regaining biological control of their planet. We will be the better for this in both ethical and scientific terms" [40].

There may also be high degrees of value associated with life that is part of terrestrial life's phylogenetic tree but that nevertheless still differs enough, especially if the transfer occurred very soon after the origin of life since evolutionary trajectories might have created life different enough from terrestrial life to take on special value. There is also the "value of place" to consider in that even if extraterrestrial life is similar to life on Earth, some may argue that its own evolution in its own unique environment is of sufficient value to preserve in its indigenous environment (especially from ecocentric perspectives which value entire ecosystems).

As an example mentioned previously, cohabitation is one long-term option to consider as a thought experiment. Ironically, Richard Taylor's comment, "Move over microbe!" might apply.[9] That is, extraterrestrial microbes might be displaced, as often happens on Earth, but they need not necessarily be harmed or destroyed. Coexistence with primitive Martian life might be possible, especially if we are proactive and ecologically careful. Would we combine into one ecosystem? Assuming we were careful, Martian life might not be destroyed. It could, however, be modified via the forces of its new ecosystem. Or maybe we will decide to preserve that life in a kind of isolated environment, perhaps with the indigenous Martian environmental conditions intact – a kind of natural conservatory – so that, to some approximation, it will be allowed to evolve as it might have otherwise.[10] This might be acceptable to many, although there will certainly be legitimate concerns about

[8] Richard Miller develops the idea of a garden narrative in: Terraforming: an ethical perspective. In *The Case for Mars VI: Making Mars an Affordable Destination, Proceedings of the 6th Case for Mars Conference, 1996* (San Diego, CA: American Astronautical Society, Univelt Incorporated, 2000).
[9] Martyn Fogg notes a radio interview with Richard Taylor in M. Fogg. *Terraforming: Engineering Planetary Environments* (Warrendale, PA: SAE International, 1995), p. 494.
[10] Robert Zubrin opens the door for such a compromise when he suggests that the polar regions will be available for indigenous life to predominate (R. Zubrin. The terraforming debate. *Mars Underground News*, **3** (1993), 3–4).

maintaining long-term cohabitation that does not adversely affect what may be a fragile indigenous ecosystem.

Frank Golley suggests that to turn away from the pursuits of colonization and terraforming of Mars would require a fundamental reorientation of our culture [9]. If devaluation of indigenous life systems is part of these pursuits, then perhaps a fundamental reorientation of our culture is warranted. This could be consistent with Robert Zubrin's and others' vision of Mars as an opportunity to conduct grand social experiments and create a new branch of human civilization with new ways of life – because with all the promise that holds, we could at the same time make respect for indigenous environments, including Martian microbial life, a fundamental principle of our exploration, settlement, and development of new worlds. The environmental and animal rights movements and the vast body of knowledge and policy that have resulted are powerful precedents on which to build. We need only learn from history and be proactive in deepening and broadening such policies if we wish to extend them to extraterrestrial environments.

Policy will ultimately be driven by the highest priority values and should not be driven by solely the likelihood of adverse effects (including extinction) on indigenous Martian life, but by the significance as well. Clearly these are difficult issues – partly because of the speculation involved, and partly because they are long-term. Nevertheless, exploring these issues now as part of long-term planning is probably wise since there is time to collect the relevant data and seek a healthy international consensus. Addressing these questions now will not be wasted if we were to indeed find a lifeless Mars because this kind of planning can only help prepare us as we move out into the rest of the solar system and universe in search of life. Perhaps we will develop a "prime directive" for non-intelligent extraterrestrial life similar to what we do for endangered species on Earth: *preserve and assist.*

Some may argue that the rational pursuit of broad ethical views and their application to policy is ultimately futile – that rationality is slave to the passions, and that self-interest is the primary motivator of human behavior. Certainly, there is truth to that. But it is also true that we can be rational and thoughtful regarding what we value and why, especially since human beings and human culture are extremely diverse and are driven by many different forces. Ultimately, through a mix of reductive, creative, and ecological thinking, as favored by Frederick Turner [41], we should, and perhaps will, pursue a reasonable balance among many diverse forces regarding the status of extraterrestrial life in our policies and worldviews. We have an opportunity to take our time and be proactive and thoughtful in exploring this delicate balance of values. And while values are important drivers of human sentiment, behavior, and policy, worldviews often give rise to values, and provide compelling contexts within which to understand values and ethics.

10.3 Worldviews

As I suggest in *The Biological Universe*, the idea of a universe full of life is more than just another theory or hypothesis; it is rather a kind of world view, which we may call the bio-physical cosmology or simply the biological universe.

(Steven Dick [42])

The Copernican principle is not established with respect to biology, culture, or ethics. In this context, the question "Are we alone?" is a deep philosophical question. That we must look beyond the Earth to address these questions is clear. In a fundamental way we seek to place the Earth, life, and most important our unique role as intelligent beings on this Earth in the broader context. In the past we have sought answers to these questions from pure logic or through spiritual revelation. Now, for the first time, we can begin a direct investigation that may construct a framework within which we may find the answers to these questions.

(Chris McKay [43])

Although the discovery of extraterrestrial life need not necessarily have any definitive effect on worldviews, there are nevertheless many plausible worldview implications worth exploring, ranging from a completely meaningless and point-less universe to a "bootstrapped universe," to a deeply meaningful and purposeful universe. In all three categories, life, intelligence, and consciousness can be seen as uniquely valuable, but for very different reasons, and with potentially quite different implications.

I think both Monod and Gould are absolutely right to perceive bleak atheism in the scenario that life and intelligence are freak accidents, unique in the cosmos. But the flip side is also true. If it turns out that life does emerge as an automatic and natural part of an ingeniously biofriendly universe, then atheism would seem less compelling and something like design more plausible.

(Paul Davies [44])

Were traces of life to be discovered elsewhere in our solar system today, it would favor the Augustinian idea that the "seeds" of life were implanted in matter from its first appearance. Such seeds could presumably come to fruition anywhere where "water and earth" provided the right environment. On the other hand, such a discovery would challenge the belief that the origin of life on Earth required a miraculous intervention on God's part. … the discovery would strengthen the case for an evolutionary origin of the first life as a consequence of the ordinary processes of nature.

(Ernan McMullin [45])

10.3.1 The accidental universe

Man at last knows that he is alone in the unfeeling immensity of the universe, out of which he has emerged only by chance.

(Jacques Monod [46])

After sufficient searching, the absence of extraterrestrial life could, particularly from a natural science perspective, be interpreted to suggest that life is a rare and accidental phenomenon in a universe that is otherwise indifferent, and possibly hostile to life. Life as a rare and random phenomenon can be interpreted to be consistent with a pointless universe worldview, one in which life and perhaps the universe itself originate only by chance and have no broader meaning or purpose in any objective sense. Coming to terms with such a reality might be one of the most important challenges to face sufficiently aware beings. In such a worldview, we might see the human condition, and more generally the biological condition, as fundamentally in tension with the nature of the universe.

In an accidental universe worldview, we might see life as a diamond in the rough, with perhaps no truly objective absolutes to guide our actions. In this worldview we are masters of the universe – it is ours to ultimately make of it what we wish. As diamonds in the rough, human beings may have more reason to turn to each other, and life more generally, with a deep reverence and respect as something truly unique and fragile in the universe. We may have to rely solely on each other, knowing we are all we have to protect and sustain us in an otherwise hostile universe.

10.3.2 The deliberate universe

The discovery of a cosmically local, independent origin of life or "second genesis" would suggest that the universe is teeming with life. Combined with a sufficient theory, perhaps one that lies closer to determinism,[11] we might think the universe not only naturally produces life, but perhaps more generally tends toward the creation of organized complex systems such as life, intelligence, and consciousness, and that as a result, we are an inevitable realization of the nature of the universe [29]. In this worldview, the "God of the gaps" has one less, very large, gap to fill – but nevertheless, life as a natural "cosmic imperative" could be seen as cosmically intrinsically valuable, and so we might see ourselves not only as truly at home in the universe, but perhaps also as occupying a special place.

10.3.3 A bootstrapped universe: "the cultural cosmos"

No meaning can be given to the idea of a view of the universe from the point of view of an observer who is outside of it or who is not a participant in it.

(Lee Smolin [47] (p. 80))

---

[11] Paul Davies writes: "I argue that on the spectrum between chance and determinism (or certainty), the closer to determinism the truth lies, the more reason we have to feel 'at home in the universe' (to borrow Stuart Kauffman's evocative phrase) and the more circumstantial evidence there would be for some sort of meaning, purpose, or design in nature" (in *Many Worlds: The New Universe, Extraterrestrial Life, and the Theological Implications*, ed. S.J. Dick (Philadelphia, PA: Templeton Foundation Press, 2000), p. 26).

But the point of your book is very simple: it is just that the whole show of the universe is so extraordinary that even the absence of God would be God enough.

(Saint Clair Cemin, commenting on Lee Smolin's *Life of the Cosmos*)[12]

The existence of even just one form of life in the universe such as *Homo sapiens* would be consistent with a kind of "bootstrapped universe" interpretation whereby the universe is seen as having fortuitously bootstrapped itself into the realm of value and meaning via the creation of beings with interests, consciousness, and culture which literally bring forth value and meaning where there may have been none prior. With cosmic evolution as the premise, this is, strictly speaking, a meaningful universe worldview – a kind of "cosmobiocentric" or "cosmocultural" worldview. The universe has created interests, value, and meaning via us, and we (along with other rational "cultural" beings) are the ultimate arbiters of that value, meaning and any purpose that might be derived thereof.

This view would also be consistent with life and intelligence existing throughout the universe, but merely as products of both chance and natural forces conspiring under the right circumstances – not necessarily because natural forces are such that the production of life is inevitable. There would not necessarily be any "deep," "deliberate," or inevitable cause (natural or otherwise) of the emergence of life and intelligence in this interpretation of a bootstrapped universe, but the universe would nevertheless be thought to possess the natural capacity or potential for life and intelligence and hence literally bootstrap itself into the realm of value, meaning, and purpose via its own potentialities that give rise to beings with such capacities. The bootstrapped universe suggests an open-ended evolution of value, meaning, and purpose, unconstrained by and perhaps independent of broader objective cosmic value, and largely in the control of rational valuing agents such as human beings who appear to be free to choose just about anything.

A "meaningless universe" is meaningless in a naturalistic vocabulary. Meaning, purpose, and values belong to the realm of human activity and thought. Even if their development might be traced back to our biological inheritance, they are nonetheless the result of the activities and mechanisms characteristic of human society. I find this view, which frees humankind from its role in the cosmological scheme of things and from its dependence on an external source of authority, to be meaningful and liberating.

(Iris Fry [48])

Lee Smolin's suggestion that our particular universe could have been "naturally selected" among many universes because its characteristics and initial conditions

---

[12] Lee Smolin quotes a comment that Saint Clair Cemin made to Smolin regarding Smolin's *Life of the Cosmos*. The quote appears in L. Smolin. Our relationship to the universe. In *Many Worlds: The New Universe, Extraterrestrial Life, and the Theological Implications*, ed. S.J. Dick (Philadelphia, PA: Templeton Foundation Press, 2000), p. 79.

allow it to persist, evolve, and create life [49] is consistent with a bootstrapped universe worldview in that it provides a potential natural mechanism for how such a universe might originate and produce life, leading to consciousness, value, meaning, and purpose.

A simple minimalist version of a bootstrapped universe worldview might suggest the universe has value and meaning merely because value and meaning exist *in* the universe in the form of valuing beings such as *Homo sapiens*, which are part and product of the universe. However, another slightly less minimalist interpretation of the bootstrapped universe could suggest the universe is meaningful because, as the source and arbiters of value and meaning, our mere declaration of a meaningful universe may be meaning enough.

10.4 Summary

This chapter covered a number of epistemological and ethical challenges and worldview perspectives related to astrobiology. A number of difficult questions were raised, and it was suggested that given what's at stake, the difficulties and limitations caused by only knowing one kind of biology suggest the need for a relatively strong precautionary principle, informed by many robotic life-detection missions, and possibly more theoretical explorations in artificial and synthetic life. Minimizing contamination, especially during a human mission, can also help reduce difficulties associated with discerning an independent origin of life.

Ethical views regarding extraterrestrial life may vary depending on a number of factors, including whether that life is an independent origin. An independent origin at least presents humanity with a novel ethical challenge for which a number of basic ethical views were examined, ranging from anthropocentrism to cosmocentrism – all of which are consistent with a cautious approach for reasons ranging from scientific value to broader cosmological value.

Given that we have some sense of these challenges well ahead of time, proactive policy work would be helpful, especially if it addressed: (a) confidence levels, (b) criteria for achieving confidence levels, (c) solar system environmental protection policies (for example international scientific preserves), (d) potential modifications to the Outer Space Treaty, and (e) assessments of a "preserve and assist" approach – including the possibility of human cohabitation with extraterrestrial life.

Worldviews were examined in part because of the interesting philosophical implications of discovering extraterrestrial life, but also because worldviews inform ethical views, which in turn inform policy. An "accidental universe" worldview might treat life as a rare and precious phenomenon, with perhaps no truly objective absolutes to guide our actions. As "diamonds in the rough," human beings may have more reason to turn to each other, and life in general, with deep reverence and respect as something truly unique and fragile in an otherwise hostile universe. A "deliberate universe" worldview, supported by a cosmically local independent origin of life (suggesting that life is a kind of "cosmic imperative") would for many mean that we are an inevitable realization of the nature of the universe, and so we might see ourselves and life as cosmically valuable, perhaps intrinsically valuable as a result of cosmological value. A bootstrapped universe, a universe that bootstraps itself into the realm of value, meaning and purpose via valuing agents and cultural evolution – a kind of "cultural cosmos" – can be for many, in the absence of more objective meaning, meaning enough.

References

[1] M. Lupisella. Using artificial life to assess the typicality of terrestrial life. *Advances in Space Research*, **33** (2004), 1318–1324.

[2] J. Szostak, D. Bartel, and L. Luisi. Synthesizing life. *Nature*, **409** (2001), 387.

[3] Unpublished email correspondence, March 2001.

[4] M. Race, M. Criswell, and J. Rummel. *Planetary Protection Issues in the Human Exploration of Mars*, Paper Number 2003–01–2523, International Conference on Environmental Systems (ICES), Vancouver, B.C. (July, 2003).

[5] R. Randolph, M. Race, and C. McKay. Reconsidering the theological and ethical implications of extraterrestrial life. *The Center for Theology and the Natural Sciences*, **17**(3) (1997), p. 6.

[6] P. Davies. Biological determinism, information theory, and the origin of life. In *Many Worlds: The New Universe, Extraterrestrial Life, and the Theological Implications*, ed. S. J. Dick (Philadelphia, PA: Templeton Foundation Press, 2000), p. 17.

[7] M. Rees. Life in our universe and others: a cosmological perspective. In *Many Worlds: The New Universe, Extraterrestrial Life, and the Theological Implications*, ed. S. J. Dick (Philadelphia, PA: Templeton Foundation Press, 2000), p. 66.

[8] R. Zubrin and R. Wagner. *The Case for Mars* (New York: Simon and Schuster, 1996). Also in personal communications.

[9] F. Golley. Environmental ethics and extraterrestrial ecosystems. In *Beyond Spaceship Earth: Environmental Ethics and the Solar System*, ed. E. C. Hargrove (San Francisco, CA: Sierra Club Books, 1986), p. 225.

[10] D. Grinspoon. Is Mars ours? The logistics and ethics of colonizing the red planet (January 7, 2004), available online at: http://slate.msn.com/id/2093579/.

[11] J. Carter McKnight. Martian landscapes and rose colored memories. *The Spacefaring Web* (January 12, 2004), available online at: http://www.spacedaily.com/news/oped-04a.html.

[12] C. Sagan. *Cosmos* (New York: Random House, 1980), p. 130.

[13] J. Baird Callicott. On the intrinsic value of nonhuman species. In *The Preservation of Species: The Value of Biological Diversity*, ed. B. Norton (Princeton, NJ: Princeton University Press, 1986), p. 163.

[14] C. McKay. Does Mars have rights? In *Moral Expertise*, ed. D. MacNiven (London: Routledge, 1990), p. 194.

[15] A. Naess. *Ecology, Community, and Lifestyle: Outline of an Ecosophy* (Cambridge: Cambridge University Press, 1989).

[16] A. Marshall. Ethics and the extraterrestrial environment. *Journal of Applied Philosophy*, **10**(2) (1993), 227–236.

[17] M. Lupisella. The rights of Martians. *Space Policy*, **13** (1997), 89–94.

[18] R. Haynes. Ecopoiesis: playing God on Mars. In *Moral Expertise*, ed. D. MacNiven (London: Routledge, 1990).

[19] R. Haynes and C. McKay. Should we implant life on Mars? *Scientific American* (December 1990), 144.

[20] D. MacNiven. Environmental ethics and planetary engineering. *Journal of the British Interplanetary Society*, **48** (1995), 442–443.

[21] M. Fogg. *Terraforming: Engineering Planetary Environments* (Warrendale, PA: SAE International, 1995), p. 490.

[22] D. MacNiven. *Creative Morality* (New York: Routledge, 1993).

[23] M. Lupisella and J. Logsdon. Do we need a cosmocentric ethic? Paper IAA-97-IAA.9.2.09, presented at the International Astronautical Federation Congress. American Institute of Aeronautics and Astronautics, Turin (1997).

[24] M. Lupisella. Astrobiology and cosmocentrism. *Bioastronomy News*, **10**(1) (1998), 1–8.

[25] J. Baird Callicott. Intrinsic value in nature: a metaethical analysis. *The Electronic Journal of Analytic Philosophy*, **3** (Spring 1995), http://ejap.louisiana.edu/EJAP/1995.spring/callicott.1995.spring.html.

[26] H. Rolston III. The preservation of natural value in the solar system. In *Beyond Spaceship Earth: Environmental Ethics and the Solar System*, ed. E.C. Hargrove (San Francisco, CA: Sierra Club Books, 1990), pp. 140–182.

[27] A. Leopold. *A Sand County Almanac* (New York: Oxford University Press, 1966), p. 262.

[28] H. Oliver. *A Relational Metaphysic* (The Hague, Netherlands: Martinus Nijhoff Publishers, 1981).

[29] P. Davies. *Are We Alone: Philosophical Implications of the Discovery of Extraterrestrial Life* (New York: Basic Books, 1995).

[30] M. Lupisella. From biophysical cosmology to cosmocentrism. Presented at *SETI in the 21st Century: Scientific and Cultural Aspects of the Search for Extraterrestrial Intelligence*, SETI Australia Centre, University of Western Sydney, Australia (January 1998).

[31] F. Dyson. *Infinite in all Directions* (New York: Harper and Row, 1998), p. 5.

[32] B.G. Norton. Environmental ethics and weak anthropocentrism. *Environmental Ethics*, **6** (1984), 131–148.

[33] A. Marshall. Ethics and the extraterrestrial environment. *Journal of Applied Philosophy*, **10**(2) (1993), 233.

[34] NASA Technical Memorandum. *Workshop on the Societal Implications of Astrobiology Final Report*, eds. A. Harrison and K. Connell (Palo Alto, CA: Ames Research Center, November, 1999).

[35] National Research Council, Space Studies Board. *Biological Contamination of Mars: Issues and Recommendations* (Washington, DC: National Academy Press, 1992).

[36] M.S. Race and R.O. Randolph. The need for operating guidelines and a decision making framework applicable to the discovery of non-intelligent extraterrestrial life. *Advances in Space Research*, **30**(6) (2002), 1583–1591.

[37] P. Sterns and L. Tennen. Preserving pristine environments: the planetary protection policy in space safety and rescue 1988–89. In *Space Safety and Rescue, Proceedings of Symposia of the International Academy of Astronautics* ed. G.W. Heath (San Diego, CA: American Astronautical Society, 1990).

[38] M. Williamson. Protection of the space environment under the outer space treaty. *IISL, Proceedings of the 40th Colloquium on the Law of Outer Space*. Paper IISL-97-IISL.4.02 presented at the 48th International Astronautical Congress, Torino, Italy (October 1997).

[39] I. Almar. What could COSPAR do to protect the planetary and space environment? *Advances in Space Research*, **30**(6) (2002), 1577–1581.

[40] C. McKay. Let's put Martian life first. *The Planetary Report*, **21** (July/August 2001), 5.

[41] F. Turner. Life on Mars: cultivating a planet and ourselves. *Harper's Magazine* (August 1989).

[42] S. Dick. Cosmotheology: theological implications of the new universe. In *Many Worlds: The New Universe, Extraterrestrial Life, and the Theological Implications*, ed. S. J. Dick (Philadelphia, PA: Templeton Foundation Press, 2000), p. 195.

[43] C. McKay. Astrobiology: the search for life beyond the Earth. In *Many Worlds: The New Universe, Extraterrestrial Life, and the Theological Implications*, ed. S. J. Dick (Philadelphia, PA: Templeton Foundation Press, 2000), p. 56.

[44] P. Davies. Biological determinism, information theory, and the origin of life. In *Many Worlds: The New Universe, Extraterrestrial Life, and the Theological Implications*, ed. S. J. Dick (Philadelphia, PA: Templeton Foundation Press, 2000), p. 27.

[45] E. McMullin. Life and intelligence far from Earth: formulating theological issues. In *Many Worlds: The New Universe, Extraterrestrial Life, and the Theological Implications*, ed. S. J. Dick (Philadelphia, PA: Templeton Foundation Press, 2000), p. 157.

[46] J. Monod. *Chance and Necessity: An Essay on the Natural Philosophy of Modern Biology* (New York: Vintage Books, 1971).

[47] L. Smolin. Our relationship to the universe. In *Many Worlds: The New Universe, Extraterrestrial Life, and the Theological Implications*, ed. S. J. Dick (Philadelphia, PA: Templeton Foundation Press, 2000).

[48] I. Fry. *The Emergence of Life On Earth: A Historical and Scientific Overview* (New Brunswick, NJ: Rutgers University Press, 2000), pp. 282–283.

[49] L. Smolin. *The Life of the Cosmos* (New York: Oxford University Press, 1997).

11

The implications of discovering extraterrestrial life: different searches, different issues

Margaret S. Race

11.1 Introduction

The search for ET life is encompassed within a broad spectrum of research efforts in the field of astrobiology [1]. In general, this multidisciplinary field seeks to understand the origin, evolution, and fate of life in the universe. While searches for ET life represent just a subset of the overall astrobiology goals, they command a disproportionate share of the public interest. To scientists it may be obvious that finding an ET microbe on Mars would be quite different than getting a message from an intelligent civilization somewhere in the Milky Way. However, to the public, the nature and implications of different types of discoveries are probably less clear. Without a systematic analysis of the varied research efforts and a consideration of the science and issues associated with them, it is impossible to get an overview of what it would mean to discover ET life. This chapter attempts such an analysis and focuses additionally on the kinds of societal issues and concerns that need to be communicated to public audiences when discussing astrobiological research and exploration efforts, particularly those centered on searching for ET life.

When searching for ET life, astrobiology uses diverse methods to identify and study potentially habitable locations, understand their environmental conditions, analyze processes that may be associated with life, and, finally, seek evidence for ET life (which may or may not be the same as finding life itself). At present, the searches for ET life can be viewed as falling into three general categories: (1) the search for extraterrestrial intelligence (SETI), which seeks signals or messages from distant technological ET civilizations; (2) detection and exploration of extrasolar and/or habitable planets beyond our solar system, which may be possible abodes of ET life; and (3) "exobiology" related research and space missions within the solar system. Each category looks in different places, using different scientific

Exploring the Origin, Extent, and Future of Life: Philosophical, Ethical, and Theological Perspectives, ed. Constance M. Bertka. Published by Cambridge University Press. © Cambridge University Press 2009.

Table 11.1 *Differences in approach for major search types*

Search type	Where	Methods used	Type of evidence
SETI	Galaxy	Radiotelescopes	Electromagnetic signals
Extrasolar planets	Galaxy	Telescopes; interferometer	Planetary transits; "wobbles"; atmospheric "signatures" with special interest in biologically related chemistry
Exobiology			
Missions (*in situ*)	Solar system (SS)	Spacecraft missions; varied scientific instruments	Physical/chemical, geological, biological (alive, dead, pieces)
Meteorite studies and cosmochemistry	Earth and SS	Lab instruments; experiments; spacecraft missions; remote data	Physical/chemical, geological, fossil, structural; molecular signatures
Origin-of-life experiments	Earth	Lab instruments and experiments (e.g. impact simulations)	Physical/chemical; molecular; synthetic replicating life forms
Extremophile research; study of analogue environments	Earth	Lab and field research to study microbial limits to life on Earth and analogue environments of relevance to ET life	Physical, chemical, geological, and biological

instruments and methods, and gathers a variety of evidence and data (Table 11.1). In addition to using different search methods in different places, each category also reflects distinct differences in the nature of presumed ET life and the scientific meaning of a successful "discovery."

Finding credible evidence for the existence of any type of ET life would be dramatic news for both the scientific community and the public at large, no doubt generating a frenzy of mass media attention. The accompanying demand for information about ET life would be a challenge for scientists, government agencies, and the mass media, regardless of when it may occur. Thus, it is prudent to consider in advance how the varied scenarios might be encountered, what they would mean to scientists, and how they would be interpreted and communicated. With these challenges in mind, this chapter begins by examining the many ways that scientists are currently searching for extraterrestrial life. It then explores the implications of each search type from a variety of perspectives, and considers the kinds of societal issues and concerns that might arise during exploration and upon discovery. The chapter also provides information on relevant policies and controls that already apply to astrobiology research and exploration. Ultimately, the distinctions among the various ET searches are important to understand because of their practical relevance in planning, risk management, decision making, and the legal and policy areas. Additionally, there are many associated issues in the ethical, philosophical, and theological realms that have begun to attract serious attention by experts in other disciplines. In short, searches for evidence of ET life and the prospect of its eventual discovery through astrobiological exploration and research activities have the potential to impact humankind, planetary environments, and ET life as well, both in the near term while we search and explore, and in the longer term post-discovery.

11.2 The searches in perspective

11.2.1 Searches for extraterrestrial intelligence (SETI)

For millennia, humans have wondered whether our species and home planet are unique. With modern advances in astronomical understanding and technology, telescopic searches are underway to scan the far reaches of space systematically for evidence associated with other life and potentially habitable planets. For several decades, searches for extraterrestrial intelligence (ETI) have been conducted using telescopes that scan stars located many light-years away and outside the solar system. The searches involve numerous efforts and many different telescopes to scan for data in the form of detectable electromagnetic "signals" conveying indirect information or possibly messages that may have been sent by distant civilizations.[1] When and if positive signals are detected, the information may be quite "old" depending on the astronomical distances involved.[2] Because the data are collected remotely and with no direct interactions with the source of the message, there are no special procedures or regulatory controls imposed during exploration, nor any specific treaties or laws that apply to ongoing SETI exploration efforts. The only issues that may arise from telescopic scans within the galaxy are perhaps those involving the telescopes themselves. Depending on the country where the telescopes are located, there may be a variety of routine environmental, health or safety (EHS) laws or regulations that apply to the construction and operation of the facilities.

The nature and scientific meaning of a potential "discovery" and the evidence associated with it are worth noting in comparison with other search types (see

---

[1] For more detailed information on SETI activities and research, see www.seti.org.
[2] Signals and light emanating from bodies or stars that are many light-years away in the galaxy will have taken tens or even hundreds of years before reaching Earth.

Table 11.2). By definition, if ETI signals are discovered, they will be presumed to emanate from an intelligent and advanced civilization, capable of deliberately devising technologies that can send detectable signals beyond their home location. Even if we detect and verify a signal, we will be unlikely to know details about the nature of the intelligent beings, their biochemistry and physiology, or even the persistence of the civilization. Even if additional studies and searches are undertaken to learn more about a particular discovery, they are unlikely to yield a full understanding about the nature of the ET life or result in direct experience or interaction with it.

11.2.2 Searches for extrasolar and habitable planets

Another search category that has grown in prominence recently is that of searches for extrasolar planets. Within the past decade, methods have been refined to detect planetary systems around other stars, and then to study them further in an attempt to determine if they might be similar to rocky, terrestrial planets that possess the features needed for habitability, based on life and biochemistry as we know it. Like SETI, these searches are conducted using telescopes that scan locations light-years away and outside the solar system (see Table 11.1). They also center on the collection of data in the form of "signals" conveying indirect or remote information about the existence of the planet and any signs of biological activity that may be linked to possible ET life. For example, astronomers can detect "wobbles" indicative of the gravitational tug between a star and an orbiting planet, or measure the dimming of light from a star as an orbiting planet transits in front of it. Other methods are used to identify the size and orbital characteristics of the newly discovered planets, their location relative to their star, and indications of whether they might be in a habitable zone by virtue of having temperatures in the range necessary for liquid water to exist. In addition, instruments are able to scan for the presence of atmospheric chemistry that is associated with life as we know it – for example indications of oxygen, carbon, or other biogenic elements or biosignatures (e.g. CO_2, H_2O). Again, like SETI, even if positive data are detected, the information may be quite "old" depending on the astronomical distances involved. Because there are no direct interactions with the target bodies, there are no treaties, laws, or special regulatory controls imposed upon these searches during exploration except those that may be associated with the establishment and operation of telescope facilities on Earth.

Like SETI, the nature and meaning of a discovery and the evidence for possible ET life on extrasolar planets are determined entirely from indirect information (Table 11.2). While positive indications for a "habitable" extrasolar planet could convey exciting information about Earth-like planets with possible biogenic atmospheres conducive to life, we are unlikely to know much with certainty about the

Table 11.2 *Scientific nature and understanding of possible discoveries*

Search type	Nature of "discovery"	Scientific meaning of discovery
SETI	Intelligent life; sentient	Advanced civilization "Aliens" of unknown biology; may be "old" (current status/ existence unknown)
Extrasolar planets	Distant planetary locations; possibly Earth-like; may detect potentially biogenic atmospheres – may be indirect evidence of metabolism?	Distant planets in other solar systems may have habitable sites; may have "signatures" of biogenic atmosphere May be "old" (current status unknown)?
Exobiology		All searches with possibilities of cross-contamination Real-time discovery; can directly study the type of life
Missions and *in situ* studies	Likely to be microbial life (may be simple or complex); evidence for ET life may be alive, dead, pieces, or biomarkers	Can determine if related to Earth life; if not, could be evidence for second genesis or multiple origins of life? Possible panspermia?
Meteorites and cosmochemistry	Probably fossil life or evidence of biological activity; biomarkers; might also find living microbes; evidence of evolving chemical systems?	Either past life or possibly extant; may be able to determine if related to Earth life; evidence of biological activity; chemical links to biology?
Origin of life	Laboratory simulations of self-replicating life; possible analogues for early Earth life or life elsewhere	Replicating chemical systems? Possibility of repeated emergence of life? Life as a cosmic imperative?
Extremophile research	Demonstrating limits to life "as we know it" from Earth analogue habitats and their microbial inhabitants	Earth life, but relevant to ET searches and studies

type of organisms or the biochemical processes, if any, that may be responsible for the observed phenomena. The uncertainty is compounded due to the fact that our interpretations are necessarily based on our singular experiences with life as we know it. In addition, even if additional studies and searches are undertaken to learn more about a particular planet, they are unlikely to yield a full understanding of its detailed geochemical nature or the source of its possible biosignatures. Considering the great distances involved, there is also little likelihood that there would be any direct experience or interaction with these newly discovered planets any time in the near future.

11.2.3 Exobiology and searches within the solar system

In contrast to SETI and extrasolar planet searches, there are a diversity of other search types and activities related to ET life and habitability that are either Earth-based or conducted in the solar system. These searches, which are grouped here into an "exobiology" category, gather real-time information relevant to ET life and solar system bodies, including Earth. The range of research and activities is impressively broad (Table 11.1) and encompasses such areas as (1) space missions and *in situ* research, (2) studies of ET materials (e.g. meteorites, cosmic dusts, lunar samples) and cosmochemistry aimed at understanding the geochemical nature of the solar system, (3) experiments and laboratory studies related to the origin of life, and (4) studies of extremophiles, which represent organisms and environments on Earth that are at the limits to life as we know it. While some of the data in this broad category are collected remotely as signals of various sorts (imaging/remote sensing, spectral analyses, etc.), considerable data are gathered in direct fashion through experiments, sample collection, roving, digging, drilling, lab experimentation, etc., thus involving situations with potentially direct impacts, interactions, and cross-contamination, unlike telescopic studies. In addition, they routinely gather evidence in the form of physical, chemical, geological, biological, structural, and molecular data, similar to routine scientific studies on Earth. Thus, unlike telescopic investigations, these activities could have potential environmental, health, and safety concerns, both on Earth and in ET environments.

In order to address the cross-contamination and environmental impact concerns, it is necessary to consider the types of exposures and discovery scenarios that might be encountered. The wide range of research types is associated with a variety of possible discovery scenarios that could be encountered within exobiology. For example, a robotic rover or lander could find evidence for microbial ET life on Mars and transmit the data to Earth; or astronauts on the Martian surface might come in direct contact with ET life sometime during a future mission. Alternatively, ET life could be discovered on Earth, in a containment laboratory examining returned Martian samples or less likely, studying meteorites or other extraterrestrial materials. Although extremophile research is focused on Earth life and environments, its discoveries of analogue environments and associated microbes are of direct relevance to many areas of astrobiological research. Because of the many possible contact scenarios and the uncertainties about ET life and contamination, regulatory and operational constraints apply to some exobiology activities, most notably in the form of planetary protection (PP) controls on missions and handling of deliberately returned ET materials on Earth.

Spacecraft and missions sent to locations in the solar system are governed by policies outlined in the Outer Space Treaty of 1967 (OST) [2], which stipulates that space exploration must be done in a manner that avoids harmful cross-contamination of celestial bodies (planets, moons, asteroids, etc.) from the introduction of extraterrestrial matter; it must also maintain the opportunity for conducting scientific exploration in both the short and long term for all humankind. In practice this means avoiding the forward contamination of target bodies by hitchhiker terrestrial microbes, and preventing the back contamination of Earth by extraterrestrial materials or samples upon return [3]. Policies and regulations to implement the Treaty are formulated and revised as necessary by an international scientific body, the Committee on Space Research (COSPAR). In the United States, the planetary protection policies are implemented through directives and requirements issued by the National Aeronautics and Space Administration (NASA) through its Planetary Protection Officer. In addition to the directives and planetary protection controls imposed by launching agencies or countries in accordance with the OST, a complex array of environmental, health and safety laws and regulations may also apply to associated research and activities conducted in Earth-based labs [4].

Unlike SETI and searches for extrasolar planets, the exobiology category builds upon a vast accumulation of direct knowledge about living systems, metabolism, and biological signatures, as well as extensive information about celestial bodies and processes in the solar system. Thus, the nature of a discovery and its scientific meaning are generally well understood (Table 11.2). Even now, it is possible to speculate about the biochemical nature of putative ET life, to consider its likely adaptation to potential habitats, and to identify promising places for other searches and/or sample collection in the solar system. Based on our current knowledge of life on Earth and conditions in the solar system, we can surmise that ET life in the solar system – or evidence for it – is likely to be microbial and may be found in places with conditions amenable for liquid water to exist, either persistently, transiently, or in the recent past. If microbial ET life is found *in situ* by robotic spacecraft or astronauts visiting new locations, more than its existence can be verified in real time. Through direct studies, it may be possible to determine whether it shares DNA and Earth-like biochemistry, and whether and how it may be related to the current three-domain view of life on Earth. Moreover, if it has a distinct biochemistry unrelated to life as we know it, we will have a direct opportunity to study the life in detail and characterize it more fully, either on Earth or in its native location.

In addition, Earth-based cosmochemical studies of meteorites and other extraterrestrial materials have also spurred research of relevance to the possible detection of ET life. For example, debates in the mid-1990s over alleged fossil life in a Martian meteorite led indirectly to extensive research on extremophiles, nanobacteria, biosignatures, the limits to life, and possible biotic and abiotic mechanisms that could explain or refute possible ET life "discoveries."[3] If an extant or dormant life form were discovered in a meteorite or other ET material, direct studies of its nature and biochemistry could be conducted to learn more about it.

Finding evidence of any type of ET life form in the solar system would transform our views of the universe, suggesting the potential for life to arise repeatedly whenever and wherever initial conditions are suitable [5]. Whether exobiological evidence for ET life is gathered on Earth or elsewhere in the solar system, a discovery could quickly generate far more direct data and understanding of the nature of extraterrestrial life and life processes than other search types. Neither searches for ETI nor extrasolar planets could conceivably yield the opportunity for such comprehensive analysis of an ET discovery. As importantly, an exobiological discovery of ET life would also prompt discussion of the broad societal implications that are immediately more complicated than those associated with search types outside the solar system.

11.3 The broad implications of discovery

Discovery and verification of the existence of ET life would be extraordinarily profound scientifically, but mere scientific knowledge about the existence of ET life *per se* is not what determines its broad meaning and implications for humankind. To understand the full impact of a possible ET discovery, we must go beyond the scientific discovery itself and consider the nature of the presumed life as well as the potential for human actions to cause impacts upon it or to realize impacts from it. For comparison, it was not the scientific knowledge about inheritable traits or the discovery of DNA's structure *per se* that had meaning or implications for humankind, but rather the uses and applications of this heritability information through deliberate human actions like genetic engineering, selective breeding and biotechnology. When and if we learn that we are not alone in the universe, we should be prepared to re-examine the nature of our plans and activities, and to communicate fully about all the implications, scientific and otherwise.

As shown in Table 11.3, the societal, policy, legal, ethical, and theological implications of a "discovery" are quite different for the various search types. For the most part, our current consideration of the impacts focuses primarily on near-term actions taken by governments and launching agencies. Even the Outer Space Treaty

---

[3] J. Billingham *et al.* Social implications of the detection of an extraterrestrial civilization. In *A Report of the Workshops on the Cultural Aspects of SETI* (Mountain View, CA: SETI Press, 1994).

Table 11.3 *Representative societal and longer term implications of discovery types*

Search types Implications:	SETI	Extrasolar	Missions and *in situ*	Cosmochemistry and meteorites	Origin of life
Policy	SETI principles (no enforcement)	None needed?	No current COSPAR or NASA policies for ET discovery; may need cosmocentric policies and laws to address non-Earth life?	No special policy needed? If extant life is found in meteorite, may need EHS regulations?	No current policies; may raise EHS and regulatory issues similar to genetic engineering or synthetic genomics
Philosophical, ethical, theological	Existence of sentient beings raises questions about creation, religious dogmas and man's place in divine plans; may raise the need for a cosmotheology?	No immediate issues unless biology is verified; then consider similar to missions	Ethical issues regarding "rights" of ET life and responsibilities toward ET environments; interference with evolutionary trajectories? Possible need to bring a cosmocentric focus to ethics and policies?	Minor or no near-term issues if fossil life; questions about how to treat ET life if alive	Multiple creations? Man's place in universe? Deliberate tinkering with natural order; "playing" God?
Individual	None	None	Lab worker safety; contamination issues; astronauts as human subjects	Lab worker safety and biohazards? Containment	Lab worker safety? responsible research and exploration?
Government actions	How should consultation be handled? Who should respond?	None immediately; in future may need to be similar to missions	Planetary protection (PP), EHS, ethical, and decision-making issues about future missions and actions; accident liability? Impact of mixing Earth and ET ecologies?	Maintain proper containment and security	EHS, PP, ethical and decision-making issues regarding future missions and actions; accidents?
Societal issues and private sector	None (although possibly some psychological questions about our place in universe?)	None	Questions about colonization, tourism, commercialization, extraction, patents, terraforming, etc.	Minor or no issues?	Issues of scale up and creation or use of "new life"? EHS and PP concerns? Regulatory controls or moratorium?

and related policies focus only on potential cross-contamination and interference with scientific exploration. There are no specific policies or statements regarding ethical considerations or the broader impacts of human activities, particularly in relation to ET life and environments. Moreover, there is no guiding framework for considering any non-scientific issues. In the longer term, it will be necessary to consider implications more broadly, including those related to the individual, the private sector, and even the extraterrestrial life types themselves. As outlined earlier, there are considerable differences in policies and oversight for the three major search types during the exploratory period. Extrapolating to the post-discovery phase, the issues get even more complicated.

When it comes to anticipating the discovery of ET life, the SETI community is alone in having conducted extensive international discussions of how to respond *if* and *when* a signal from ETI is detected. A set of "SETI Principles" were developed over a period of years and approved by the SETI Committee of the International Academy of Astronautics (IAA) in 1989. The Declaration of Principles is not legally binding and has no enforcement provisions, but has been endorsed by numerous major organizations [6]. Rather than presuming anything about ET life itself, the SETI Principles focus instead on the human response anticipated in the face of a discovery scenario, providing step-by-step operational guidelines for verifying the signal, sharing information openly, and consulting broadly and internationally prior to making contact in the form of a return message. Already researchers have identified a number of anticipated implementation problems likely to arise upon discovery, despite the existence of the SETI Principles. These include problems with the level and type of organizational readiness [7], the limited discovery scenarios considered [8], and a lack of preparations for mass media communications in situations involving breaking news and incomplete scientific information [9, 10].

Despite the shortcoming of the SETI Principles, they represent the only organized attempt to codify guidelines and policies about what to do upon discovery. For extrasolar planet searches, there are no current or anticipated policies related to the discovery of ET biogenic atmospheres or potentially habitable locations. Even so, it is unlikely that any are needed because the indirect nature of contact and the extensive distances involved preclude significant practical concerns, except perhaps the sharing of scientific information. The same cannot be said of exobiological searches within the solar system. There are currently no policies or recommendations applicable upon the discovery of non-intelligent ET life, whether it is found in extraterrestrial material examined in a laboratory, or on other solar system bodies. Put simply, there is no clear guidance on what to do in either the short or long term if and when microbial, non-intelligent ET life is found. This policy and procedural gap persists despite the magnitude of organized and individual research efforts, and the fact that a discovery could occur at any time and would prompt a need for immediate response from multiple levels – scientifically, governmentally, and societally. Science and technology are clearly ahead of policies in a number of areas, resulting in potentially serious gaps in knowing what will or should be done upon discovery. Ideally, this gap should be addressed as part of astrobiological exploration planning, and the public should be involved in discussions of the broad uncertainties involved.

11.4 Communicating about exploration and discovery

We already know that communicating about an ET discovery is likely to be complicated by public attitudes, misperceptions, Hollywood-style science fiction, national interests, and ethical/theological considerations. The communication process is likely to be frantic, with input from scientific and space community interests as well as an array of international institutions. Already there is need for education and information in advance of a discovery to allow public audiences to put information into appropriate contexts. This means that planning for communication about a discovery must consider at an early stage how to deal with potential conflicts, gaps, misunderstanding, and debate, whether they center around scientific topics or otherwise. Prominent among the anticipated concerns are likely to be varied questions of theological and ethical implications [11, 12], potential risks to Earth's biota and environment [13] and the long-term advisability and impacts of continued exploration and interaction. To the extent that humans are directly involved in a discovery (e.g. handling or collecting samples, analyzing alien life in labs etc.), serious questions arise about laboratory worker and/or astronaut safety [14, 15], both on Earth and in extraterrestrial environments. Additional concerns include the rights of ET life and responsibilities toward it [16, 17], extraterrestrial property rights and environmental ethics [18, 19], and future actions by either governments or the private sector with the potential for large-scale or global impacts (e.g. colonization, commercialization, extractive industries, tourism, terraforming, pollution, etc.) [16, 20]. For the most part, discussions of issues to date have taken the perspective of the developed world, emphasizing "modern" science and the technological capabilities of current space-faring nations. In considering the implications for all humankind, it will be important to integrate the perspectives of indigenous peoples and others, whose concerns, diverse views, and knowledge are likewise valid, yet often overlooked [21]. Finally, since all policies, laws, religions, and ethics on Earth are based upon life as we know it, some have even suggested the need for comprehensive overhauls from a cosmocentric perspective if ET life is discovered [12, 19, 22, 23]. Recently, a National Research Council report recommended a significant updating of forward contamination requirements for Mars in light of the uncertainties about both liquid water and special regions on Mars where microbial life could exist. The report also asserted the need to consider whether planetary protection policies for Mars should be extended beyond "protecting the science" to include "protecting the planet" [24]. The report specifically recommended that an international workshop be convened to consider the ethical implications and responsibility to explore Mars in a manner that minimizes harmful impacts on potential indigenous biospheres (whether suspected or known to be extant), and to determine whether revisions to current policies and the Outer Space Treaty are necessary.

The discovery and verification of ET life will undoubtedly result in intense public interest and mass media scrutiny, providing an opportunity to educate the public about the science and technology behind the discovery as well as the potential use of scientific information in risk analyses and decision making, and the longer-term societal implications that draw meaning from equally important non-scientific disciplines. Undoubtedly, there will be a need to speculate about the possible responses and prepare in advance for handling the full range of information and perspectives anticipated. Table 11.4 is an example of a preliminary and far from comprehensive attempt to begin organizing the exploration of these issues.

In addition to communicating with the public, it may advisable to plan targeted pre-discovery communication efforts aimed specifically at the scientific and space communities. Astrobiology is a diverse, multidisciplinary field in which professionals are likely to focus on their single discipline or search type, and perhaps not recognize the diversity of issues and implications behind current searches. Ideally, plans for communication should involve the education of professional audiences in order to inform them about the diversity of issues and engage them in systematic analysis and public discussion of all aspects of the search and discovery scenarios.

Looking ahead, another complication may arise based on which type of ET life is found first. It is not totally clear whether the discovery of one particular type of ET life would enhance or adversely impact other search efforts underway or planned. For example, if ET life were discovered by a rover on Mars or in a containment glove-box on Earth, what would be the impact on plans for future sample return or human missions? What additional controls and policies would apply after ET life is verified? Will there be added oversight of laboratory studies and experiments aimed at characterizing ET life and understanding its biochemistry, or could there even be calls for a moratorium on specific types of research? Would a verified SETI signal somehow affect ongoing exobiological research in the solar system? Who among scientific, governmental, or theological institutions should be involved in making decisions for humankind about the advisability of further contact or interactions with ET life, particularly if it is "just" microbial? Would it be

Table 11.4 *Categorical view of information type and issues by search type*

Search type	Science and technology facts	Risks/decisions?	Societal meaning/actions
SETI	Presumed existence of intelligent, sentient beings; technological civilization; no likely details on form, function or persistence of ET life	Send a return message? Psychological impact?	We are not alone. Societal interactions unlikely unless significant advances in technology. Raises questions about second genesis and theological implications, etc.
Extrasolar planets	Existence of planetary locations; taxonomy of solar systems and planet types; information on atmospheric composition/processes; evidence of possible metabolism and habitability? Old signal?	None. Knowledge that terrestrial planets are more common could raise interest and concern about other ET searches?	There may be other habitable locations – with or without life. Unlikely to prompt societal actions or generate major ethical or theological questions beyond those associated with the possible existence of ET life *per se*
Exobiology *In situ* missions (orbiters, rovers, sample return, humans)	Physical, chemical, biological evidence for microbial ET life; characterize and understand habitat, biochemistry, features, etc. May extend taxonomy of life in the universe	Mission-related issues (e.g. sample return to Earth; biocontainment; astronaut interactions; mixing ecologies; impacts on Martian ecology and evolutionary trajectories caused by human actions, etc.)	ET life exists now; need to determine if/how related to Earth life; multiple or second genesis? New concerns about cross-contamination; possible mixing of ecologies; ethical "rights" of ET life? Obligations towards ET life and environments? Altering "natural order"? Need for cosmocentric policies and ethics? Possible impacts on planned space activities?
Meteorites and cosmochemistry	Microbial life (simple/complex; alive, dead, pieces; fossil)? Biomarkers, etc. Possible evidence for evolving chemical systems?	None, unless living microbe detected	ET life existed in recent past; may have broader implications if live ET forms are detected (see above)
Origin of life	Laboratory studies to mimic/synthesize life and replicate process of its emergence or early formation; model of precursor conditions and initial chemistry for life; comparative understanding of types of biochemistry; terrestrial, interstellar and interplanetary life; life may be a cosmic imperative	Concerns similar to genetic engineering? Laboratory and environmental controls over man-made life? Limits or moratorium on creation or applications? Altering "natural order"? Concerns about cross-contamination, safety, etc.	Life may be widespread and simple to emerge. Theological, ethical and legal implications of man-made life and artificial replicating systems. Possible impacts on future research and development

advisable to have different institutions involved for deliberations about intelligent vs. microbial ET life? How would we know whether planned actions involving ET life could be mitigated or reversed if necessary?

Citizens worldwide deserve to be informed and educated about the facts about any presumed ET discovery and involved in deliberations about planned subsequent actions. This means that communication, education and outreach regarding the discovery of ET life must be planned with a multidisciplinary and long-term view – integrating the scientific, technological, and societal aspects in the information that will be conveyed to the public, whether via mass media coverage or through educational outreach channels. The complex ramifications of discovery and the possible future actions by space-faring nations and the astrobiology community compel us to think broadly about the meaning of life, its future on both our home planet and beyond, and the evolutionary trajectories of humankind and other life forms that may cohabit the universe with us. How we respond in the short or long term to the discovery of ET life has significant repercussions for current and future generations on planet Earth. It likewise may have significant impacts on the ET life as well. As we plan to communicate about scientific efforts and possible successes in the ongoing search for ET life, we must deliberately take steps to consider them in the broadest context – that of responsible exploration for all. That is precisely the intention behind this chapter.

References

[1] NASA Astrobiology Roadmap (2003), available online at: http://astrobiology.arc.nasa.gov/roadmap.
[2] United Nations Treaty On Principles Governing The Activities Of States In The Exploration And Use Of Outer Space, Including The Moon And Other Celestial Bodies, UN Doc A /Res /2222(XXI); TIAS #6347, NY, 1967.
[3] J. D. Rummel. Planetary exploration in the time of astrobiology: protecting against biological contamination. *Proceedings of the National Academy of Sciences*, **98** (2001), 2128031.
[4] M. S. Race. Planetary protection, legal ambiguity and the decision making process for Mars sample return. *Advances in Space Research*, **18**(1/2) (1996), 345–350.
[5] National Research Council. *Size Limits of Very Small Organisms: Proceedings of a Workshop* (Washington, DC: National Academy Press, 1999), available online at: www.nap.edu.
[6] S. J. Dick. *Life On Other Worlds: The Twentieth Century Extraterrestrial Debate* (London: Cambridge University Press, 1998).
[7] J. E. Tarter. Security consideration in signal detection. Invited paper number IAA-97-IAA.9.2.05, International Astronautical Federation, Paris, 1997.
[8] P. Schenkel. Legal frameworks for two contact scenarios. *Journal of the British Interplanetary Society*, **50** (1997), 258–262.
[9] S. Shostak. Media reaction to a SETI success. International Astronautical Congress, Turin, Italy (Paris: International Astronautical Federation, 1997).

[10] C. Oliver, *et al.* The case of EQ Peg: challenge and response. International Astronautical Congress (Paris: International Astronautical Federation, 1999).

[11] M. S. Race and R. O. Randolph. The need for operating guidelines and a decision making framework applicable to the discovery of non-intelligent extraterrestrial life. *Advances in Space Research*, **30**(6) (2002), 1583–1591.

[12] P. Davies. E.T. and God. *The Atlantic Monthly* (September 2003), available online at: www.theatlantic.com/issues/2003/09/davies.htm.

[13] National Research Council. *Mars Sample Return: Issues and Recommendations* (Washington, DC: National Academy Press, 1997), available online at: www.nap.edu.

[14] J. D. Rummel, *et al.* (eds.). A draft test protocol for detecting possible biohazards in martian samples returned to Earth, NASA/CP-2002–211842 (Washington, DC, 2002).

[15] M. S. Race, *et al.* Planetary protection and humans on Mars: NASA/ESA workshop results. *Advances in Space Research*, **42**(6) (2008), 1128–1138.

[16] C.P. McKay. Does Mars have rights? An approach to the environmental ethics of planetary engineering. In *Moral Expertise*, ed. D. MacNiven (New York: Routledge, 1990), pp. 184–197.

[17] M. Lupisella. The rights of Martians. *Space Policy*, **13**(2) (1997), 84–94.

[18] E. Hargrove (ed.). *Beyond Spaceship Earth: Environmental Ethics and the Solar System* (San Francisco, CA: Sierra Club Books, 1986).

[19] R. Randolph, *et al.* Reconsidering the ethical and theological implications of extraterrestrial life. *Center for Theology and Natural Sciences Bulletin* **17**(3) (1997), 1–8. Available online at: www.CTNS.org.

[20] I. Almár. What could COSPAR do for the protection of the planetary and space environment? *Advances in Space Research*, **30** (2002), 1577–1581.

[21] R. von Thater-Braan. The six directions: a pattern for understanding Native American educational values, diversity and the need for cognitive pluralism, SECME Summer Institute, University of Arizona, Tuscon, 2001.

[22] M. Lupisella and J. Logsdon. Do we need a cosmocentric ethic? International Astronautical Congress (Turin, Italy), Invited Paper Number IAA-97-IAA.9.2.09. International Astronautical Federation, Paris, 1997.

[23] S. Dick (ed.). *Many Worlds: The New Universe, Extraterrestrial Life and the Theological Implications* (Philadelphia, PA: Templeton Foundation Press, 2002).

[24] National Research Council. *Preventing the Forward Contamination of Mars*, C.F. Chyba (chair) (Washington, DC: National Research Council, National Academy Press, 2006), available online at: www.nap.edu.

12

God, evolution, and astrobiology

Cynthia S. W. Crysdale

There is a *Family Circle* cartoon that shows the father of the family escorting his daughter home from school. The caption is a single question that the little girl asks of her father: "If we send astronauts to Mars, do they hafta drive past Heaven?" While this strikes us as funny, it illustrates the double world in which many of us live. Few educated adults would ask the question in such simplistic terms. Nevertheless, many people live in a bifurcated world in which they have accepted the results of science, and presume the reasonable world of scientific endeavor. However, when it comes to thinking about God, some people still have a somewhat childish, antiquated, or rudimentary worldview. In the scientific world everything is open to question and results are only as valid as the evidence that supports them. With regard to God, however, both believers and non-believers often assume that religious issues can only be settled by reverting to a kind of mythic fideism.

In the contemporary Christian world of North America, the notion that God might have created other worlds with novel emergent life forms can threaten this basic religious intuition. The Genesis accounts of creation presume that the Earth is central to God's purposes, most notably that of establishing a special relationship with mankind. The discovery of extraterrestrial life in any form would seem to contradict this core assumption of the Judeo-Christian tradition. Furthermore, the theological challenges raised by the discovery of such life, or its artifacts, are related to still unresolved tensions over the theory of evolution. To the degree that evolution is assumed to imply a randomly unfolding universe with no direction or purpose, some Christians reject the scientific worldview in favor of belief in a Creator God who has instilled a purposeful orientation into creation.[1]

---

[1] Central to the Judeo-Christian tradition is the idea that humans are created in the image of God and thus have a special place in the unfolding of creation and history. Genesis 1:26–28 are the verses most often cited here, where God creates male and female in God's image and gives them the world for their sustenance. Thus, even

Exploring the Origin, Extent, and Future of Life: Philosophical, Ethical, and Theological Perspectives, ed. Constance M. Bertka. Published by Cambridge University Press. © Cambridge University Press 2009.

The discovery of extraterrestrial life would challenge this narrative of a God who intentionally ordered the cosmos towards the emergence of human life. Such a discovery would further convince some that the Genesis creation narratives are indeed primitive myths to be jettisoned. For others it would simply expand their notions of God to include a much grander view of God's purposes in creation. For many Christians it would raise the stakes in the perceived divergence between a theological and a scientific view of the universe.

My purpose in this article is to examine how our views of God might change if we find evidence of extraterrestrial life, even and especially microbial life.[2] I am working from within the Judeo-Christian tradition, and will assume that the God questions will confront the presumed omniscience, omnipotence, and unchanging nature of the classical Christian understanding of God. Since the current, ongoing, renewed debates over God's role in evolution are intimately tied to any resolution of questions about God and astrobiology, the bulk of this article will focus on questions of emergence in relation to models of God's agency. Questions about how God is related to an emerging world are at the heart of current religious politics in North America, and confusions in this regard will only be exacerbated with the discovery of extraterrestrial life.

12.1 Chance versus determinism

Let us begin by reviewing the commonly assumed dichotomy between a worldview in which there is purpose and one in which the universe has emerged according to mere chance. Paul Davies does this in an article in *The Atlantic* on "E.T. and God" [1]. Here he cites Jacques Monod as illustrating the classic position of the scientist who believes that evolution has now shown the universe to be meaningless: "Man at last knows he is alone in the unfeeling immensity of the universe, out of which he has emerged only by chance" [1].[3] In contrast, Davies alludes to scientists who believe they have found order and purpose in the universe. Many such scientists refer to the unlikely coincidence of events that would be necessary for the emergence of life on Earth. As Davies puts it:

if one does not interpret the first chapter of Genesis in a literal way, one still has a profound theology of the relationship between God, humans, and the rest of the created world. This tradition is thus "anthropocentric" – centered on human life – and it is this anthropocentrism that will be challenged if other life forms are found in the universe.

[2] Note that, while many researchers are actively involved in the Search for Extra-Terrestrial Intelligence (SETI), it is most likely that the first discovery of extraterrestrial life will involve microbial life, or evidence of the existence of microbial life in the past.

[3] Davies is quoting Jacques Monod. *Chance and Necessity* (New York: Vintage Books, 1972), p. 180. See also Ian Barbour's use of this quote, and discussion, in *Religion and Science: The Historical and Contemporary Issues* (San Francisco, CA: Harper Collins, 1997), pp. 79–80.

Many scientists believe that life is not a freakish phenomenon (the odds of life's starting by chance, the British cosmologist Fred Hoyle once suggested, are comparable to the odds of a whirlwind's blowing through a junkyard and assembling a functioning Boeing 747) but instead is written into the laws of nature [1] (p. 114).

Davies goes on to add the theological component in here. Generally, those who would agree with Monod about the purposelessness of the universe exclude the notion of a Creator God – hence the "alone in the universe" mantra. Those who might agree with Hoyle about order being written into the universe such that life is destined to emerge tend to be more sympathetic to a theological worldview. Davies contrasts "sheer chance" with "lawlike certitude":

The theological battle line in relation to the formation of life is not, therefore, between the natural and the miraculous, but between sheer chance and lawlike certitude. Atheists tend to take the first side, and theists line up behind the second; but these divisions are general and are by no means absolute [1] (p. 114).

Let us examine the notion of determinism more closely. In general, determinism is the idea that the laws of nature are set from the beginning, and what emerges in the universe is merely an unfolding of what was already ingrained from the start. With regard to the emergence of life, a deterministic approach assumes that the appearance of life on Earth, especially human life, is written into the laws of nature. World process is teleological, meaning that it is headed towards an "end" or purpose. This view assumes that there is purpose in the universe in that it is ordered to certain outcomes. The broad thrust of evolutionary changes is innately pre-determined, particularly when it comes to the emergence of life.[4]

Let me make a further distinction between "mechanistic determinism" and what I call "statistical determinism." In mechanistic determinism one presumes a closed universe. There is little room for contingency or probabilities in this perspective. One assumes a direct causality among all events. Everything can be explained in terms of "A" causing "B," and "B" then causing "C," and so on. If we run up against something that we can't explain, or seems to involve constant change, it is

---

[4] Generally, evolutionary scientists reject any kind of determinism as thus described. Ernst Mayr, one of the elder statesmen of evolutionary biology, in his book *What Evolution Is* (New York: Basic Books, 2001), rejects any notion of *finalism*, defined as "belief in an inherent trend in the natural world toward some pre-ordained final goal or purpose, such as the attainment of perfection" (p. 286). He rejects in turn a notion held in the decades after Darwin, called *orthogenesis* and defined as "the refuted hypothesis that rectilinear trends in evolution are caused by an intrinsic finalistic principle" (p. 288). His definition of teleology is "the study of final causes; the belief in the existence of direction-giving forces" (p. 291). He is insistent that natural selection is in no way directed toward some pre-determined end: "Selection is not teleological (goal-directed). Indeed, how could an elimination process be teleological? Selection does not have a long-term goal" (p. 121). Notions of adaptation that include this directedness are erroneous: "One must always remember that adaptation is not a teleological process, but the a posteriori result of an elimination (or of sexual selection)" (p. 149). The determinism that is not supported by empirical findings is one in which a specified goal is determined from the beginning of the process.

simply a matter of ignorance. Once we understand the phenomenon further, it is presumed, we will be able to see the direct "mechanistic" relations to explain the phenomenon. Eventually, the entire universe can be explained according to these directly causal relations.

This view, often attributed to Sir Isaac Newton, has lost favor in recent centuries, once the role of probabilities in explanatory science has come to be acknowledged, whether in evolutionary theory or in quantum physics. Nevertheless, with regard to the emergence of life, some scientists continue to hold a statistical determinism, assuming that outcomes are in some way set from the beginning. A statistical determinist acknowledges the role of probabilities in emergence. World process is not a closed set of direct causalities but open to change, full of potential for novelties to emerge. What makes this position "determinist" is the idea that eventual outcomes are set in the initial conditions. Paul Davies cites Christian De Duve as holding such a position:

Christian De Duve does not deny that the fine details of evolutionary history depend on happenstance, but he believes that the broad thrust of evolutionary change is somehow innately predetermined – that plants and animals were almost destined to emerge amid a general advance in complexity [1] (p. 116).

Where does God fit in to such a view? A deterministic worldview does not require a Creator God. However, it lends itself to a belief in a Creator who either wrote the laws of the universe into a closed system or set the initial conditions in an open system. In such a view God is the one who determines outcomes, who gives a direction to emergence and sets the goals towards which the unfolding of the universe is headed. Such a divine persona is not needed in order to hold a deterministic worldview, but the notion of direction or purpose raises the question of a divine intentionality, to which theists respond affirmatively.[5]

Now let's take the other side of the coin and consider "chance." A person who ascribes to chance as the principle by which the universe unfolds would acknowledge the occurrence of regular and universal laws. But the emergence of such laws, for example the law of gravity, has no necessary permanence and no directionality. While the ongoing functioning of such laws reflects a regularity that mimics "necessity," the emergence of these laws is purely random. With regard to the emergence of life on Earth, its appearance and development is a happy accident and not the result of any kind of purpose or design. Once again, Paul Davies captures this position:

---

[5] Note that the understanding of God that parallels this determinism is that of Deism, a rationalistic understanding of God that grew out of the Enlightenment and Newton's mechanics. God is thus envisioned as a divine power that established the order of the universe and then let it "unwind" on its own. Such a deistic understanding of God is not to be mistaken for a more robust and complex theology of God that has its source in a medieval synthesis of Biblical and philosophical thought, what I will call here, "classical theism."

Most biologists regard a "progressive evolution" with human beings its implied goal, as preposterous. Stephen Jay Gould once described the very notion as "noxious." After all, the essence of Darwinism is that nature is blind. It cannot look ahead. Random chance is the driving force of evolution and randomness by definition has no directionality [1] (p. 116).

Is there room for God in such a perspective? Many assume that a universe governed by pure chance excludes God. Evolution and world process have no design, no purpose or principles written into their unfolding. No Creator acts to determine or even influence the outcomes, the direction of what emerges and what does not, what survives emergence, and what atrophies.

On the other hand, there are those who cite chance and the low probabilities of life's emergence as grounds for belief in the Divine. Fred Hoyle, as already mentioned, notes how seemingly impossible it is to imagine life emerging merely on the basis of pure chance. The chance of this happy accident occurring is akin to the chances of a whirlwind blowing through a junkyard resulting in a Boeing 747. In this case, the low probabilities of such a happy accident lead some to postulate the intervention of a God of purpose and design.

This review of the divergence between determinism and pure chance is a caricature of positions held by various scientists, philosophers of science, and religious believers. Nevertheless, it depicts a range of views that underlie many debates over an evolving universe and the role of God in evolution, particularly in the emergence of life on Earth. The dichotomy, however, gives an overly simplistic view of the options here. In fact, a number of theologians have been searching for new models of divine action in light of evolutionary theory. The question, thus, of how our notions of God will change should we discover life on another planet, depends on what our notion of God's agency and purpose is in the first place. How we understand God to act or to have acted in the unfolding of life on this planet will determine our response should we need to include extraterrestrial life in the narrative of a creating God.

12.2 Divine action in an emergent universe

If one is to move beyond a closed system to acceptance of an emergent universe, three components must be addressed: order, novelty, and telos. If one wants to incorporate God into such an emergent worldview, one must account for God's role in establishing order, allowing or creating novelty, and intending or enacting purpose. With regard to God and the element of chance in the unfolding of the universe, Ian Barbour outlines three options: God controls events that appear to be random; God designed a system of law and chance; and God influences events without controlling them [2] (pp. 239–240).

The first view assumes that events are determined by God even though we are unable to see the order in this determination. What seem to us to be chance events are really hidden divine actions guiding the flow of emergence. This view would be akin to what I have called statistical determinism above and what Barbour labels "God as Determiner of Indeterminacies" [2] (pp. 305ff). The problem with this approach is that while it seems to allow for novelty, at root it is still deterministic and does not allow for true novelty. It leaves one trying to imagine every detail of evolutionary history as subject to God's direct intervention. Further, it tends to be reductionistic in assuming God enters the process at the lowest levels (in atomic or molecular details).[6]

The second option includes the action and purpose of God but at a more remote level. The design or purpose of God is in the whole system of interaction between law and chance. This view is most concerned with the probabilities for the emergence of life. Those who hold this view see God as setting up the systematic conditions that made life and consciousness possible.[7] Thus we have Fred Hoyle's allusion to a whirlwind, a junkyard, and a Boeing 747; the clear improbability of life's emergence is cited as an indication that God must have had a hand, even if indirectly, in setting up conditions such that life as we know it unfolded. Design is identified with the whole system of regularity and probability, making higher life forms possible. Objections to this view are that it seems to limit God's activity to originating and sustaining natural processes [2] (p. 240).

The third view is that God continues to be involved in what emerges but not through direct control. Rather, chance, lawful causes, and God are involved in the novel constitution of each event. This view rejects any determinism, as latent in the first view, yet allows for a more direct divine involvement in evolution than is apparent in the second position above. Continuing creation is an experiment of trial and error. Waste, suffering, and blind alleys can and do occur, but these in turn have allowed for infinite variety and beauty in what has emerged. The God involved here is one who, out of love, allows for risk, who acts by means of persuasive love rather than coercive power. This is a God who influences and is influenced by the world. Theologians who adopt this view are often informed by the process philosophy of Alfred North Whitehead.

---

[6] Barbour here cites those who insist that God works "from the bottom up" in determining events at the most rudimentary levels: W. Pollard. *Chance and Providence* (New York: Charles Scribner's Sons, 1958), chapter 3; D. McKay. *Science, Chance, and Providence* (Oxford: Oxford University Press, 1978); P.T. Geach. *Providence and Evil* (Cambridge: Cambridge University Press, 1977). Note, however, that it is possible to envision God affecting probabilities at higher levels, for example, even in terms of human freedom. Some versions of predestination in the vein of Calvinist thought would view God as determining individual salvation, in spite of appearances of human freedom.

[7] These ideas are often discussed under the label of the "anthropic principle," which is the idea, first suggested by an astrophysicist in 1973, that the universe is fine-tuned with exactly the right factors to allow for the highly improbable emergence of life as we know it. For an example of this thinking, particularly with its religious overtones, see http://ourworld.compuserve.com/homepages/rossuk/c-anthro.htm. Barbour discusses this in *Religion and Science: The Historical and Contemporary Issues* (San Francisco, CA: Harper Collins, 1997), pp. 204–206.

A fourth alternative emerges from the distinction between *natural theology* and a *theology of nature* [2] (pp. 246–249). The above three positions fall loosely into a theology of nature: they all assume a distinction between belief in God (religion) and theories of emergence (science). Given this distinction they try to explain how their religious views of God are compatible with the probability shaped nature of evolution. Natural theology, by contrast, claims that theistic conclusions can be drawn from scientific data. This view harkens back to William Paley who in 1802, well before Darwin, insisted that the properties of each living species reveal the designing hand of God [3]. Modern defenders of natural theology insist on the inadequacies of Darwinian and neo-Darwinian theories of evolution. While they accept the general thrust of evolutionary theory (unlike Biblical creationists who reject evolution altogether), they believe that chance and natural selection are not sufficient to explain the complexity of what has emerged. Instead, the hand of a designer must be invoked as a scientific explanation for certain moments in the evolutionary narrative, most notably in the emergence of life and of human life in particular. The most recent advocates of this position come from the Intelligent Design movement, which has gained a following in the last decade and has sought to affect what is taught in science classrooms across the USA.[8]

All of this needs to be understood within the drama of history as it unfolded from the medieval world to the modern scientific milieu [2] (pp. 281–284). First, classical theism, in the Christian tradition, comes from the medieval view of nature that combined Greek and Biblical ideas. It is most widely known through Thomas Aquinas's synthesis of Aristotelian philosophy and Christian theology, though the Protestant Reformers retained the Monarchical view of God inherent in this position. In this medieval synthesis, nature is understood to have a fixed order, though chance occurrences are not overlooked. Not only is there order but there is purpose – everything comes from God and is returning to God. Everything has a final cause – built-in goals towards which each creature is oriented. Most importantly, the model of God is as a governor, using providence to manage creation's unfolding. As *omnipotent*, God governs with providential wisdom. God is also *omniscient* – knowing everything about everything, past, present and future. God's *eternity* is foremost – having a permanence beyond the vicissitudes of earthly existence. Most central, and of greatest challenge to our modern view, God is *immutable* and *impassible* – God never changes.

---

[8] See P.E. Johnson. *Darwin on Trial* (Washington, DC: Regnery Gateway, 1991); and Michael Behe. *Darwin's Black Box: The Biochemical Challenge to Evolution* (New York: The Free Press, 1996). The Discovery Institute supports this approach: see http://www.discovery.org/csc/. For a critique of this movement, see B. Forrest and P.R. Gross. *Creationism's Trojan Horse: The Wedge of Intelligent Design* (Oxford: Oxford University Press, 2004).

With the advent of Newtonian science, assumptions about both nature and God shifted. The Newtonian view gives greater scope to change, but change is seen only as the re-arrangement of the fundamental particles of nature. This world is deterministic in its teleology. That is to say, God's intentions are written into creation from the beginning and then unfold mechanistically. Mechanical causes rather than "ends" or purposes are what is to be explored by science – and the future could be perfectly predicted given full knowledge of past causes. While God is still very much in the scope of this worldview, he is seen as the creator who sets the mechanism in order and then lets it run its course – the watchmaker God. This is the Deism of Benjamin Franklin and Thomas Jefferson – it incorporates a generic God who is to be revered but who never intervenes in worldly affairs.

Our modern worldview differs significantly from either of these perspectives. The most dramatic change has to do with our perspectives on time, history, and change. While the role of chance has been discussed by philosophers down through the ages, Darwin made it constitutive to the explanation of natural history. Freud unveiled the unconscious, revealing that the narratives we tell may not be as obvious as we first think. Marx dissected political interests to show that our social systems have hidden and, at times, malicious meanings. So the order apparent in the medieval synthesis and assumed in Newtonian science has been upended: nature, human nature, and social life are all subject to development, change, and unseen or unexpected meanings.

Instead of order, or the re-arrangement of fundamental units, we now have a world that is evolutionary, dynamic, and emergent. Instead of determined purpose we have an ever-shifting interaction of law and chance. The future is unpredictable both in principle and in practice. Nature is seen to be an interdependent set of relations that is subject to demise as well as development.

With such a changed understanding of the world, our notion of God has come under severe scrutiny. Most in jeopardy is the immutability of God and God's relationship to time. Can one retain a view of God as unchanging yet engaged with a world that is ever changing? How can God's purposes be implemented in a world subject to law and chance? While some insist that our changed worldview calls for the jettisoning of the notion of God altogether – omniscient, omnipotent, eternal, and immutable – others, as we have seen, are attempting to modify the concept of God in a way that accords with our new understanding of an emergent universe.

Above we reviewed several general trends in the attempt to incorporate God into an ever emerging universe. Let me go further here to present specific models of God's relation to and activity within an evolving world.[9]

---

[9] These are selected and abridged from Barbour, chapter 12. Another good resource on the options here is *God's Activity in the World*, ed. O. Thomas (Chico, CA: Scholars Press, 1983). See also C. Southgate. *God, Humanity and the Cosmos*, 2nd edn. (New York: T & T Clark International, 2005).

12.2.1 A revised classical theism

In recent years neo-Thomist scholars have sought to retrieve the notion of God's omnipotence in an emergent world without succumbing to the inactive God of Deism. They have done this by renewing the distinction Thomas Aquinas made between primary and secondary causes. God is the primary cause in conferring on all things their distinctive forms and powers. God's action in nature is primarily in conservation: without God's continual presence all of reality would cease to exist. But God also works through secondary causes, and these are the proper functioning of things at the level at which they were created – atoms in being atoms, stars in their orbits, and humans in their choices. God is present in these created things and relations, not through direct intervention but secondarily. The image invoked is often that of toolmaker and tool: God as wielder of the axe cuts the wood but the axe serves as a secondary cause in its own right. This metaphor lags in that it does not convey the dynamism involved in secondary causality. Most important is the idea that God has given each being an inclination, which both expresses the essence of its being at the same time that it expresses God's purposes.[10] God's influence lies in that attraction to the good and, while there is room here for the miraculous intervention of God on occasion, God's action lies primarily in the power of love drawing all things towards himself.[11]

12.2.2 God as determiner of indeterminacies

We have alluded to this possibility above. It lies in claiming that the very elements of chance that atheists cite as evidence of God's absence may be the point at which God acts. God's providential wisdom may be manifest in controlling events that appear to us as chance. This does not mean that God intervenes in every movement

---

[10] Some scholars note that classical theism involves an essentialist ontology in which "a thing becomes what it is" rather than an existentialist ontology, more compatible with an evolutionary worldview in which "a thing is what it becomes." It is true that classical theism, with its roots in Aristotle, envisioned an understanding of being in which "things" were understood to be an integration of matter and form, with a final cause "built in" as it were. The issue, to my mind, is not so much whether "things" have or become "essences" but whether the notion of being itself can be transposed from a closed worldview to a dynamic one. I am not convinced that a metaphysics of being, in which the notion of essence is central, is to be totally jettisoned in favor of a metaphysics of becoming in which essence or formal cause is considered suspect. This issue is probably the most central in a divergence between classical theism and modern process thought. There is not room here to engage in this debate, but my own work leans towards retrieving and transposing a classical notion of being into a dynamic worldview rather than assuming that contemporary evolutionary thought makes this impossible. In fact, this is what Bernard J.F. Lonergan (discussed below) attempts to do in chapters 14–17 of *Insight: A Study in Human Understanding*, ed. F.E. Crowe and R.M. Doran (Toronto: University of Toronto Press, 1992 [1957]).

[11] The neo-Thomists that Barbour cites include Étienne Gilson. The corporeal world and the efficacy of second causes. In *God's Activity in the World*, ed. O. Thomas (Chico, CA: Scholars Press, 1983), pp. 213–230; and R. Garrigou-Lagrange. *God: His Existence and His Nature* (St. Louis: Herder, 1934). He also includes in this category of thought the works of Austin Farrer and Karl Barth. See A. Farrer. *A Science of God?* (London: Geoffrey Bles, 1966); and K. Barth. *Church Dogmatics*, vol. 3, pt. 3 (Edinburgh: T & T Clark, 1958).

of every atom in an exceedingly large number of directly causal events. Instead, God actualizes one of the potentialities already existent. God acts in the combination of lawlike order and chance events to determine in the long run just what emerges or occurs. This view tends to see God's causal influence as coming from the bottom up, and can be subject to a certain reductionism, though most authors do allow for God's involvement at higher levels, thus exerting a top-down effect on lower levels. This approach fits in with chaos theory and the recognition that an infinitesimal initial change can have profound effects at higher levels of existence. It involves a determinism of sorts, what I have called statistical determinism above, yet avoids both the distant God of Deism and the mechanistic determinism that fails to recognize the element of chance in world process. While quite distinct from the neo-Thomism discussed above, this approach can be compatible with both the models discussed below.[12]

12.2.3 God as communicator of information

Increasingly, the communication of information has become the cornerstone of understanding organic processes. Evolutionary history in particular reveals the unfolding of the capacity of organisms to encode, process, and respond to information internally and from the environment. Clearly, the role of DNA as the essential factor in the origin and sustenance of life shows the vital importance of information storage and retrieval in the existence of life on Earth. Both John Polkinghorne and Arthur Peacocke have appealed to the image of God's action in the world as that of the input of information.[13] This approach fits in with the notion of God as the determiner of indeterminacies but specifies the mode of God's presence as the communication of information that evokes the occurrence of one event over another. The evolutionary narrative can be understood as the communication of an agent who expresses intentions but does not dictate a pre-determined plan. Other

---

[12] The work of Nancey Murphy and Robert John Russell can be cited here as examples of this approach. See N. Murphy. Divine action in the natural order: Buridan's ass and Schrodinger's cat. In *Chaos and Complexity*, eds. R.J. Russell, N. Murphy, and A.R. Peacocke (Rome: Vatican Observatory, and Berkeley, CA: Center for Theology and the Natural Sciences, 1995), pp. 325–357; N. Murphy and G.F.R. Ellis. *On the Moral Nature of the Universe: Theology, Cosmology and Ethics* (Minneapolis, MN: Fortress Press, 1996).

[13] Barbour discusses these two thinkers on pp. 314 and 315. He cites the following as examples of this point of view: John Polkinghorne. *Reason and Reality* (Philadelphia, PA: Trinity International Press, 1991), chapter 3; idem. *The Faith of a Physicist* (Princeton, NJ: Princeton University Press, 1994), pp. 77–78; and idem. The metaphysics of divine action. In *Chaos and Complexity*, eds. R.J. Russell, N. Murphy, and A.R. Peacocke (Rome: Vatican Observatory, and Berkeley, CA: Center for Theology and the Natural Sciences, 1995), pp. 147–156; A.R. Peacocke. *Creation and the World of Science* (Oxford: Clarendon Press, 1979), chapter 3; idem. *Theology for a Scientific Age* (Minneapolis, MN: Fortress Press, 1993), chapter 9; and idem. God's interaction with the world. In *Chaos and Complexity*, eds. R.J. Russell, N. Murphy, and A.R. Peacocke (Rome: Vatican Observatory, and Berkeley: Center for Theology and the Natural Sciences, 1995), pp. 263–288. See also John F. Haught. *God After Darwin: A Theology of Evolution* (Boulder, CO: Westview Press, 2000), pp. 69–80.

images are that of the choreographer who sets up the dance but leaves the precise movements up to the dancers as the dance unfolds. Likewise, one can imagine an unfinished symphony in which improvisation on a theme is the *modus operandi*. The early Christian appropriation of the Greek notion of a divine *logos* (word) in interpreting Christ as the Word of God parallels this image of God as communicator – both creation and redemption are understood as the expression of a loving God. At the human level God's purposes are communicated to self-conscious persons who can respond in kind. But this communication can be extrapolated to the whole of emergence in the universe.

12.2.4 Process theism

Process theology, in which Christian theologians have adapted the philosophical work of Alfred North Whitehead, is probably the most thoroughly developed of the theologies that have tried to integrate modern science with Christian theism.[14] The process view accepts the dynamic openness of creation. It assumes a community of interaction at all levels of existence – a social and ecological model rather than a monarchy or a machine. While God is the primordial ground of order, God is also the ground of novelty. God lures the world through attraction to ongoing development. God acts primarily through persuasion but, most importantly, God himself is influenced by the unfolding of world process. While God is eternal and does not perish, nevertheless God's ongoing presence in the world is stressed. God is omnipresent and intimately involved in a unique and direct relationship with each member of the cosmic community. This view allows for particular divine initiatives, with the focus being on the involvement of God in the ongoing interaction between order and novelty. While the primordial nature of God is immutable, the power and providence of God entails a God of persuasion, whose actions in the world are always in response to the choices of created entities, at all levels. This is in contrast to an omniscient God who knows what will occur, when the future, including God's future, is radically open. Process theologians reject the metaphysics of being from classical theism in favor of a metaphysics of becoming. Rather than reconcile classical tenets about God with the modern recognition of contingency and change, they have revised their notion of God to embody time and development.

---

[14] Ian Barbour would count himself within the category of Process Theism. He devotes an entire chapter in *Religion and Science* (chapter 11) to a discussion of process thought as well as including it in his discussion of contemporary models of God, where he gives an assessment and critique of process thought (pp. 322–328). He cites as the basic resources for understanding process thought, A. North Whitehead. *Science and the Modern World* (New York: Macmillan, 1925); and idem. *Process and Reality* (New York: Macmillan, 1929). On process theology see J.B. Cobb, Jr. and D.R. Griffin. *Process Theology: An Introduction* (Philadelphia, PA: Westminster Press, 1976). A more recent Christian theologian who has incorporated process thought into his work on evolution is John Haught. See *Deeper Than Darwin: The Prospect for Religion in an Age of Evolution* (Boulder, CO: Westview Press, 2003) and *God After Darwin: A Theology of Evolution* (Boulder, CO: Westview Press, 2000).

In sum, there are a number of responses to the modern worldview and its evolutionary character. Some reject this worldview outright and cling to a literal Biblical account of creation. Others ascribe to a natural theology that accepts the general thrust of evolutionary theory but sees God as intervening at crucial moments to introduce complexity where otherwise it would not emerge. Finally, there is a broad spectrum of theologians who develop a theology of nature in which they adapt notions of God, order, chance, and purpose to accommodate an ever-emerging universe. These innovations vary to the degree that they let go of the tenets of classical theism and/or in the way they understand God's purposes to interact with the dialectic of law and probability.[15]

12.3 Emergent probability

Having outlined several options in the modern reconciliation of God and emergence, let me develop one particular approach that I believe resolves to a large degree the polarity of chance and determinism. The work of Canadian Jesuit, Bernard J. F. Lonergan (1904–1984) falls within the category of a revised classical theism. Lonergan was a Roman Catholic philosopher schooled in the work of Thomas Aquinas, whose life spanned the entrance of the Catholic church into the modern world in the twentieth century. His ultimate interest was in developing a method for a renewed theology [4], but he believed that he first needed to understand the rise of modern science and its methods. He fits within the classical mode of theology in that he relies on the Aristotelian synthesis of Thomas Aquinas. On the other hand, he addresses the issues of modernity and incorporates into his Thomism an emergent view of world process.[16]

---

[15] While the bulk of this article has addressed ways in which theologians have adapted their ideas of God to evolution, the fact of the matter is that in the American populace at large the theory of evolution is often rejected in favor of some kind of interventionist divine action. A Gallup poll in February of 2001 revealed that 45% of responding US adults agreed that "God created human beings pretty much in their present form at one time within the last 10,000 years or so." Only 37% accepted the idea of allowing both God and evolution as their understanding of how the world unfolded. Gallup polls over the years have shown little change in this – the view that God alone and not evolution produced humans has never drawn a response of less than 40%. See D. Quammen. Was Darwin wrong? *National Geographic* (November, 2004), 6.

[16] Lonergan probably would not be considered, first and foremost, a philosopher of science. You would not find him in a collection of articles about science, philosophy, and religion. His ultimate goal, and achievement, was to develop a method in theology. But he knew that the epistemological issues of modernity had to be tackled in order to re-invent theology for the modern age. So he approached his massive study *Insight: A Study of Human Understanding* by advocating "self-appropriation" – considering what operations everyone uses in knowing. This leads to a consideration of how this everyday set of operations becomes methodological in scientific or scholarly work. Thus his philosophy of science is *heuristic* – laying out what scientists do when they seek to understand phenomena in the world. Such a heuristic does not determine the content of what is known by the scientist but explains the processes by which the scientist will have come to his/her conclusions when they occur. Likewise, emergent probability is a *heuristic* device in the sense that it does not tell us what has or will emerge, but provides a structure by which whatever does emerge will emerge. Among Lonergan scholars there are a number who are involved in the science and religion dialogue, such as Patrick Byrne and Charles Hefling at Boston College. Unfortunately, there are not many scientists or philosophers

Lonergan's seminal work of 1957 is entitled *Insight: A Study of Human Understanding*, in which he develops an epistemology of "critical realism" [5]. In the process of examining how we know what we know, Lonergan elucidates two different kinds of laws that scientists discover in the world: the laws of *classical science* that explain regularity and universality and the laws of *statistical science* that help us understand variability and the likelihood of certain events occurring in the world. The interaction of these two kinds of laws accounts for *emergent probability* and it is this explanation of world process that I think can be instrumental in sorting out order, novelty, and telos in relation to God and the unfolding of the universe. Let us explore emergent probability before returning to the question of God's action in the world and, particularly, God's relation to astrobiology.

The most basic thing to understand is that the laws that govern the world as we know it are of two distinct types. In speaking of "laws of nature" we refer, most often, to the regularities that humans have come to understand as recurrently operative in the world around us. But such explanations, abstracted from concrete instances, come in two kinds. There are the classical laws of science by which we understand principles that work with regularity and universality across a wide range of similar circumstances, for example the law of gravity. But there are also statistical laws that explain if and when something is likely to occur. These are laws of probability, well known by weathermen, casino owners, and quantum physicists alike.

A closer examination of the difference between these is called for. First, in determining classical laws, scientists use a long series of instances of some phenomenon in order to discover the common principles governing such events. So Galileo sought to grasp the nature of the free fall by observing similar behaviors of numerous objects dropped from various heights. By cataloging and classifying behaviors common to these many instances, he was able to discover certain patterns. Eventually he was able to determine a co-relation between spatial extension and temporal duration, an invariant relationship that could explain all similar phenomena, all other things being equal. This invariance is what is presumed by science of the classical type, which seeks to explain the systematic aspects of the world.[17]

Note another characteristic of these kinds of laws. They abstract from the particular places and times that verify the law in question. Many aspects of these concrete circumstances are left unexamined because they are not relevant to the

---

of science who are familiar with Lonergan's work. However, the important issue is not knowing Lonergan's work but grasping the fundamental concepts he proposes and determining whether in fact they do have heuristic explanatory value.

[17] Note that "classical" as used here is to be distinguished from "classicist." The "classicist worldview" refers to a pre-modern understanding of both human nature and the world, and it is to be contrasted with the modern worldview that incorporates historical consciousness. While it can generally be said that the classicist worldview conceives of science as predominantly that of discovering classical laws, classical science continues within the purview of historical consciousness. What is new is the recognition of statistical as well as classical types of investigations.

question at hand. For example, febrile seizures in children are explained by a failure of the brain's electrical signals due to the sudden onset of a high fever. This explanation, and the concrete insight that yields a diagnosis in the hospital emergency room, ignores the colors of the clothes the child is wearing, whether her hair is in a ponytail, and/or how tall she is.[18] Thus, classical laws, by their very nature, are abstract and leave unexplained many other aspects of a situation.

While classical laws explain the aspects of the world that occur with regularity, statistical science deals with the variable aspects of the world. Classical laws are based on the proviso "all other things being equal," but statistical investigation recognizes first and foremost that all things are not always equal. The latter concerns itself, not with the *invariance* of certain phenomena, but with the *frequency of the occurrence* of such phenomena. Classical laws explain the laws of motion by which a coin tossed in the air moves through space, but statistical science deals with the likelihood of such a coin landing with heads up. Galileo explained the nature of a free fall, but his theory could not determine whether or how often objects fall off buildings.

Statistical science thus deals with the frequencies of events. Like classical scientists, statisticians aim at generalizations, but in this case they are seeking *ideal frequencies* rather than principles of regularity. One specifies the event to be counted and sets parameters of time and geography in which to count these events. For example, one might count the number of homicides in Washington DC in 2007. Out of n crimes reported, how many were murders? Or, for every 1000 residents in the city, how many murders took place? One can break this down into neighborhoods and compare the ratio of murders to crimes reported in various parts of the city. Ideal frequencies are numerical ratios that are the mean (the average murder rate in the city) around which actual frequencies (the actual number of murders in a given neighborhood) diverge non-systematically.[19] One can then compare the number of homicides over a number of years, or compare one city with another. Ideal frequencies thus can be used to determine probabilities: the likelihood of dying by homicide this year compared to last year, or in Washington DC as compared to Baltimore.

---

[18] There are some factors that *are* relevant, such as the age of the child and the family history of such a disorder. Clearly, one of the important aspects of gaining insight into the systematic factors that define and explain a phenomenon such as this is determining which factors are systematically related to the phenomenon and which are *merely* coincidental. For example, while racial origin or sexual activity may be merely coincidental elements in relation to the explanation of one disease, they may be significant factors in explaining another disease.

[19] These non-systematic divergences from an ideal frequency (an average) illustrate a proper understanding of the term "random." That is, a particular event is random in relation to an aggregate of events. There is no such thing as a single random event. Events are random relative to a pattern that emerges in a group of events. Thus, the probability of rolling a dice and getting a six is one in six. This is the ideal frequency, the average. Any particular throw of a dice (the actual frequency) is random relative to this expected pattern. The important point is that to talk about the world's unfolding as "completely random" makes no sense, since one can only determine any event or set of events as random if one has first found a pattern, or regularity with which to compare it.

Still, while statistical scientists seek such generalizations, the work of statistical science differs from that of classical science in that one must often calculate one's findings anew, as situations being studied shift and change. While classical investigation often reaches a term at which a theory about a systematic relationship is confirmed beyond a reasonable doubt, statistical science is subject to ongoing changes. So, while the search for the cause of diabetes has come to a term with the discovery of insulin, determining the likelihood of being murdered in Washington DC requires the recurrent gathering of information in each subsequent year. While classical science determines "the nature of" some phenomenon (uterine cancer), statistical science sets out to elucidate "the state of" – the weather, the economy, public health, vehicle safety, and so on. To the degree that these phenomena are ever changing, the statistician must engage in recurrent studies.[20]

While I have been at pains to explain the difference between these two endeavors, I do not mean to suggest that we live in a bifurcated world. It is not the case that some aspects of our lives can be explained by regular laws of the classical type, while in other arenas we live by laws of probability. Rather, both are different ways of explaining the same world – by attending to different kinds of data and by asking different kinds of questions.

To illustrate this, let us use the world of baseball. Consider the case of a baseball hero. The data about Johnny Damon's past success at hitting home runs answer the question "How often?" and give us the information we need to calculate the likelihood of his hitting home runs in the new season – such is the role of statistical science. Classical science can explain the laws of motion whereby the ball moves through the air if and when such a home run is hit. But classical science cannot indicate whether and when such home runs are likely to occur. So it is that batting averages are calculated, and calculated under a variety of conditions: Johnny Damon's home run average in the post-season, against this particular pitcher, against left-handed pitchers, and so on.

To give another example, the classical laws of biology explain what occurs when a sperm fertilizes an egg and conception takes place. But the biological definition and explanation of conception cannot, in and of themselves, determine fertility rates. To determine these one must count and calculate, considering a range of variables, such as age, education, health, and frequency of intercourse, among couples within a certain geographic location.

So both classical and statistical laws seek to explain the same world. But they do so in very different but complementary ways. Let us move one step further

---

[20] Note that what is being studied (e.g. the incidence of disease in bees) may show little change and hence reveal great stability, but unless these phenomena are studied in a variety of places over a determinate timeframe, one does not know that the stability exists. Since the nature of statistical science is to determine the conditions under which certain events occur, one has to keep counting these events under a variety of conditions in order to arrive at one's goal.

to show how these two kinds of laws operate in the ongoing development of the world.

The first point to grasp in this case is the notion of a scheme of recurrence. A scheme of recurrence occurs when a series of conditions for an event coil around in a circle, such that event "A" fulfills the conditions for the occurrence of event "B," which in turn fulfills conditions for "C" to occur, which then satisfies the conditions for "A" to recur. A recurrent cycle emerges that has a certain stability to it. Further, defensive mechanisms can develop so that any intervening event that threatens the cycle is offset by a second cycle designed to eliminate the intruder. Examples of schemes of recurrence include the planetary system, the circulation of water over the face of the Earth, the digestive system of mammals, the nitrogen cycle that keeps plants alive.[21] Examples of defensive systems would be the body's immune system or the compensatory reactions of an environment when the ecological balance is disturbed.

Further, not only are there single schemes, there are conditioned *series* of schemes of recurrence. So it is that the circulation of water over the face of the Earth is a scheme that itself is a condition for the possibility of the nitrogen cycle of plant life to occur. And the nitrogen cycle of plant life is a scheme that is itself a condition for the possibility of the digestive system of animal life to occur. So individual cycles themselves form a conditioned, recurrent series of schemes.[22]

Note that these schemes of recurrence are conditioned and not inevitable. Thus, though the scheme itself is a combination of classical laws that function with regularity, "schemes begin, continue, and cease to function in accord with statistical probabilities" [5] (p. 141). At any stage of history, then, there are probabilities for the emergence and probabilities for the survival of various schemes of recurrence. The emergence of new schemes depends on a coincidental manifold of underlying events that produce the conditions for such an emergence. An example would be the random genetic mutations that provide the opportunity, over time, for adaptive traits in a population to emerge and stabilize, perhaps eventually leading to a new species or subspecies. The survival of schemes of recurrence depends on the continued survival of the underlying conditional schemes. So it is that an ecosystem, when changed, leads to the extinction of species. So it is that a metabolism, when its fundamental processes fail to work, breaks down to yield the death of an organism.[23]

---

[21] Two ideas emerging in the scientific literature reflect this idea of schemes of recurrence. These are *autopoiesis* and *autocatalysis*. On the former, see http://en.wikipedia.org/wiki/Autopoiesis. On the latter, see R.E. Ulanowicz. Ecosystem dynamics: a natural middle. *Theology and Science*, **2** (2004), 231–250 and idem. *Ecology, The Ascendent Perspective* (New York: Columbia University Press, 1997). See also the discussion of autocatalytic cycles in D.J. Depew and B.H. Weber. *Darwinism Evolving: Systems Dynamics and the Genealogy of Natural Selection* (Cambridge, MA: The MIT Press, 1995), pp. 409–411; 466–475.

[22] Global warming provides a most pertinent and publicly discussed example of the interlocking sets of schemes of recurrence in the natural world today.

[23] Note that "probabilities for the emergence and survival of schemes of recurrence" is not the same as talking about probabilities for the emergence of species. First, the definition and boundaries of a species are still

Emergent probability, then, is a heuristic explanation of world process that contradicts a determinism by which all of creation is governed according to classical laws.[24] While classical laws explain the systematic and recurrent aspects of the world, the emergence and survival of these systems depend on underlying conditions. And these underlying conditions occur according to probabilities.[25] This worldview incorporates an indeterminacy that is not mere chance and which has its own intelligibility (that is, one can make sense of it), even though it is not the intelligibility of automatic progress or of a totally determined system.[26]

Is there purpose or direction here? Lonergan retrieves the classical notion of "final cause" by referring to "finality" as operative in the unfolding of world process. Finality is the "upwardly but not determinately directed dynamism" of world process [5] (p. 497). This is teleology but with a difference. There is dynamism, not merely a mechanism playing out pre-determined goals. The dynamism generates ever more complex systems, even though the direction of such development is not determined. So there is a tendency for physical elements to gather into the higher integrations of chemistry. And physical chemistry yields organic chemistry. Organic chemistry makes sentient beings sentient while the neurological make-up of the human brain is oriented to ever greater understanding. Thus the dynamism is "upwardly directed."[27] Yet, most importantly, it is "upwardly but not determinately" directed. There is great openness and flexibility here – while the dynamism of world process is toward higher integrations of lower schemes of recurrence, just what exactly will emerge and what will atrophy is not determined. While the

matters for debate in biological circles. Second, a species is a population of organisms that is subject to an immense number of schemes of recurrence, from metabolic to environmental schemes, including, in humans, schemes of value and meaning. So changes in the incidence of certain schemes of recurrence can have a profound effect on the nature of a population and its ability to adapt to its surroundings. Such shifts in schemes of recurrence is what adaptability and natural selection are all about. If the schemes of recurrence shift within a population in such a way that the members of this population are no longer able to mate with members of an earlier related population, a new species has emerged. Thus the probabilities of the emergence and survival of a species are dependent on the emergence and survival of a complex set of schemes of recurrence.

[24] What does it mean to say that emergent probability is a *heuristic* explanation? The point of a heuristic method or explanation is to direct one's attention in learning or discovery. Without determining the answer to a question one can work out what the parameters of the answer will be once it is reached, and thus aid in the problem-solving process. See http://en.wikipedia.org/wiki/Heuristic.

[25] Lonergan's definition of emergent probability is as follows: "Emergent probability is the successive realization in accord with successive schedules of probability of a conditioned series of schemes of recurrence" (B.J.F. Lonergan. *Insight: A Study of Human Understanding*, eds. F.E. Crowe and R.M. Doran (Toronto: University of Toronto Press, 1992 [1957]), pp. 125–126).

[26] The third section of chapter 4 of *Insight* deals with a "clarification by contrast," in which Lonergan discusses the Aristotelian, Galilean, and Darwinian worldviews, as well as that of Indeterminism. See B.J.F. Lonergan. *Insight: A Study of Human Understanding*, eds. F. E. Crowe and R. M. Doran (Toronto: University of Toronto Press, 1992 [1957]), pp. 151–161.

[27] The language of "upwardly directed," of course, raises the question of a hierarchy in the natural world, an idea that lost favor along with earlier notions of teleology and the demise of the "great chain of being." The issue of how to conceive of increased complexity as "higher" but not necessarily "better" continues. See E. Mayr. *What Evolution Is* (New York: Basic Books, 2001), p. 278; and D.J. Depew and B.H. Weber. *Darwinism Evolving: Systems Dynamics and the Genealogy of Natural Selection* (Cambridge, MA: The MIT Press, 1995), p. 495.

human mind is oriented toward asking and answering questions, just which answers will emerge, whether one grasps the answers correctly or makes mistakes, is not pre-determined. What emerges is subject to probabilities.

The point of this exposition is that the dichotomy between an unfolding universe that is subject to determinism – "A" always and everywhere causes "B" – and a universe that unfolds according to "mere chance" is misguided. Appealing to *determinism* versus *chance* as the only two possible engines that drive the world's unfolding is simply a false dichotomy. There are some sets of events that can be explained by systematic patterns of relations that occur regularly, all things being equal. But whether and when such events occur is a matter of underlying conditions. In both cases there is some intelligibility to be grasped – appeals to chance occurrences are not appeals to nonsensical processes. Thus, the processes by which the world has emerged, and continues to unfold, are twofold, and an adequate account of this unfolding must involve accounting for both regularity and probability, both chance and necessity, in interaction with each other.

Why is this theologically important? At the very least it is important in rejecting the idea that there can only be purpose – and therefore divine intention or initiative – if the universe unfolds deterministically. The God tied to determinism is the Deistic God that came along with Newton's mechanics. This God has gone by the wayside, rightly so, in the wake of our understanding of probabilities as constitutive of world process. But there is still plenty of room for a transcendent God who works out his/her purposes through the interaction of regularity and probabilities.

Here it is important to distinguish two types of "contingency" – i.e. dependence or conditionality. Classical theism has always held that God is a (the only) *necessary* being, which is to say, God is not dependent on anything else for God's existence. God is the "pure act of being": in Aquinas God is the only self-subsisting being (ST I q2 and q3), in Lonergan's terms, God is an unrestricted act of understanding [5] (pp. 667–692). There is no "potentiality" in God. God already is what God is: God is not subject to development, emergence, or evolution.

In contrast, the world is wholly *contingent*, which is to say, it need not be. But we need to distinguish two ways in which our world is contingent. *Within* world process as it unfolds there are *necessary* (or *determinate*) relations – the regularities discovered by classical laws – which are nevertheless subject to *contingencies* – the conditions making the concrete operations of such laws possible – which are studied by statistical scientists. But in addition to this internal contingency, there is simply the brute fact of existence itself. This second kind of contingency has to do with the fact that there is something rather than nothing. In other words, God did not have to create anything. As a fully self-subsistent being God chose freely to create something, and the world God created happened to be a world in which regularity and conditionality, necessity and contingency, work themselves out over time. (God could have created a totally determined world but chose not to.) So there is the contingency that is an aspect of the world God created and there is the contingency of created being itself.

The importance of highlighting this distinction is to retain the independence of God from the world he/she creates. That God created something is a matter of God's free choice. That God created a world subject to the interaction of necessity and contingency, regularities subject to probabilities, is also a matter of God's free choice. Neither kind of contingency has an impact on the transcendence and necessity (God's self-subsistence) of God. Our supposedly modern discovery of chance need not impinge on our conception of God as wholly other and independent of his/her creation. What has been challenged is a deterministic worldview in which God's necessity and knowledge are presumed to yield pre-destined outcomes. It is the God of the Newtonian settlement that has been dethroned by the theory of evolution and its discovery of chance as constitutive of world process.[28] In a properly understood classical theism, God's omniscience and omnipotence and freedom are the ground of anything existing at all.

12.4 God, astrobiology, and the future of life

We have taken a long journey to come back to the question of God and astrobiology. How will our notion of God change if we find evidence of extraterrestrial life? The short answer is: not a whole lot. But that is presuming a view of emergence that allows for both novelty and continuity, and that presumes a Creator God who is not the Deist God of mechanistic determinism. My point throughout is that what is at stake in the discovery of extraterrestrial life is not our notion of God but our understanding of God, order, novelty, and purpose. There is indeed a major theological challenge for our time – but it has to do not with God and astrobiology but whether we can reconcile an evolving, emerging world with a God who is present yet transcendent, who created a cosmos with an order and orientation but without determining specific outcomes. If that is the case, and I have given a number of

---

[28] On the shift away from this clear distinction regarding two types of contingency in the years after Aquinas, see N. Ormerod. Chance and necessity, providence and God. *The Irish Theological Quarterly*, **70** (2005), 263–278, at 270; E. McMullin. Natural science and belief in a creator: historical notes. In *Physics, Philosophy and Theology*, eds. R.J. Russell, W.R. Stoeger, and G.V. Coyne (Vatican City: Vatican Observatory, 1988), pp. 49–79, at 60; and G.B. Deason. Reformation theology and the mechanistic conception of nature. In *God and Nature: Historical Essays on the Encounter Between Christianity and Science*, eds. D.C. Lindberg and R.L. Numbers (Berkeley, CA: University of California Press, 1986), pp. 167–191. On the Newtonian worldview and its effects on culture, see M.J. Buckley. The Newtonian settlement and the origins of atheism. In *Physics, Philosophy and Theology*, pp. 81–102; and M.C. Jacob. Christianity and the Newtonian worldview. In *God and Nature: Historical Essays on the Encounter Between Christianity and Science*, eds. D.C. Lindberg and R.L. Numbers (Berkeley, CA: University of California Press, 1986), pp. 238–255.

examples of contemporary approaches in Christian theology in which this is the case, discoveries of new forms of life will simply illustrate the breadth of God's reach in interacting with law and chance as the universe unfolds.

So my conclusion is that the theological challenges that astrobiology raises are challenging but not new. Now the question of *religious politics* is another matter. Not all Christians have accepted the need for a new synthesis in understanding God's relation to the world. Many accept a dichotomy between "chance" and "purpose" and believe that a Biblical literalism or, at best, an interventionist God of miracles is the only God who can combat the supposed meaninglessness of a world governed by probabilities. The lines drawn over issues of evolution will be writ even larger if novelty and emergence are discovered elsewhere in the universe.

My contention is that such a discovery will not challenge our notion of God as much as it will challenge our notions of ourselves. We have faced this dilemma before – Copernicus and Galileo dethroned the human; Darwin made us incidental products in the history of life. Slowly the human race is discovering that we are not the center of the universe, that both space and time are so vast that we are mere blips on the screen. *This* has archetypal impact and will not go down lightly. I see this as a critical moment to learn humility. But to the degree that the religious public has a certain narrative about how the world has unfolded, a narrative that includes divine determinism and/or intervention, and especially one that sees human life and salvation as the apex of meaning in this unfolding, the discovery of life elsewhere will cause huge problems.

Let me conclude with a final set of comments about the future of life. Just as the notion of emergent probability affects the way we think about God's agency, so too does it have implications for our understanding of human agency. On the one hand, we can celebrate the openness of the universe and discard as simply erroneous ideas that human interventions are in some way breaking unchanging laws determined by God. The human quest for understanding and for creating are examples of emergent probability, and are to be highly regarded and pursued. On the other hand, we ought not to deceive ourselves into thinking that *our* agency and creativity is a deterministic one. We are not agents of direct causality. We ought not to mistakenly think of our actions as a matter simply of action "A" necessarily causing outcome "B." We too are subject to the vicissitudes of emergent probability, and thus can initiate actions with unintended consequences. This means that the best we can do is to set up the conditions of possibility whereby the outcomes we hope for might emerge and have a good chance of surviving. Whether what we expect to occur does occur is a matter of probability, not of certainty. There may be a very high probability of a certain outcome, but such "certainty" is just one measure along a spectrum of probabilities.

We need to think of ourselves as living within an ethic of risk not an ethic of control. I say this in direct reference to the actions we take in "terraforming" or colonizing or exploring other planets. My caution is to point out that the conditions of possibility that we establish in the hopes of one outcome may *at the same time* establish conditions under which totally unforeseen schemes of recurrence become established. And since new schemes can yield great leaps in the probability of still further – perhaps unwelcome – novelty, we need to proceed with great caution. Thus my explanation of emergent probability yields an ethic of risk that grounds in turn an ecological ethics – no action yields only a single outcome. We need instead to think in terms of an ecology of moral action – a systemic network of outcomes (schemes of recurrence) resulting from best laid plans.

Finally, let me draw these comments out from a religious perspective. Acceptance of an ethic of risk may produce a caution with regard to our actions. But such cautions may not yield true humility. Indeed, military strategists, imperialists, dictators, and narcissists know all about risk, they just expect to manage it in their favor. What a religious perspective contributes here is an "otherworldly falling-in-love" [4] (pp. 104–107). Such falling in love with "that of which nothing greater can be thought" adds to an ethic of risk, an ethic of gratitude.[29] Calculating risks, managing probabilities, no matter how benign, is radically changed if one believes one is playing with gifts that have a sacred origin. To begin with love, gratitude, and humility and then consider manipulating probabilities yields a distinct deliberative process. To believe that one can encounter the Giver as well as make use of the gifts is what makes a worldview religious. This is not to deny the oppressive or dysfunctional use of religion in trying to control the world. Nor is it to insist that there cannot be moral agents or moral agencies (policy makers) without an explicitly religious approach. It is merely – in good Judeo-Christian fashion – to recognize ourselves as creatures rather than creators and to at least acknowledge that the question of a Creator or Giver of gifts is a valid question as one makes plans to affect the future of life in the universe.

References

[1] P. Davies. E. T. and God. *The Atlantic Monthly* (September 2003), pp. 112–118, available online at: www.theatlantic.com/issues/2003/09/davies.htm.

[2] I. Barbour. *Religion and Science: The Historical and Contemporary Issues* (San Francisco, CA: Harper Collins, 1997).

---

[29] The notion of God as "that of which nothing greater can be thought" comes from St. Anselm (d. 1109). For more on the relationship between risk and gratitude, see C.S.W. Crysdale. Risk, gratitude, and love: grounding authentic moral deliberation. In *The Importance of Insight: Essays in Honour of Michael Vertin*, eds. D. Liptay and J. Liptay (Toronto: University of Toronto Press, 2007), pp. 151–171.

[3] W. Paley. *Natural Theology, or Evidences of the Existence and Attributes of the Deity, Collected from the Appearances of Nature* (Boston, MA: Gould and Lincoln, 1851).

[4] B.J.F. Lonergan. *Method in Theology* (Toronto: University of Toronto Press, 2001 [1971]).

[5] B.J.F. Lonergan. *Insight: A Study of Human Understanding*, eds. F.E. Crowe and R.M. Doran (Toronto: University of Toronto Press, 1992 [1957]).

Part III

Future of life

13

Planetary ecosynthesis on Mars: restoration ecology and environmental ethics

Christopher P. McKay

13.1 Introduction

There has been a lively discussion recently about the science and ethics of "terraforming" Mars. The high level of interest is a result of spacecraft discoveries about Mars combined with the realization that humans are effectively warming the Earth and wondering if they can, and should, do the same on Mars. I suggest that terraforming is more appropriately called planetary ecosynthesis, and in this chapter I review the scientific studies of planetary ecosynthesis and the environmental ethics associated with instigating such global change on another planet.

Mars today is a cold, dry, frozen desert world on which not even the most hardy of Earth life could survive. Temperatures average $-60\,°C$ and the pressure averages $0.6\,kPa$, over one hundred times less than atmospheric pressure at the surface of the Earth. As a result of the low pressure, and secondarily the low temperature, water is not liquid on the surface of Mars at any location or season. Strong solar ultraviolet radiation reaches the surface of Mars to complete the deadly mix of hostile environmental conditions.

But Mars has not always been this harsh. There is compelling evidence that early in its history Mars had stable liquid water on its surface [1, 2]. Presumably this phase of liquid water was associated with a higher pressure and somewhat warmer atmosphere. This evidence for liquid water on Mars – originally from the Mariner 9 and Viking missions and now confirmed by recent missions (see Figure 13.1) – is the central motivation for the search for past life on Mars [33, 34].

The fact that Mars once supported widespread liquid water and possibly life on its surface opens the question of the feasibility of restoring such conditions on Mars by artificial means. The fundamental challenge of restoring habitable conditions on Mars is to warm up the planet from its current $-60\,°C$ to over $0\,°C$, and perhaps

Exploring the Origin, Extent, and Future of Life: Philosophical, Ethical, and Theological Perspectives, ed. Constance M. Bertka. Published by Cambridge University Press. © Cambridge University Press 2009.

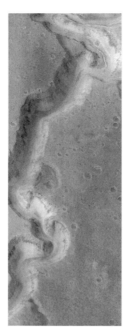

Figure 13.1 Liquid water in the past on Mars. Mars Global Surveyor image showing Nanedi vallis in the Xanthe Terra region of Mars. Image covers an area 9.8 km by 18.5 km; the canyon is about 2.5 km wide. This image is the best evidence we have that some of the fluvial features on Mars were carved by liquid water in stable flow on the surface for an extended interval. Photo from NASA/ Malin Space Sciences.

as warm as +15 °C – reaching parity with Earth. Humans have demonstrated, and implemented, the technology to warm planets with Earth as our first target. The level of human-induced warming on Earth is debated but is probably on the order of a few degrees. On Mars the warming needed would be tens of degrees – many times larger than on Earth – but the extrapolation from Earth to Mars is conceptually straightforward. Energy balance calculations suggest that warming Mars might be achieved in 100 years or less [3, 4]. Producing an oxygen-rich atmosphere would take more than 100,000 years. Thus, warming Mars is within current technology and this fact frames the discussion about Mars in a fundamentally different way than planetary scale environmental alteration on any other world of the solar system. Because the question of "can we" has been tentatively answered for Mars in the affirmative, the question of "should we" and "will we" warrant consideration.

The scientific issues associated with planetary ecosysnthesis on Mars have been discussed in the scientific literature for decades. Initial discussions were limited [5, 6, 7, 8] until "Making Mars Habitable" was featured on the cover of *Nature*

Table 13.1 *Habitability (adapted from McKay et al. 1991 [3])*

Parameter	Limits	Note
Global temperature	0–30 °C	Earth = 15 °C
Composition for plants, algae, microorganisms		
Total pressure	>1 kPa	Water vapor pressure plus O_2, N_2, CO_2
CO_2	>0.015 kPa	Lower limit set by photosynthesis No clear upper limit
N_2	>0.1–1 kPa	Nitrogen fixation
O_2	>0.1 kPa	Plant respiration
Composition for breathable air		
Total pressure:		
Pure O_2	>25 kPa	Lung water vapor plus CO_2, O_2
Air mixture	>50 kPa	Based upon high elevation
	<500 kPa	Buffer gas narcosis
CO_2	<1 kPa	Set by toxicity
N_2	>30 kPa	Buffer gas
O_2	>13 kPa	Lower limit set by hypoxia
	<30 kPa	Upper limit set by flammability

magazine and the field was rightly considered to have entered mainstream scientific discussion [3]. There are currently two international science journals, *Astrobiology* and the *International Journal of Astrobiology*, that explicitly consider planetary ecosynthesis as part of their content.

13.2 Conditions needed for habitability

The first step in considering ecosynthesis on Mars is the delineation of the requirements for habitability. We tend to think of the present Earth as the only model for a habitable world. However, there are two alternative possibilities for life supporting states for Mars, one with oxygen and one without. These two alternative states are listed in Table 13.1, adapted from McKay *et al.* [3]. Life could survive on Mars in an atmosphere composed of carbon dioxide with moderate levels of nitrogen and low levels of oxygen. Such an atmosphere is thought to have prevailed on Earth before the rise of oxygen 2 Gyr ago. It is also the likely composition of a thick early atmosphere on Mars. Many bacteria, some plants and even a few animals can survive in low oxygen atmospheres. Humans would require a source of oxygen and could not breathe the high carbon dioxide.

Table 13.2 *Comparison of Mars and Earth*

Parameter	Mars	Earth
Mass	0.107	1
Surface pressure	0.5 to 1 kPa	101.3 kPa
Average temperature	$-60\,°C$	$+15\,°C$
Temperature range	$-120\,°C$ to $+25\,°C$	$-80\,°C$ to $+50\,°C$
Atmosphere composition	95% CO_2	78% N_2
	2.7% N_2	21% O_2
	1.6% Ar	1% Ar
Incident sunlight	$149\,W\,m^{-2}$	$344\,W\,m^{-2}$
Surface gravity	$3.73\,m\,s^{-2}$	$9.80\,m\,s^{-2}$
Solar day	24 h 39 m 35.238 s ("sol")	24 h
Sidereal year	687 days, 668.6 sols	365.26 days
Obliquity of axis	25 deg	23.5 deg
Eccentricity	0.0934	0.0167
Mean distance to Sun	1.52 AU	1 AU (1.49×10^8 km)

A second habitable state to consider is an oxygen-rich atmosphere that is essentially the same as on the present Earth. As we discuss below it appears feasible to create a habitable state on Mars based on a carbon dioxide-rich atmosphere but it is not feasible to create an oxygen-rich atmosphere.

13.3 What went wrong with Mars?

There is compelling evidence that Mars had more habitable conditions in the past. However, it is not clear what happened to the atmosphere and hydrosphere of early Mars. It is generally thought that Mars lost its carbon dioxide atmosphere through a combination of processes all related to its small size (Mars is 1/10 the mass of the Earth, Mars and Earth are compared in Table 13.2). These processes include the formation of carbonates, loss to space due to solar wind sputtering, and atmospheric erosion due to impacts of comets and asteroids. The relative importance of these processes is being debated [35] but these processes all occur on Earth as well. They are more pronounced on Mars because it is so much smaller than the Earth.

Earth is large enough that its internal heat flow can drive plate tectonics. Mars has a single thick plate. As a result Mars does not have the recycling of material that results from the subduction of one plate under another and the ejection of gases in the resulting arc volcanoes. The gases emitted from arc volcanoes such as Mt. St. Helens in the Cascade Range represent the recycling of material into the atmosphere by plate tectonics. Mars is too small to have plate tectonics and hence has no way to recycle materials such as carbonates. The lack of recycling and high loss rates due to lower gravity and no magnetic field are thought to be responsible for the loss of most of the carbon dioxide of the early Martian atmosphere. The amount of CO_2 still present on Mars is unknown.

There may have also been some loss of the initial water on Mars but a large part of it is still present on Mars – frozen into the ground in the polar regions. Based on the morphology of the craters in the polar regions of Mars, it was deduced that the ground there is rich in ice [10]. Direct confirmation came from the gamma ray and neutron detectors on the Mars Odyssey mission, which indicated the presence of ice-rich ground in the upper two meters for latitudes poleward of about 60° in both hemispheres [11]. Radar results indicate that this ice-rich ground extends down to 2 km or more in the northern hemisphere [12]. A veritable frozen ocean of water is still present on Mars.

The factors that resulted in Mars' loss of habitability are related to its small size and resulting lack of plate tectonics. These are not factors that can be changed with foreseeable technologies. Therefore, one possible objection to ecosynthesis on Mars is that it would be doomed to fail over geological time due to the same factors that doomed an initial habitable environment on Mars. The logic of this argument is correct: Mars newly restored to habitability would only have a finite lifetime. This lifetime would be approximately given by the timescale of the removal of the atmosphere due to carbonate formation, about 10 to 100 million years. This is a short time compared to the age of Mars – 4.5 billion years – but a long time compared to human timescales. It is relevant here to note that Earth will not remain habitable for much longer than this. Current estimates suggest that Earth will become uninhabitable in 500 million years or less as it becomes Venus-like due to the progressive brightening of the Sun [13]. Thus, a habitable Mars that persists for 10 to 100 million years is "Earth-like" in terms of its life expectancy. No solutions for infinite lifetime exist for Mars or Earth. Nothing lasts forever, not even the Earth and sky (despite the popular song by the rock group Kansas to the contrary).

As discussed above, it appears certain that Mars still has a vast amount of water, albeit frozen in the ground. The total CO_2 amount on Mars, as carbonate, absorbed gas in the soil, or frozen in the polar caps is unknown. There is only about 0.6 kPa of CO_2 in the atmosphere at the present time but estimates of the total CO_2 in the soil, atmosphere, and cap range from as low as a few kPa to as high as 100 kPa [14].

A biosphere requires large amounts of CO_2 and H_2O but also N_2. Nitrogen gas is essential for a breathable atmosphere (see Table 13.1) and nitrogen is needed by life as an essential macronutrient. The only known supply of nitrogen on Mars is in the atmosphere at a level of 0.016 kPa, a tiny amount compared to the 80 kPa of N_2 in Earth's atmosphere. If this nitrogen were entirely converted to biological material it would form a layer only 1 cm thick. Mars cannot support a biosphere if the current atmospheric N_2 is the total nitrogen available on the planet. Unfortunately, there are no data on the amount of nitrogen in the soil of Mars as nitrate. Theoretical arguments suggest that lightning and meteorites should have produced nitrates on Mars and there may be up to 30 kPa present [14, 15]. The question of the nitrogen supply is probably the key question in terms of the feasibility of ecosynthesis on Mars using near-term technologies.

Mars does not have a planetary magnetic field but probably does not need one to be habitable. There are two primary reasons that a magnetic field is sometimes proposed as required for habitability: (1) to provide shielding against radiation and (2) to prevent solar wind erosion of a thick Martian atmosphere.

13.4 Radiation protection

The Earth's magnetic field does not deflect galactic cosmic rays because these particles are much too energetic. These particles are primarily stopped by the mass of the Earth's atmosphere, which is equivalent to 1 kg cm^{-2}. The Earth's magnetic field deflects solar protons, channeling these particles to the polar regions, creating the aurora. However, even without the magnetic field these particles would not penetrate the Earth's atmosphere and would not reach the surface.

If Mars had an Earth-like surface pressure of 1 atm, its atmospheric mass would be 2.6 kg cm^{-2} due to the lower gravity (to reach the same surface pressure with a lower gravity, $0.38\,g$, requires a more massive atmosphere). Thus, the radiation shielding effects of the Martian atmosphere would exceed those for the Earth and a magnetic field is not essential for radiation protection. Furthermore, Earth occasionally loses its strong dipole field during field reversals. These events are not correlated with any increases in extinctions in the fossil record.

13.5 Atmospheric erosion

Because Mars (and Venus) do not have magnetic fields, the solar wind impacts directly on the upper atmospheres of these planets. This does result in a small rate of atmospheric loss at the present time. However, the loss rate would not increase if we increased the surface pressure of the Martian atmosphere. This is due to the fact that conditions at the top of a thicker atmosphere would be similar to the conditions at the top of the present atmosphere only raised by a small elevation. For example, if the surface pressure on Mars were to increase to one atmosphere, the low pressure regions of the atmosphere would be raised in altitude. We can estimate the height change by computing the scale height in a warm Earth-like Martian atmosphere (because scale height is inverse with gravity and Mars' gravity is 0.38 times Earth's, and inverse with mean molecular weight; Mars 44, Earth 29, the scale height on Mars would be 14 km, compared to 8 km on Earth). To increase the pressure on Mars from 0.6 kPa to 100 kPa requires a pressure increase of 166 or

5.1 scale heights ($e^{5.1} = 164$) resulting in an altitude gain of 71 km for the upper atmosphere. This is a tiny increment compared to the radius of the planet. Thus, the top of the atmosphere would feel essentially the same gravity as it does today and would feel the solar wind at the same intensity. The net result is that the erosion of gases from the Martian atmosphere by the solar wind would remain unchanged. The current loss rate is not significant; for example the loss rate of water on Mars today corresponds to the loss of a layer of water two meters thick over 4 billion years (for example, see [36]).

13.6 Energy and time requirements

The discussion above shows that the necessary materials to construct a biosphere are likely to be present on Mars and that, in addition, the fundamental physical aspects of Mars that would be virtually impossible to alter, such as axial tilt, rotation rate, and eccentricity, are similar to the corresponding values for Earth (see Table 13.2). The one exception is the surface gravity, which is 0.38 of the Earth value. It is typically assumed that life from Earth can accommodate this lower gravity but this has not yet been tested.

If the physical materials are present, the next question in determining the feasibility of planetary ecosynthesis on Mars is to compute the energy and time required to affect the desired change. The problem naturally divides into two phases [3, 4]. Phase 1 is warming Mars from the present cold state and restoring the thick atmosphere of CO_2. Phase 2 is the production of O_2 in sufficient quantities to be breathable by humans. The energy requirements of each phase are listed in Table 13.3 and are also expressed in terms of the sunlight incident on Mars. Trapping and using sunlight is the only plausible energy source for changing the environment on a global scale. From Table 13.3 we can see that if the sunlight incident on Mars could be utilized with 100% efficiency it would take only ~10 years to warm Mars and restore the thick CO_2 atmosphere. Clearly 100% efficiency is an overestimate but atmospheric supergreenhouse gases, as discussed in the next section, can effectively alter the energy balance of a planet and efficiencies of 10% are plausible. Thus, the timescale for warming Mars is ~100 years.

To produce a level of breathable O_2 from CO_2 on Mars would require the energy equivalent of 17 years of Martian sunlight. While the energy level here is comparable to the energy to warm Mars, the efficiency is much lower. Because of basic thermodynamic constraints, the efficiencies for warming are much higher than the efficiencies for causing chemical reactions. Sunlight will not spontaneously produce O_2 from CO_2: a mechanism to drive the reaction is required. The only known mechanism that can operate on a global scale and use sunlight to convert CO_2 to O_2 is a biosphere. This conversion is precisely the reaction that produces biomass in

Table 13.3 *Energy requirements for terraforming Mars*

Initial state	Final state	Amount	Energy $[J\,m^{-2}]$	Solar energy[a] [years]	Time [years]
Surface warming					
$CO_2(s)$ at $-125\,°C$	$CO_2(g)$ at $15\,°C$	$200\,kPa$; $5.4 \times 10^4\,kg\,m^{-2}$	3.7×10^{10}	7.9	
Dirt at $-60\,°C$	Dirt at $15\,°C$	$\sim 10\,m$; $2 \times 10^4\,kg\,m^{-2}$	1.2×10^9	0.3	
$H_2O(s)$ at $-60\,°C$	$H_2O(l)$ at $15\,°C$	$10\,m$; $1 \times 10^4\,kg\,m^{-2}$	5.5×10^9	1.2	
$H_2O(s)$ at $-60\,°C$	$H_2O(g)$ at $15\,°C$	$2\,kPa$; $5.4 \times 10^2\,kg\,m^{-2}$	1.6×10^9	0.33	
			Total:	10	100
Deep warming					
$H_2O(s)$ at $-60\,°C$	$H_2O(l)$ at $15\,°C$	$500\,m$; $5 \times 10^5\,kg\,m^{-2}$	2.8×10^{11}	56	500
Making O_2					
$CO_2(g) + H_2O$	$CH_2O + O_2(g)$	$20\,kPa$; $5.4 \times 10^3\,kg\,m^{-2}$	8×10^{10}	17	100 000

[a] Energy divided by the total solar energy reaching Mars in a year, $4.68 \times 10^9\,J\,m^{-2}\,yr^{-1}$. Adapted from McKay *et al.* [3].

the Earth's biosphere. Given that the Earth has an extensive biosphere that has been utilizing this reaction for billions of years, the efficiency of the Earth's biosphere is a plausible upper limit to the efficiency of a Martian biosphere. The production of biomass on Earth corresponds to the energy equivalent of 0.01% of the energy incident on the Earth as sunlight. Assuming this optimistic efficiency it would take over 100,000 years to produce the minimal breathable O_2 levels on Mars, as defined in Table 13.1. Clearly this efficiency for Earth is an average over all of the Earth's biomes with widely different individual efficiencies, from dry deserts to lush rain forests. If a planet could be entirely covered with rain forests then clearly the efficiency would go up, maybe as much as a factor of 10. However, a global biosphere on Mars is likely to have a range of ecosystems just as does the Earth. The long timescale indicated by this calculation is not a precise prediction but it is a robust indicator that producing a breathable O_2 atmosphere on Mars will take many thousands, or hundreds of thousands, of years.

Energetic considerations indicate that warming Mars and restoring a thick CO_2 atmosphere could be accomplished over human timescales (~ 100 years). Altering the atmosphere to make it breathable to humans ($\sim 20\%\ O_2$) is not possible with present technologies for all intents and purposes.

13.7 Warming Mars

There have been many proposed methods for warming Mars [17] but the only one that is clearly rooted in demonstrated technology is based on supergreenhouse gases. Originally suggested by Lovelock and Allaby [16] and worked out in detail by Marinova *et al.* [9] the basic idea is to use supergreenhouse gases to increase the greenhouse effect on Mars. Marinova *et al.* [9] considered the production on Mars of supergreenhouse gases composed only of F, S, C, and H. Supergreenhouse gases containing Cl and Br were specifically excluded from consideration because of the deleterious effect these chemical species have on ozone. Climate calculations have shown that if there is CO_2 ice present in the polar regions of Mars, then a warming of 20 °C will cause the complete evaporation of that ice through a positive feedback mechanism [3, 16]. The results of the atmospheric energy balance calculations of Marinova *et al.* [9] indicate that the production of fluorine-based gases at levels of 0.1 to 1 Pa is a possible way to increase the mean Mars temperature to the levels necessary to cause the outgassing and evaporation of all available CO_2 ice on Mars. Figure 13.2 shows the temperature increase for the present Mars for several gases. Of the gases considered, C_3F_8 was the most potent artificial greenhouse gas for use on the present Mars. Less than 1 Pa of C_3F_8 (a few ppm in an Earth-like atmosphere) would result in sufficient warming of Mars to cause complete CO_2 outgassing.

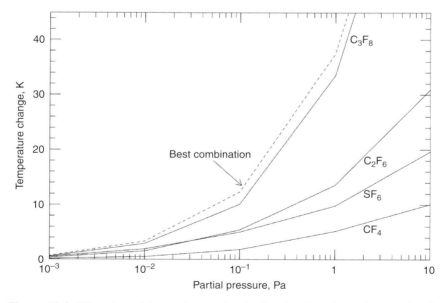

Figure 13.2 Warming of Mars due to simple fluorine-based gas independently and the best gases combination (dashed line) for the given total greenhouse gas amounts ($pCO_2 = 0.6$ kPa) from Marinova *et al.* [9].

Releasing CO_2 is a key step in warming Mars because CO_2 contributes to the greenhouse effect and thus a positive feedback on further CO_2 release. Furthermore, CO_2 is the most readily available gas on Mars to bring the pressure up high enough that liquid water can be stable.

13.8 Ethics

In the previous sections I have briefly summarized the state of our scientific understanding of the possibilities of planetary ecosynthesis on Mars. More detailed reviews can be found in McKay *et al.* [3], Fogg [17], Marinova *et al.* [9] and Graham [18]. I suggest the following conclusions based on the scientific literature:

1. It is likely that Mars has adequate amounts of CO_2, H_2O, and N_2 for the construction of a biosphere.
2. The timescale for warming Mars and restoring a thick CO_2 atmosphere is relatively short (~100 years). However, it is not practical to create a breathable (~20% O_2) atmosphere on Mars.
3. If restored to habitability, Mars would maintain its habitable state for 10 to 100 million years.
4. Supergreenhouse gases based on F, S, C, and H could be produced on Mars and used to warm the planet. Human-produced supergreenhouse gases are currently warming the Earth. Concentrations 100 to 1000 times higher than the anthropogenic levels in Earth's atmosphere are needed on Mars.
5. The scientific community considers planetary ecosynthesis on Mars as a serious topic in space research.

The fact that we are altering the Earth's global climate and the possibility that using the same methods we "can" alter the habitability of Mars implies that the question of "should" we conduct planetary ecosynthesis is a timely and relevant one. The environmental ethics of planetary ecosynthesis on Mars is therefore a subject for consideration and in the rest of the chapter I provide some preliminary observations.

Environmental ethics has developed on Earth and mostly in response to environmental crises. It is perhaps not surprising that extrapolating environmental ethics to an apparently lifeless world is not straightforward. The universal yet unexamined assumption of environmental ethics is that nature is equivalent to life. On Earth this equivalence is obviously true. No aspect of the environment on Earth can be considered separate from biological effects. Most systems of environmental ethics have not attempted to make any meaningful distinction between nature and life. Consider as an example the "Land Ethic" articulated by Aldo Leopold [19]. "A thing is right when it tends to preserve the integrity, stability, and beauty of the biotic community. It is wrong when it tends otherwise." As I have discussed before [21] it is unclear how to apply this to Mars. Indeed as we leave the Earth we see that the equivalence of life and nature is, as far as we now know, only applicable to Earth. Everywhere else in our solar system we encounter nature that is profoundly devoid of life. The fundamental challenge in applying environmental ethics beyond the Earth is to delve one level deeper than has been necessary on Earth and examine the difference between the environmental ethics of nature and the environmental ethics of life.

I have argued [21] that to understand the difference between nature and life in a system of environmental ethics requires a clarification of the fundamental assumptions – the normative axioms – on which the system is based. Systems of environmental ethics are based on varying combinations of three normative axioms, which I identify as:

1. Preservationism – the fundamental principle that nature is not to be altered by human beings.
2. Wise stewardship – the fundamental principle that the measure of all things is utility to humans, in the broadest and wisest sense of utility.
3. Intrinsic worth – the fundamental principle that there exist sets of objects which have intrinsic worth regardless of their instrumental value to humans.

These are principles on which systems of environmental ethics are based and not categories into which systems of environmental ethics are grouped. Furthermore, they are not mutually exclusive. Indeed, virtually all systems of environmental ethics are based on some combination of these principles in varying degrees. Anti-humanism as expressed by Ehrenfeld [20] is an extreme form of preservationism in which all human actions are to be suppressed. Anti-humanism was the original label used by McKay [21]. Rolston [22] and Hargrove [23] provide a more balanced view of preservationism and suggest the term now used here.

Wise stewardship is certainly the most common and influential fundamental principle in environmental ethics. It is so pervasive that it is often assumed without statement. The most common argument in environmental ethics is that we humans jeopardize our own interests when we degrade the environment.

The third fundamental principle listed above is relatively recent and not fully developed in the environmental ethics literature. The principle of intrinsic worth posits that there are sets of objects which have value of themselves independent of any instrumental or intellectual utility to other beings. The principle itself has wide adherence if the set of intrinsic objects is restricted to human beings. Historically the restrictions were more severe, but over time the sphere of consideration has expanded along the dimensions of race, gender, disability, and ethnicity so that now virtually everyone would agree that all humans are in the set of objects with intrinsic worth. There have been serious arguments presented to expand the sphere of consideration to non-human animals [24], to ecosystems [25], and to all life itself [26, 27]. In all cases the discussion has been in the context of life forms of intrinsic worth pre-existing within the state of nature. No choice between nature and life forms of intrinsic worth has been considered. However, this is just the choice we face on Mars. The state of nature is lifeless and we have before us a choice to replace that state with a state with life forms of intrinsic worth. If richness and diversity in life forms is a value in itself, then planetary ecosynthesis on Mars is a good thing.

13.9 Utilitarian motivations for planetary ecosynthesis on Mars

Given the importance of the utilitarian principle in environmental ethics it is instructive to consider how planetary ecosynthesis on Mars is – or is not – useful to humans.

Mars is not useful as a "lifeboat" to which humanity flees after having destroyed the Earth. This is a common idea in science fiction but has no technical basis. Past migrations of people have occurred in which a significant fraction of the population relocates over a great distance. An example is the Irish migration to North America. However, this example does not hold for space travel. Any foreseeable technology for space travel will involve only a very small number of travelers.

The advantages of a small number of humans and assorted life forms from Earth present on Mars as a genetic "Noah's ark" are dubious as well. There are certainly possible scenarios, such as a large asteroid strike or unchecked plague, in which large fractions of the human population on Earth are killed, but even if the survival rate on Earth was one in a million the residual population would be huge compared to any plausible population on Mars. Events that could literally sterilize the Earth – such as a nearby supernova and the Sun entering the red giant stage – would also sterilize Mars.

Finally the idea that a human colony present on Mars or the prospect of making Mars habitable would allow us, or cause us, to disregard environmental principles on Earth is absurd.

A more plausible utilitarian value that can be obtained from planetary ecosynthesis on Mars is the knowledge that we, human society, would derive from such activities. Humans have a global effect on the environment of Earth and our effects are growing as our population and technology increase. It is therefore inevitable that humans will assume some management responsibility for the Earth's environment. The view that the environment is self-regulating and does not require human intervention is not tenable if humans continue to have such a growing impact. Understanding the Earth well enough to manage it is a daunting task. Clearly the knowledge needed for this task will primarily be obtained from studies of the

Earth itself. However, studies of other planets can provide important comparison points and context for the Earth. Along these lines, studies of planetary ecosynthesis on Mars could provide important lessons as to how a biosphere can work. The words left on Richard Feynman's blackboard on his last day of work "What I cannot create I do not understand" express this point. An essential task in understanding, and managing the biosphere on Earth, may be the creation of a biosphere on Mars.

The potential knowledge to be gained by studying ecology on Mars is greatly increased if Mars has indigenous life that represents a separate origin from life on Earth [28]. Throughout this discussion I have assumed that Mars is a lifeless planet. However, it is possible that Mars has life either in secluded habitable zones or frozen dormant in the polar ices. If there is life on Mars, or was life, it may share a common origin with life on Earth or it may represent an independent origin of life. As recently as a decade ago it was assumed that if there was life on Mars it would have to represent an independent origin since it was present on a different planet. However, we now know that rocks from Mars have reached Earth intact and furthermore that the temperatures in these rocks would not reach sterilizing levels during the trip [29]. These rocks are the result of impacts on Mars ejecting material into space. Early in the history of the solar system the impact, and hence transfer rate, would have been much higher. Thus, life from Mars could have been carried to Earth inside one of these rocks. Presumably the converse is also true – rocks from Earth could have carried life from Earth to Mars. Because of this interplanetary rock transfer it is no longer assumed that life on Mars would necessarily be of a separate origin. Indeed most researchers would consider that the most likely case would be a common origin for life on Earth and Mars and convincing evidence to the contrary would be required before it was concluded that Mars had an independent origin of life.

If there has never been life on Mars then there are minimal implications for planetary ecosynthesis. If there was life on Mars and it is now extinct beyond recovery, then planetary ecosynthesis can be viewed as a type of "restoration ecology." If there is life on Mars, or recoverable life, but it shares a common ancestor with life on Earth then it seems plausible that planetary ecosynthesis can proceed using Earth life forms as needed.

Perhaps the most interesting and challenging case is that in which Mars has, or had, life and this life represents a distinct and second genesis [30, 31]. The discovery of a second genesis of life has profound scientific, as well as philosophical and ethical importance. Philosophically, the discovery would directly address the question of life in the universe, and would strongly support the idea that life is a naturally emergent phenomenon and is widespread and diverse in the universe. Scientifically, having another example of life expands the scope of biology from one to two. There may well be significant advances in medicine, agriculture, pest control, and many other fields of biological inquiry, from having a second type of life to study. I would argue that if there is a second genesis of life on Mars, its enormous potential for practical benefit to humans in terms of knowledge should motivate us to preserve it and to enhance conditions for its growth. Observations of Mars show that currently there is no global biosphere on that planet and if life is present it is in isolated refugia or dormant. It is possible that life present on Mars today is at risk of extinction if we do not alter the Martian environment so as to enhance its global habitability.

An appreciation for the potential utility and value of the restoration of a Martian biota does not depend on the assignment of intrinsic value to alternative life forms. The creation of a second biosphere using a second genesis of life could be of great utilitarian value for humans in terms of the knowledge derived ranging from basic biology to global ecology. And a case can be made [30] that its value exceeds the opportunity cost of not establishing human settlements on Mars.

The utilitarian arguments presented above indicate that we should alter Mars to allow any indigenous life to expand and form a global biosphere even if the resulting biosphere is never a natural home for life from Earth or humans. If there is no indigenous life, these utilitarian arguments indicate that we should alter Mars to support life from Earth even if this never results in a biosphere that can be a natural home for humans. The point is only a theoretical one since our current understanding of Mars and planetary physics suggests that it is not possible using foreseeable technology to make Mars into a world that can be an Earth-like home for humans. Humans would require some sort of O_2 source to move around the planet – a significant improvement in habitability compared to the current state.

This discussion has implications for near-term exploration of Mars by robots and humans. Until we know the nature of life on Mars and its relationship – if any – to life on Earth, we must explore Mars in a way that keeps our options open with respect to future life. I have argued elsewhere [31, 32] that this means that we must explore Mars in a way that is biologically reversible. Exploration is biologically reversible if it is possible and practical to remove all life forms carried to Mars by that exploration. Because of the high UV and oxidizing conditions on Mars, biological reversibility is achievable. Previous missions to Mars, such as the Pathfinder mission and the two MER rovers, have carried microorganisms to the Martian surface where they remain dormant as long as shielded from ultraviolet radiation. To reverse this contamination already present on Mars, it would be necessary to collect all metal objects within which microbes could remain viable. Furthermore, the soil at crash sites and in the vicinity of landers that had come into contact with the spacecraft would have to be thrown up into the atmosphere where it would be exposed to sterilizing ultraviolet radiation. A similar approach can be used to reverse the contamination from human bases.

13.10 Summary

Planetary ecosynthesis on Mars is being seriously discussed within the field of planetary science. It appears that restoring a thick atmosphere on Mars and the recreation of an environment habitable to many forms of life is possible. It is important now to consider if it "should" be done. To do this takes us into new and interesting territory in environmental ethics but both utilitarian and intrinsic worth arguments support the notion of planetary ecosynthesis. Strict preservationism arguments do not. It is important to have the long-term view of life on Mars and the possibilities of planetary ecosynthesis. This affects how we explore Mars now. Mars may well be our first step out into the biological universe, it is a step we should take carefully.

References

[1] M. H. Carr. *Water on Mars* (Oxford: Oxford University Press, 1996).

[2] M. H. Carr and M. C. Malin. Meter-scale characteristics of Martian channels and valleys. *Icarus*, **146** (2000), 366–386.

[3] C. P. McKay, O. B. Toon, and J. F. Kasting. Making Mars habitable. *Nature*, **352** (1991), 489–496.

[4] C. P. McKay and M. M. Marinova. The physics, biology, and environmental ethics of making Mars habitable, *Astrobiology*, **1** (2001), 89–109.

[5] C. Sagan. Planetary engineering on Mars. *Icarus*, **20** (1973), 513–514.

[6] J. A. Burns and M. Harwit. Towards a more habitable Mars – or – the coming Martian Spring. *Icarus*, **19** (1973), 126–130.

[7] M. M. Averner and R. D. MacElroy (eds.). *On the Habitability of Mars* (NASA-Ames Research Center, Moffett Field, CA: NASA Special Publication, 1976), p. 414.

[8] C. P. McKay. Terraforming Mars. *Journal of the British Interplanetary Society*, **35** (1982), 427–433.

[9] M. M. Marinova, C. P. McKay, and H. Hashimoto. Radiative-convective model of warming Mars with artificial greenhouse gases. *Journal of Geophysical Research*, E03002, doi:10.1029/2004JE002306 (2005), p. 110.

[10] S. W. Squyres and M. H. Carr. Geomorphic evidence for the distribution of ground ice on Mars. *Science*, **231** (1986), 249–252.

[11] W. C. Feldman, W. V. Boynton, R. L. Tokar *et al.* Global distribution of neutrons from Mars: results from Mars Odyssey. *Science*, **297** (2002), 75–78.

[12] G. Picardi *et al.* Radar soundings of the subsurface of Mars. *Science*, **310** (2005), 1925–1928.

[13] K. Caldeira and J. F. Kasting. The life span of the biosphere revisited. *Nature*, **360** (1992), 721–723.

[14] C. P. McKay and C. R. Stoker. The early environment and its evolution on Mars: implications for life. *Reviews of Geophysics*, **27** (1989), 189–214.

[15] C. V. Manning, C. P. McKay, and K. J. Zahnle. The nitrogen cycle on Mars: impact decomposition of near-surface nitrates as a source for a nitrogen steady state. *Icarus*, **197** (2008), 60–64.

[16] J. E. Lovelock and M. Allaby. *The Greening of Mars* (New York: Warner Books, Incorporated, 1984).

[17] M. J. Fogg. *Terraforming: Engineering Planetary Environments* (Warrendale, PA: SAE International, 1995).

[18] J. M. Graham. The biological terraforming of Mars: planetary ecosynthesis as ecological succession on a global scale. *Astrobiology*, **4** (2004), 168–195.

[19] A. Leopold. *A Sand County Almanac: With Essays on Conservation from Round River* (New York: Ballantine Books, 1970 [1949]).

[20] D. W. Ehrenfeld. *The Arrogance of Humanism* (New York: Oxford University Press, 1978).

[21] C. P. McKay. Does Mars have rights? An approach to the environmental ethics of planetary engineering. In *Moral Expertise*, ed. D. MacNiven (New York: Routledge, 1990), pp. 184–197.

[22] H. Rolston III. The preservation of natural value in the solar system. In *Beyond Spaceship Earth: Environmental Ethics and the Solar System*, ed. E. C. Hargrove (San Francisco, CA: Sierra Club Books, 1986), pp. 140–182.

[23] E. Hargrove. *Foundations of Environmental Ethics* (Englewood Cliffs, NJ: Prentice-Hall, 1989).

[24] P. Singer. *Practical Ethics* (Cambridge: Cambridge University Press, 1979).

[25] C. D. Stone. *Should Trees Have Standing?* (Los Altos, CA: William Haufmann, 1974).

[26] A. Naess. The shallow and the deep, long-range ecology movement: a summary. *Inquiry*, **16** (1973), 95–100.

[27] A. Naess. A defense of the deep ecology movements. *Environmental Ethics*, **6** (1984), 265–270.

[28] C. P. McKay. The search for extraterrestrial biochemistry on Mars and Europa. In *Exobiology: Matter, Energy, and Information in the Origin and Evolution of Life in the Universe*, eds. J. Chela-Flores and F. Raulin (Dordrecht: Kluwer, 1998), pp. 219–227.

[29] B. P. Weiss, J. L. Kirschvink, F. J. Baudenbacher, *et al.* A low temperature transfer of ALH84001 from Mars to Earth. *Science*, **290** (2000), 791–795.

[30] C. P. McKay. Let's put Martian life first. *The Planetary Report,* **21** (July/August 2001), 4–5.

[31] C. P. McKay. Hard life for microbes and humans on the Red Planet. *Ad Astra,* **19**(3) (Summer 2007), 30–31.

[32] C. P. McKay. Biologically reversible exploration. *Science*, **323** (2009), 718.

[33] C. P. McKay. Exobiology and future Mars missions: the search for Mars' earliest biosphere. *Advances in Space Research*, **6**(12) (1986), 269–285.

[34] C. P. McKay. The search for life on Mars. *Origins of Life and Evolution of the Biosphere*, **27** (1997), 263–289.

[35] C. V. Manning, C. P. McKay, and K. J. Zahnle. Thick and thin models of the evolution of carbon dioxide on Mars. *Icarus*, **180** (2006), 38–59.

[36] E. Carlsson *et al.* Mass composition of the escaping plasma at Mars. *Icarus*, **182** (2006), 320–328.

14

The trouble with intrinsic value: an ethical primer for astrobiology

Kelly C. Smith

14.1 Introductory comments

A philosopher always needs to know his audience. In this case, I am going to assume that most readers of this piece will be scientists and engineers working in the space sciences – that is, non-philosophers who have relatively little familiarity with ethics as a field of study. Certainly we all know about ethics in the sense that we all know it has to do with rules about our behavior, especially our behavior towards other people. In this sense, ethics is rather similar to evolution – people think they know what it is because they understand the basic idea. However, there is much more to ethics than can be summarized in a few rules of behavior, just as there is much more to evolution than can be captured in a paragraph explaining natural selection.

My goal here is thus not so much to break new ground as to discuss some basic concepts in ethics which are often misunderstood in such a way as to block fruitful ethical discussion. Indeed, it is quite common for people to adhere to views which actually undercut the whole ethical enterprise without even realizing this. So I will begin this article by discussing some extremely common confusions about what ethics is and isn't. Then I will discuss a few basic ways of assigning moral value, trying to show the disadvantages of intrinsic value assignments as they are often used in environmental ethics contexts such as astrobiology. I end by arguing for a ratio-centric approach to ethical value and show how it can be applied to the most dramatic question in astrobiology: whether we should Terraform Mars.

14.2 Some basic ethical concepts

14.2.1 Ethics, science, and reason

Ethics is a field of study which attempts to identify and systematically justify principles governing how we should treat others. Ethicists attempt to formulate

Exploring the Origin, Extent, and Future of Life: Philosophical, Ethical, and Theological Perspectives, ed. Constance M. Bertka. Published by Cambridge University Press. © Cambridge University Press 2009.

general ethical principles which cover as wide a range of situations as possible. They spend a lot of time making these principles very clear (e.g. distinguishing them from similar principles, clarifying when they do and do not apply, etc.). Then they justify the principles using general considerations of epistemology, reason, psychology, etc. Finally, they elucidate the practical implications of the general theories.

Ethical reasoning is really not so different from scientific reasoning. In both ethics and science, the emphasis is on explaining what happens (or what should happen) in a particular case by recourse to more general principles or laws. We think we have *explained* (science) or *justified* (ethics) an action when we can show how it follows (or should follow) from a general theory. Both scientists and philosophers have to clarify and support the theory as best they can, even when the general theory is not directly testable, contains vague elements in need of clarification, etc. Both disciplines place great emphasis on logical argument and terminological clarity to minimize irrelevancies and confusions. The bottom line is that both scientists and ethicists are in the business of generating systematic justification of their claims.

A scientist may have a personal opinion about something, but he should be careful to differentiate this from what science tells us. Similarly, we all have personal opinions about what constitutes ethical behavior and what does not. Unfortunately, when it comes to ethical issues, we are *not* all as careful to distinguish personal opinion from systematic ethics. This is the first important confusion about ethics – it is not merely a collection of personal opinions. Ethical answers are the products of a general normative theory of human action. I freely grant, of course, that the question of *which* ethical theory one should rely on is a very complicated debate. However, that is a separate question from whether ethical questions have defensible answers, just as the debate about which possible explanation of quantum gravity is the best is distinct from an explanation of why physics is a science.

Part of the reason people tend to think of ethics as "just opinion" is a distressing tendency to approach complex issues in a dichotomous, black or white fashion. Children are taught in elementary school to differentiate between "statements of fact" and "statements of opinion" as if this were an easy thing to do. We get used to thinking this way and thus easily classify science as dealing with facts and ethics as dealing with mere opinions.[1] But if "facts" are taken to mean statements whose truth we are certain of beyond the possibility of error, then science does not have any. If "opinions" are taken to be statements for which we have no evidence at all, then ethics is not about them.

---

[1] This is why, for example, the claim that evolution is "just a theory" is such an effective tool in the hands of the creationists. Evolution is certainly a theory, but adding the word "just" implies that theories are merely opinions serving as stopgaps while we await the discovery of irrefutable facts.

14.2.2 Got facts?

One important difference between science and ethics is that science attempts to generate true *descriptive* statements about the universe. Its job is primarily to correctly describe how things actually work. Ethics, on the other hand, is not about describing the world, but about *prescribing* how it should be. This is a simple point, but it's also a point people routinely lose sight of in the course of an ethical argument. This is perhaps especially true of scientists, who are so accustomed to thinking about descriptive statements and their empirical support that they instinctively seek shelter in them when things get too far from what they can measure directly.

For example, trying to determine the ethically correct procedures for protecting Martian life is very difficult. It is neither empirically tractable nor subject to easy consensus. The rules NASA has written to govern such procedures are another matter, however, as anyone who knows where to look can read these and figure out how to apply them. The tendency to dodge difficult normative questions about what we should do to protect Martian life with descriptive answers about what NASA rules currently say we should do is thus understandable – but it's also misguided.

After all, what are NASA's rules based on? More to the point, what *should* they be based on? Would we think they provided an appropriate answer to questions about how to manage Martian exploration missions if we knew they represented merely the personal opinion of a particular NASA administrator? I think not. Rather, they have the force they do because we believe (or at least are willing to assume when we are frustrated and overworked) that they are based on legitimate ethical principles. Therefore, rules of behavior, whether we are talking about NASA regulations or federal law, must be based on ethical principles, not vice versa. The underlying ethical principle, not the regulation itself, is the appropriate source of ethical justification.

A slightly different way in which people blur the line between ethical and descriptive analysis is to try to derive ethical principles from descriptive ones (a move philosophers refer to as *the naturalistic fallacy*). Suppose we are discussing whether we should terraform Mars for the benefit, not of terrestrial life forms like us, but of native Martian organisms. A scientist or engineer might well point out that this proposal isn't going to get very far politically because people are selfish and they aren't going to vote to spend money on things which do not directly benefit them. This is usually put in such a way that it implies the discussion is over and we *should* act in accord with how people *do* act. Whatever you may think of the proposal's merits, this argument is fundamentally wrongheaded. Even if we suppose, without argument, that people really *do* only vote to support things which directly benefit them, we still need to know what bearing this has on the question, "What *should* we do on Mars?" The latter is a normative question which can't, in principle, be answered by a description of human psychology.

14.2.3 *It's all relative?*

"Who is to say what is right and what is wrong?" This is one of the most common sorts of responses people give when engaged in what seems to be a lengthy and frustrating ethical dispute. Ultimately, it's an expression of a position philosophers call *ethical relativism* and one of the most difficult challenges facing anyone who wishes to have a serious ethical discussion is the widespread and entirely uncritical acceptance of ethical relativism today, especially here in the USA. Basically, an ethical relativist is claiming that there are no truths about ethics, and thus what any one person or society *considers* ethical is entirely relative. One of the reasons this view is so popular is that it sounds so intuitively correct – until and unless you examine its consequences. The bottom line is that relativism is a far, far more radical position than most of its casual adherents realize.

The basic relativist claim is like saying that all ethical judgments are ultimately the same sorts of things as subjective expressions of taste. Thus, "It is immoral to kill innocents" is in the same essential basket as "I don't like North Carolina style barbeque." Let's suppose briefly that this is true and consider where it leads. For one thing, if ethical judgments are really just subjective matters of taste, ethical arguments are, quite literally, meaningless. There is no way I can convince another that killing innocents is wrong any more than I can argue them into believing that NC style BBQ doesn't taste good. It's just not a rational process.

The problem is, very few people who espouse ethical relativism are willing to accept its consequences. "Relativists" thus continue to debate at least some ethical points, despite holding a view which should make this an exercise in futility. Not only do they debate such points, they continue to hold strong ethical views about morality which they believe *others* really truly should adopt – I assume, for example, that very few relativists believe that killing innocent babies because it is fun should be allowed. The logical consequence of this kind of thinking then is, quite literally, a might makes right view of ethics – it's ultimately all opinion, but sometimes I just have to force others to accept my point of view, not because it's objectively right (since there is no such thing) but because I feel strongly about it and have the power to force compliance.

If you find it too radical to believe that ethical discussions are silly, that people who stand up for ethical principles are chumps, and that might really does make right, then you need to reconsider any attraction you might have towards ethical relativism. We must believe that there is "something like truth" out there and that we can, however haltingly, make progress towards it by applying the same basic tools of reason used in other disciplines like science.

I want to make one final point along these lines: believing in "something like truth" is not necessarily the same as believing that there is a uniquely correct answer to any particular question. All we really need to believe is that not all answers to ethical questions are equally good. This allows us to reject many bad answers to ethical questions for very good (if not definitive) reasons, and thus makes ethical progress of an important sort possible. We might wish the situation were clearer still, but if physics can live with the particle/wave duality of light, surely ethics can live with its own type of pluralism.

14.3 Building an ethical framework

14.3.1 On valuing

Now that we've dealt with the general relationship between truth and ethics, we can move on to the more specific question of how to justify particular ethical positions. One of the most important points of debate in modern ethics has to do with the "moral community question." That is, to whom precisely do we have moral obligations? In the good old days, this question did not arise much since it was so obvious that only healthy, wealthy, European males had a serious claim to moral standing. The demise of this certitude may have made the world a better place in many ways, but it greatly complicates ethical debate: Do we have obligations to future generations not yet born? Do non-human animals have ethical standing? Do rivers and mountains have rights we should consider?

Now, things can be ethically valuable in two crucially different ways. On the one hand, I can value something because it serves my needs. If I want to julienne a fifty pound sack of carrots, a food processor might be seen as quite valuable. However, the value of a food processor is entirely dependent on its utility as a means to my own ends. If I were to find myself stranded on a desert island with only a food processor for company, most of us would allow that its value is very close to zero because I have no real use for it. Philosophers call this sort of valuation *instrumental*, since the value of the object is a function of the way in which I can use it as an instrument to meet my own goals.

On the other hand, most of us believe that we should value other human beings, not because of the uses to which they can be put, but because they are the sorts of things which have value in and of themselves. This sort of value is called *intrinsic* because it has to do with the intrinsic nature of a thing, not its relation to other things or its utility. Thus, while it's true that I may value a plumber in part because I need my toilet fixed, this is not the most important reason to value her. If the plumber were stranded on a desert island all by herself with no plumbing to fix, we do not think this reduces her ethical value at all. It would be immoral to kill her even if doing so would not directly inconvenience any other person because her value is a consequence of the sort of entity she is, not the uses to which she can be put.

Traditionally, philosophers think of intrinsic value as a kind of line in the sand. Those things with intrinsic value are the fundamental units of ethical analysis.[2] Instrumental value is viewed as either irrelevant or at least of decidedly secondary importance. One reason this is enormously important is that the values of items which have a merely instrumental value are subject to the forces of the market and thus fluctuate greatly depending on external circumstances. We certainly do not want the ethical value of humans to vary with the availability of plumbing or the whims of those who surround us at any particular time. So it's vital to forestall any instrumental calculation of human value by showing how it is completely beside the point from an ethical point of view – humans have instrumental value, to be sure, but that's not why we should treat them as morally important. As Immanuel Kant puts it, "People have a dignity, not a price" [9].

Of course, this immediately raises the thorny question: What exactly is it about humans that accounts for their intrinsic moral value? The answer to this question has been debated for a very long time indeed, but anyone who uses the intrinsic value account must necessarily provide some answer to this question, on pain of holding that everything in existence is equally valuable from a moral standpoint. In other words, we can argue about where the line in the sand should be located, but not about whether there needs to be a clear line.

14.3.2 Inclusive intrinsic value

The entire field of environmental ethics can be characterized, without too much oversimplification, as an attempt to expand the boundaries of our moral community beyond human beings. If we accept that humans are special from a moral standpoint, but refuse to accept that they are *uniquely* special, then we have to consider whether other entities should be included in the community of moral players. So the environmental ethics movement can be seen as an attempt to move the line in the sand further and further from its original location where only humans were seen as possessing intrinsic value. The further you move the line, the more inclusive your account of intrinsic value becomes.

To be sure, inclusive intrinsic value accounts have been a much needed corrective to thousands of years of unthinking assumptions of human superiority. In particular, environmental ethics has done an excellent job of critiquing *anthropocentrism*, which is basically the view that humans are special simply because they are human. If we wish to avoid begging the question of moral valuation, there must be some attribute humans possess which accounts for their status. Once this is

---

[2] Thus one common response to the food processor example above would be to say that the food processor really has no *ethical* value at all.

established, it becomes an open question whether humans are the only entities that possess this attribute.

So there is nothing wrong with asking whether intrinsic value extends to entities other than human beings. On the other hand, it's my view that the environmental ethics movement has historically suffered from a kind of selection bias that affects its character in an odd way. The fact of the matter is that most people who specialize in environmental ethics have a deep interest in, even a reverence for, natural things. As a result, there's an understandable push to establish the significance of non-human "things" in the environment: animals, plants, ecosystems, mountains, rivers, etc. The basic idea is this: "The well being of non-human life on Earth has value in itself. This value is independent of any instrumental usefulness for limited human purposes" [1]. Indeed, such luminaries as Holmes Rolston are willing to extend intrinsic value well past living things [2]. There are many different positions that can and have been taken on the exact nature of intrinsic value, who/what possesses it, how they/it possess it, and how we can know that they possess it, but the intrinsic value justification is certainly central to modern environmental ethics. As John O'Neill puts it: "To hold an environmental ethic is to hold that non-human beings and states of affairs in the natural world have intrinsic value" [3].

One could argue that a more inclusive view of intrinsic value is a much needed corrective to a long history of unthinking anthropocentrism. As an exercise in the sociology of academia, this certainly makes some sense. However, it does mean that the treatments of these questions in the environmental ethics textbooks that non-experts consult for authoritative opinions tend to have a definite point of view that is not necessarily representative of the ethics community more generally, which has for a very long time indeed tended to the view that reason is morally special (e.g. Aristotle, Descartes, Kant, etc.). As a result, non-experts often take relatively radical ethical claims as uncontroversial.

I first became aware of this problem when I started presenting my own ratio-centric account to astrobiological audiences. When I told people that I thought only creatures with the capacity to reason have ethical standing in the strict sense, the response I often got was something like, "Yes, but what's your *ethical* position?" The clear implication is that this sort of view has long since been discredited and is currently considered something of a fringe position. In fact, whatever else might be said about the merits of a position like ratio-centrism, it is most certainly *not* an outlier within the community of professional ethicists.

Indeed, some version of ratio-centrism is probably the position adopted by the vast majority of professional ethicists today. It is instructive to note the kind of reaction I get from my colleagues when I describe some of the more extreme positions that have been taken in astrobiology. For example, Chris McKay's claim that the inherent rights of any Martian life we discover would "confer upon us the obligation to assist it in obtaining global diversity and stability" [4] is routinely met with expressions of incredulity. The typical reaction is first to question whether that is *truly* what he meant, and when it becomes clear that it is, to dismiss it with a rueful shake of the head as simply absurd. My point here is certainly not to imply that the popularity of a position is evidence of its truth, since a deep ecologist would certainly argue the reaction of the professional ethics community is simply perpetuating antiquated and indefensible theories, etc. On the other hand, the existence of a very wide gap between the way ethical theory is portrayed by many in the environmental movement and how it's viewed by a more general sample of the ethics community should give non-experts pause.

For whatever reason, many non-experts have gotten the idea that the best way to insure the ethical status of non-human life and the like is to simply extend the protection of intrinsic value to them. The advantages of intrinsic valuation are clear – once established, it functions as a trump card against any attempt to argue that we should ignore the welfare of such an entity in favor of some instrumental use. The problem is that they rarely realize the implications of such a move. In particular, the strength of an intrinsic value approach is inversely proportional to its exclusiveness – the more entities are said to have intrinsic value, the more difficult it becomes to make any ethical decision, since instrumental tradeoffs are not allowed. Intrinsic value is a powerful ethical tool, but it must be used sparingly.

Suppose, for example, we hold that a dog has merely instrumental value. In that case, I can only argue weakly for its protection, since there are many potentially useful things I can do with a dog, only some of which are in its own interests. Moreover, under the instrumental view, nothing I do to a dog is ever "simply wrong" because dogs are not the kinds of thing which really have ethical status in their own right. We could easily argue that the greater good is served by allowing a dog to be used for medical research or stuffed and mounted for people's amusement, etc. In order to decide what to do with a dog, I need to objectively weigh all the different uses to which it might be put and this calculation might well endorse actions that are hardly in the dog's best interests. If, on the other hand, I believe the dog has intrinsic moral value, then I can use this to scotch utility arguments of this sort. Something with intrinsic value simply should not be treated in certain ways, irrespective of the possible benefits to others of doing so. Just as we would not accept an argument that the greater good would be served by using a human being for medical experiments against their will, I have a clear duty to protect a dog from such exploitation if I believe it is intrinsically valuable.

What sets ethics apart from mere personal intuition is the attempt to justify ethical positions in the context of a general ethical theory. For our purposes, this has two important implications. First, we can't simply *assert* the intrinsic value of non-human entities, rather we must put forward a principle of division which draws a line between those things which have intrinsic value and those which have merely instrumental value. To be sure, there are a great many ways one might argue the principle of division should work. Throughout much of Western philosophy it was thought that the possession of reason is the *sine qua non* of intrinsic ethical value. More recently, alternative concepts have been put forward, including the ability to feel pleasure/pain or self-awareness or being alive or a certain level of organized complexity, etc. [2, 5, 6, 7].

Second, we have to show that our principle of division is not simply an ad hoc move that we favor simply because it results in a self-serving division. Both in science and in ethics, the best way to guard against ad hoc moves is to require them to flow from a more general theory. If we want to view a particular principle of division as general rather than ad hoc, we must look carefully at the implications of applying it in a wide variety of cases, not just those of immediate interest. It's in this second aspect of assessment that the uncritical use of intrinsic value assignments falls apart.

Let me illustrate what I mean. If we cast the net of intrinsic moral value widely, we always generate uncomfortable cases philosophers sometimes classify as *lifeboat cases*. Suppose, for example, that I am stranded in a lifeboat with my son, Sam, and my dog, Cleo. Suppose further that we are all on the verge of dying of dehydration and it seems rescue cannot possibly arrive in time to save us. What should we do? If we reject something like reason as the principle for determining who has intrinsic moral value, then all three occupants of the boat are probably going to be seen as having intrinsic moral status, since it seems pretty clear that Cleo is alive, can feel pleasure/pain, is a complex entity entitled to a proper name, etc. We might like this sort of move in some contexts because it gives us compelling moral reasons not to exploit Cleo. However, in this case we have to make a difficult decision. This situation pits Cleo's moral status against that of myself and my son – someone is going to lose, the only question is who.[3]

Now, someone who wants to defend Cleo's intrinsic moral status has a real problem. She has two basic options. On the one hand, she could stick to her guns and insist that Cleo has intrinsic moral value and that such value trumps any considerations of Cleo's utility as a source of water for the humans in the boat. Traditionally, intrinsic value is thought of as something which does not come in degrees. Something either has intrinsic value or it doesn't. Thus, Immanuel Kant argues that humans have intrinsic value because they are rational, but he does not think that smarter humans have more value than dumb ones. This is important because viewing intrinsic value

---

[3] This is, of course, an unlikely scenario designed to elicit a strong intuitive reaction. However, conflicts between individuals with intrinsic moral value will be commonplace and, the wider you cast the net of intrinsic value, the more often you will be faced with such dilemmas (albeit usually less poignant than this one).

this way prevents a different type of haggling over who is more valuable, etc. Thus, presumably the moral recommendation from a defender of Cleo's intrinsic value to the boat's occupants would either be to do nothing and pray for rescue or to devise some fair and impartial way of deciding who will give up their water for the sake of the others. I think it's safe to say that the vast majority of people would find it exceedingly difficult to accept these consequences and in fact the word "absurd" probably occurs to most people at this point.

Another option would be to argue that intrinsic value comes in degrees. In principle, there is no reason we can't argue that Cleo does have intrinsic moral value, yet still allow that her value is not as great as mine or my son's. This would open the door to sacrificing Cleo for the sake of the humans. In effect we are saying that, though all creatures with intrinsic moral value are equal, some are more equal than others. This sort of move is intuitively very appealing, because it solves the immediate problem in the lifeboat in a way we find convenient. The problem here is that such a move robs the intrinsic value account of its original motivation. The whole point of distinguishing intrinsic from instrumental value in the first place was to prevent any kind of haggling over who is more important than whom in different circumstances. In the eyes of a traditional defender of intrinsic moral value like Kant, to even enter into a discussion about which intrinsically valuable entities are more valuable than others is to have missed the whole point of moral value in a fundamental way. The solution is not to more precisely determine the circumstances which set the moral value, but to reject the very idea that intrinsic values depend on circumstance at all. Dignity is non-negotiable.

To put it another way, it may make us feel that we are being more fair to say that Cleo has intrinsic moral value. However, is that really changing anything if every time Cleo's supposedly intrinsic value is pitted against significant human interests, she loses out? We are still drawing an implicit line in the sand that separates humans from non-humans, we are just not being clear about that up front (which creates other problems). As John Stuart Mill [8] observed long ago, what good does it do to recognize moral rights in the abstract if they have no application in practical circumstances? I submit that if we really want to make this move, then the argument over whether moral value is intrinsic or instrumental becomes purely academic, since it ultimately does not affect the practical decisions moral theory must make.

Actually, there is at least one other option open to the defender of Cleo's intrinsic value. It's not really a position so much as a refusal to take a position, but it is nevertheless a common way to "solve" the problem. One could always maintain that Cleo's moral status is intrinsic just like that of the human castaways, but then refuse to take any clear position on how we should adjudicate the conflict in the lifeboat. Typically this is not put as baldly as I state it, but rather by the less obvious route of issuing very vague ethical guidance such as "we must respect the rights of all concerned." This sounds good if we don't think about it very critically, but ultimately we have an obligation to state at least something about what it means to "respect rights," including some clear mechanism for how we can decide when we should respect the rights of some entities more than others. If we don't do that, then we have not really dealt with the problem at all. Using ethics to argue for a position but then failing to provide clear guidance precisely when it is most needed – when the decisions are difficult and hard choices must be made – gives professional ethicists a deservedly bad name.

14.4 The ratio-centric account

14.4.1 The account outlined

So if we are serious about actually getting to the truth of the matter in ethics, we must insist on some basic requirements for any ethical theory. For our purposes, the two most important are that: (a) the theory must unflinchingly explore its implications in a wide variety of cases and (b) the theory must be specific enough to be of practical utility in difficult situations. If a theory makes no effort to be general or does not deal with difficult situations, it should not even be seriously considered. Only if it does these things can we move on to the further question of whether it's the best alternative amongst competing theories.

I would now like to offer for the reader's consideration an account which avoids the sort of problems I have discussed above while also allowing for a real environmental ethic. It is true that this account also, depending on your ethical intuitions, might have some counter-intuitive consequences, but at least it does not buy respectability at the price of clarity. I must emphasize that this is merely a sketch, however, since ratio-centrism draws on such an enormous body of prior work that doing it justice is far, far beyond the scope of this chapter.

The position I want to sketch is called *ratio-centrism* because it views the possession of reason as the essential feature dividing those things with intrinsic moral value from those things with merely instrumental value. By "reason" here I don't mean intelligence, but a much more general trait. To be rational is to have a conception of oneself and one's environment, to be capable of symbolic thinking, to be aware of one's own interests, and to have some conception of the nature of a rule of conduct. This is not a perfectly clear criterion, of course, but it does make some stark distinctions since human beings are clearly rational and bacteria are clearly not.[4]

---

[4] Of course, any attempt to draw clear lines is going to have a difficult time defending their precise location against all possible alternatives. There are two points here that it is crucial to keep in mind, however. First, the

Why should reason in this sense be the *sine qua non* of ethical value? I would argue that the ultimate point of ethical systems in general is to provide the means for the smooth functioning of societies. The basic idea is that ethics are a product of cultural evolution in complex societies and their function is to stabilize and enrich such societies.[5] I will not defend this claim here in detail, though I will note that part of the reason for finding it attractive is the paucity of alternatives if on the one hand you reject any kind of supernatural basis for ethical objectivity while on the other hand recognizing the dangers of ethical relativism.

In any event, if the purpose of ethics really is to provide for the smooth functioning of societies, then it seems to follow that its primary objects are those entities capable of functioning in such a society. In other words, the primary moral community is composed of those with the requisite abilities to participate in a society. Some level of rationality seems a prerequisite for social functioning[6] since, at a minimum, participation in a society implies the ability to conceive of rules of behavior, to communicate with each other concerning those rules, and then to constrain one's behavior accordingly. Entities which have these abilities can participate in a moral community and thus the community must recognize their moral worth as a necessary requirement for its formation. Entities which cannot participate in the moral community may certainly be valued – indeed, their value in some circumstances can be quite high – but this value is always of a fundamentally different character than that of rational creatures because they are external to the community.

From an ethical point of view, then, a ratio-centrist world is populated with three basic classes of things. First, there are entities which clearly possess reason sufficient for participation in a moral community and have intrinsic moral value precisely because of this (e.g. humans, space-faring aliens, God). Second, there are entities which clearly do not possess reason and thus have merely instrumental value (e.g. mountains, rivers, bacteria, sponges). Finally, there are entities which possess what seems to be a limited form of reason that falls short of what would be needed for full participation in a moral community (e.g. dolphins, octopi, my dog Cleo, other primates). Another way of putting it would be to say that these entities

---

fact that the precise location of a line is not clear does not indicate that there is no meaningful distinction to be made, it simply means that we will have to accept some degree of fuzziness about the boundary between concepts we wish to use. Fuzzy boundaries are actually quite common, as when, to take an astrobiological example, one tries to distinguish between living and non-living systems. Second, this is a problem faced by *any* account which imposes distinctions. We must always keep in mind that ethical theories (and scientific ones, for that matter) represent choices between imperfect alternatives. The question we must always focus on, therefore, is not whether a theory can avoid all criticism, ambiguity, etc., but rather whether it fares better than any available alternative.

[5] Honesty compels me to point out that, unlike ratio-centrism in general, this is not a popular view amongst my fellow ethicists.

[6] I am also assuming a certain notion of society here, namely one which is capable of producing culture. This rules out insect "societies," but might well include some non-human primate groups.

can participate in a community in some ways (e.g. sharing affection), but not in others (e.g. formulating rules).

The exact moral status of an entity in this last class is unclear. However, it's unclear for largely epistemological, not metaphysical, reasons. That is, the basic criterion for inclusion in the class of intrinsic value entities is relatively clear, the problem with the last class of entities is that we don't know enough to know for sure just how rational they are. This does nothing to impugn the ratio-centric principle, it just shows that there remain some open questions about membership in the community it defines. I am not taking a position on their ultimate ethical status and certainly continued research and debate is appropriate here. Even with this uncertainty, the line in the sand can be agreed upon. Moreover, it does significant moral work – on Earth, for example, the vast majority of life forms (and all non-life) are quite easily and uncontroversially placed in the non-rational category.

14.4.2 Anthropocentrism

Now I want to briefly consider some of the more common objections to ratio-centrism. The most obvious criticism is that this account is unacceptably anthropocentric, or in other words, it rather too conveniently arrives at the same "humans are the best" view of ethics that has been so destructive in the past. The first thing to say in my defense here is that the account is not anthropocentric, but ratio-centric. This is more than a semantic game. An anthropocentric account would accord humans special status simply because they are human. This is clearly indefensible, but not because it results in a bias in favor of humans. Rather, it's because the principle itself is indefensible. Why should membership in a particular species, in and of itself, have any bearing on moral standing at all?

Ratio-centrism is quite different, however. True, the only organisms on Earth which seem clearly to have the kind of reason required for intrinsic moral status are humans and perhaps this is prima facie reason to be suspicious. The key questions that need to be asked are: (1) is the basic principle used defensible and (2) is membership open, in principle, to entities other than humans? Certainly there is some debate over whether reason is the best principle one might use to draw moral lines, but it's certainly defensible as evidenced not only by its long history but by the way even those who reject it in the abstract tend to adopt it in extreme situations. And certainly membership in the moral community thus defined is open to other entities – it's not about being human, it's about being rational, and if one takes the project of astrobiology seriously, this is unlikely to be a uniquely human characteristic in the universe.

To put it another way, a biased outcome is not necessarily evidence of a morally problematic selection process. Suppose, for example, that I hire a woman for a job in preference to a man. Suppose I do this all the time, so that the only people I have ever hired for this job in the past 30 years are all women. It is tempting to accuse me of sexism immediately, and certainly you would be justified in taking a closer look at what is going on, but it's entirely possible I am doing nothing wrong. The question we need to answer is whether the principle of division I am using is justifiable, and we don't know this simply from the fact that it results in a skewed pattern of hiring. If the job for which I am hiring is that of a wet nurse, for instance, we readily admit that there has been no morally problematic discrimination. So, privileging humans simply because they are human is one thing, privileging them because they are rational is another matter entirely.

14.4.3 *"Mere" instrumental value*

I think one reason people react so viscerally towards ratio-centrism is that they feel it won't be possible to accord non-human entities any serious weight in our moral deliberations. In one sense, of course, this is true – non-rational entities have merely instrumental value. However, this is far from saying they are not valuable at all. Modern science has revealed much about how the interests of rational humans are intimately tied up with the interests of non-human entities in the environment. Thus, a sensible ratio-centrist should argue quite forcefully for the preservation of the world's remaining tropical rainforests, even when this means significant inconvenience for humans who would wish to farm the land, etc. The reason for this has nothing to do with the intrinsic moral value of rainforests, however. Rather, it's simply not very smart to destroy that which provides you with what you need to stay alive, even when doing so might meet some short-term goals. Killing the goose that lays the golden eggs because you are feeling momentarily hungry is just not smart. Thus, a thoughtful and scientifically literate ratio-centrist will agree on many points with someone who wishes to accord non-rational entities intrinsic value, though for very different reasons. In fact, what makes ratio-centrism so powerful is that it not only provides a principle of division, it also establishes a common system of measuring moral worth and thus a means for adjudicating ethical disputes. Thus, not only can I grant that non-rational entities can be valuable, I have the means to assess that value and weigh it against other things of value in order to make an informed decision. More inclusive intrinsic value theories, as discussed above, cannot do this well at all.

14.4.4 *Abuse*

A slightly different criticism of ratio-centrism is that such a ratio-centric account is likely to encourage people to do horrible things to non-human entities. To some extent, I agree. It's undeniable that the past history of ratio-centrism is nothing to be especially proud of. We have tended to take the most naïve, most short-sighted, and most narrow possible view of the value of anything other than humans. People often think that any time human interests conflict with the interests of non-human entities, humans should win, regardless of the circumstances. It may even be that people who are likely to adopt these simplistic analyses tend to flock to ratio-centrism because it seems to provide them with what they want. Note, however, that the problem here lies not with the theory of ratio-centrism, but with its application. People who think this way are either sloppy ratio-centrists who place far too much value on short-term rather than long-term analysis or they fail to understand what we have learned about the importance of ecological interactions. In any event, the way to show that these positions are indefensible is to adopt a good general ethical theory. If we use ratio-centrism, we can easily show people that in many cases what they *think* is in their best interests actually isn't. Far from being the problem, a proper version of ratio-centrism is the cure.

Moreover, if we are honest we have to admit that any ethical position can be abused and probably that the likelihood of abuse is proportional to its popularity. Certainly inclusive intrinsic value accounts in environmental ethics have been used to support extremely radical activities such as spiking trees. No doubt the list of abuses attributable to people who subscribe to inclusive intrinsic value accounts will grow as those positions become more popular and dominant. In general, though, it's quite unfair to criticize any theory for the abuses of its most radical defenders.

14.5 Issues astrobiological

14.5.1 The insights of astrobiology

We can at long last turn to the application of all this theory to ethical issues in astrobiology, especially to the question of whether it's ethical to terraform Mars. Before we do that, I do want to talk briefly about how astrobiology enriches ethical theory. One thing I find especially fascinating about astrobiology is that it throws old questions in ethics into a new light. For one thing, when you are talking about what sorts of ethical obligations we might have to extraterrestrial life, it forces you to take the distinction between ratio-centrism and anthropocentrism very seriously indeed. In more traditional discussions of ethics, we really only knew of one case of rational creatures – humans. True, God is often put forward as another possible rational entity, but this seems a very different thing and, in any event, we are not sure of His existence. If, however, you are addressing a group of people whose goal is to actually find life beyond Earth, you are forced to explicitly consider the possibility that some of this life may be fully rational and thus have intrinsic moral value.

The distinction between intrinsic and instrumental value is also much more stark in astrobiological contexts. One problem you have in terrestrial contexts is that different people often stake out the same position for very different reasons. Because they are ultimately defending the same position, they have the luxury of being less than clear about precisely what their reasons are and how they differ from alternative justifications. As I pointed out above, a ratio-centrist might be a vociferous advocate of rainforest conservation, but for completely different reasons than some of his fellow travelers who believe trees have moral rights. The bottom line is that Earth is such an interconnected biosphere that fiddling with any part of it is also going to affect other parts, including ourselves. Thus, there is always rational self-interest at stake on *both* sides of terrestrial environmental issues. This is why, for example, environmental publications often freely mix very different kinds of ethical justifications – here an appeal to economic considerations, there a plea for common moral standing. If we do find life beyond Earth, however, we will not have to worry about a shared biosphere. There will still be great instrumental value to extraterrestrial life even if it's obviously non-rational, but it's very unlikely we will be in the same ecological lifeboat, so to speak, in any meaningful sense and thus we can disentangle our interests more easily from the aliens'.

Finally, at least for the next few hundred years, there are quite likely to be fewer problem cases in astrobiology than on Earth. That is, there will be fewer cases like dolphins – entities which seem to possess some but not all aspects of reason and thus present real challenges to any general principle of division. Part of the reason for this is that we will not share a common evolutionary history (presumably) with extraterrestrial life and thus the differences between them and us will be much more pronounced. There's also this, through: if we think of the search for extraterrestrial life as a random walk though the possibility space of reason, we are far more likely to encounter entities with either very high or very low levels of reason relative to us than entities which are similar to us. This means that the vast majority of ethical disputes in astrobiology will be much less messy than on Earth. They will either involve interactions between two rational species, in which case both sides clearly have intrinsic moral value and we can treat potential conflicts of interest the same general way we would treat ethical disputes between humans, or they will involve interactions between rational creatures and entities which clearly have no reason at all and we can treat them as straightforward questions of how best to serve the interests of those within the moral community.

14.5.2 *On terraforming Mars*

So, what does a ratio-centrist have to say about a major manipulation of the environment beyond Earth for human purposes, such as Terraforming Mars? Should we attempt to remake Mars in the image of Earth so that we can more efficiently exploit its resources, or should we leave it as an unspoiled wilderness?

The first question, of course, is whether there is life on Mars. If there is no life, then there are no rational entities and how we should decide what to do is clear. I do not mean by this that there will be no disputes over how best to utilize Mars, but simply that Mars should be looked at as a resource for rational creatures. This doesn't necessarily mean we should turn the planet into a giant strip mine. True, once the technology to do this is actually in hand, I am sure there will be no shortage of people calling for precisely this sort of destructive policy, but they are not really ratio-centrists, or at any rate not very thoughtful ones. In fact, it's not clear they have an ethical position at all as opposed to a very short-sighted kind of self-interest. If nothing else, Mars would be a truly *unique* resource that we would be well advised to deal with carefully.

At the other extreme, let's suppose that Mars does have life and, moreover, that the life is clearly rational. A ratio-centrist will have to allow that these creatures are our moral equals, just like the American Indians were the moral equals of the early European colonists in the new world. I use this example on purpose because, while many would be quick to point out there that we might very well treat Martian creatures much the same way we treated the American Indians, the point is that this is clearly an *immoral* way to proceed. Far from being a justification of such exploitation, ratio-centrism provides the theoretical means to see just why we shouldn't have used the Indians and shouldn't use the Martians as mere means to our own selfish ends. Fortunately, the odds of discovering rational life on Mars seem to be pretty close to zero, which simplifies the moral calculation greatly.

If there is life on Mars at all, it is almost certainly in the form of unicellular microbes living beneath the Martian surface. In such a case, there would presumably be no question of its non-rationality. As such, their ethical value would be entirely instrumental. Thus, what should be done to them is a function of the best interests of truly rational creatures on Earth, namely us. It's crucial to note, though, that the rational creatures of Earth do not speak with one voice about what is in their interests. Some of them want strip malls and parking lots, some want areas of pristine wilderness to contemplate, some want scientific investigation of the unknown, etc. Even when we grant that a question about Terraforming Mars is exclusively about serving human interests, we are still pulled in many different directions. In fact, it's entirely possible, even likely, that well-intentioned, thoughtful ratio-centrists will disagree about how we should behave. Even so, there are some things we can readily agree on.

The first point of agreement is that Mars is a treasure trove of scientific information. This is true even if there has never been a single speck of life anywhere on the planet, but the scientific value increases exponentially if there is life, even more if the life is of an independent evolutionary origin, yet greater still if it is organized in complex ecosystems, etc. Indeed, we will not even *know* what Mars contains, living or otherwise, until a significant piece of science has been conducted there. It does not require a view that Mars or its potential inhabitants have intrinsic moral value to see that we should not destroy, willy nilly, that which we do not even understand. Therefore, we should all be able to agree that a strong conservative approach to exploration is in order initially. We should not do things to Mars, especially things which we can never reverse, until we have a much better picture of what is at stake. This is going to require an extended period of scientific exploration without large-scale intervention.

The second point of agreement is that some rational creatures on Earth will wish us to leave Mars in pristine condition. To the ratio-centrist, it doesn't matter in some sense whether their reasons are correct. All that matters is that they desire this. Even though a ratio-centrist feels that the Martian landscape and non-rational Martian life have no intrinsic ethical value, he cannot deny his obligations to his fellow rational creatures on Earth. Failing to take their feelings into account, even if I don't share them, will harm them and I have to care about this. Immanuel Kant, for example, clearly feels that animals have no intrinsic moral status because they are not rational:

> … every rational being, exists as an end in himself and not merely as a means to be arbitrarily used by this or that will. … Beings whose existence depends not on our will but on nature have, nevertheless, if they are not rational beings, only a relative value as means and are therefore called things [9].

This does not mean that I can do whatever I wish with non-rational animals, however. Although I have no direct moral obligation to your cat, I do have an obligation to you. So Kant would argue that killing your cat (without very good reason) is immoral, not because it harms the *cat* but because it harms *you*.

For these kinds of reasons, we should be cautious about Terraforming Mars. We need to take adequate time to develop a clear picture, both of what Mars has to offer and of how rational creatures on Earth really feel about the use of Mars. It is important to emphasize when rational convergence occurs, and here we have convergence: both an inclusive intrinsic value theorist and a ratio-centrist will agree that we should leave off large-scale manipulations of Mars until we are better able to answer these questions. Of course, they will immediately get into an argument about how long we must wait and when precisely we will have answered the questions adequately, but certainly a period of 50 years or so is not at all unreasonable, particularly if we need to investigate a subsurface ecosystem.

On the other hand, we can't lose sight of the fact that Mars does offer very enticing possibilities. Sooner or later, the ratio-centrist will argue that the scales have tipped and we know enough about Mars' potential to end the exploration phase. Assuming it's technologically feasible, terraforming and colonizing Mars would offer enormous advantages to billions of humans. To begin with, it would establish a second home base for the human race, thus significantly reducing the likelihood of our perishing in some disaster or other. It's hard to calculate the value of something like this, but certainly avoiding the destruction of our species has to be one of the most important considerations any ratio-centrist could possibly contemplate. A large colony on Mars would certainly offer access to all sorts of resources which we could use to fuel our economies. Money itself is of no moral value, of course, but efficient and thriving economies improve standards of living for rational creatures and this is certainly important. Finally, expanding human habitation to Mars would allow us to maintain larger populations of humans. We tend to think of larger populations as a bad thing because on Earth at present they are seen as stretching limited resources. However, a larger human population coupled with an enormous increase in natural resources would offer real advantages such as a faster rate of progress in science and technology (via larger absolute numbers of scientists), a more diverse and rich culture, and larger economies of scale. To those who still remain unconvinced of the utility, I ask them to consider questions like this:

1. What is the relative ethical value of contemplating a pristine Mars versus contemplating a new human culture on another planet?
2. Is there no way short of blocking Terraforming efforts altogether to maintain the native Martian environment, at least in part?

Now I grant that things are not as simple as I paint them. Terraforming may not be feasible for a very long time, determining when we have learned enough to progress to active measures on Mars will be controversial, etc., etc. However, consider the advantages of ratio-centrism. First, it provides a common yardstick of value by which we can measure competing plans of action. Second, we are unlikely to encounter the kind of problem cases we have on Earth where it is sometimes unclear to what extent creatures are able to participate in a moral community. Third, and most importantly, though there will certainly be disputes, there is a clear mechanism for making difficult moral decisions when people cannot agree. A choice of moral theory is, like most other choices, a question of picking the best available alternative. Certainly a ratio-centric ethic is not perfect and cannot avoid all controversy, but then it's also much more practical than an inclusive version of intrinsic value.

References

[1] A. Naess. A defense of the deep ecology movement. *Environmental Ethics*, **6** (1984), 266.

[2] H. Rolston. *Conserving Natural Value* (New York: Columbia University Press, 1994).

[3] J. O'Neill. The varieties of intrinsic value. In *Environmental Ethics*, eds. A. Light and H. Rolton (Oxford: Blackwell Publishing, 2003), p. 131.

[4] C. McKay. Does Mars have rights? In *Moral Expertise*, ed. D. MacNiven (London: Routledge, 1990), p. 194.

[5] P. Singer. *Animal Liberation* (New York: Harper Perennial, 1975).

[6] T. Regan. *The Case for Animal Rights* (New York: Routledge, 1984).

[7] J. Rachels. *Created from Animals: The Moral Implications of Darwinism* (Oxford: Oxford University Press, 1991).

[8] J. S. Mill. *On Liberty* (New York: Penguin Classics, 1869).

[9] I. Kant. *The Groundwork for the Metaphysics of Morals*. Available at http://www.swanseapolitics.org.uk/texts/kant/kanta.htm, Hypertext Library electronic text of the translation by Thomas Kingsmill Abbott, 1785.

15

God's preferential option for life: a Christian perspective on astrobiology

Richard O. Randolph

15.1 Why should we care about Christian ethical perspectives on astrobiology?

Why should we care about a Christian ethical perspective on astrobiology? After all, the world is becoming an increasingly pluralistic society, with a rich plethora of different religious and philosophical perspectives. Why should we care about what Christians or any other religious group thinks about the ethics of space exploration? For that matter, we may question whether there should be any ethical concerns about astrobiology or the ethics of space exploration at all.

We stand on the brink of an exciting new age in space exploration and science. Already, we have discovered extrasolar planets in our galaxy, launched several very successful robotic missions to Mars, and lifted our gaze above the clouds of Earth through the Hubble and Spitzer space telescopes. This promises to be just the beginning, a foretaste of what is yet to come! In the next twenty years, we could well discover evidence of simple, microbial extraterrestrial life in our solar system – if it exists. In addition, NASA has enterprising plans for the exploration of space beyond our own solar system. The discovery of advanced, intelligent extraterrestrial life remains an intriguing possibility. At the heart of NASA's research strategy is astrobiology, a newly emerging science that will study the origin, extent, and future of life – both in space and on planet Earth. At last, through astrobiology humans are ready to make that gigantic leap off our small planet and become a genuine space-faring species. How could an ethical analysis possibly contribute to the birth of this new science? Further, among all of the diverse religious and philosophic ethical perspectives, why should we concern ourselves with a Christian perspective, especially given the vocal anti-science sentiments voiced by many Christian clergy and laity?

Exploring the Origin, Extent, and Future of Life: Philosophical, Ethical, and Theological Perspectives, ed. Constance M. Bertka. Published by Cambridge University Press. © Cambridge University Press 2009.

My first response to the skepticism outlined above is to argue that part of astrobiology's maturation process as an authentic science is to develop an ethical framework that guides its research and exploration. Just as researchers in other scientific disciplines must meet the ethical requirements mandated by their Institutional Review Boards before proceeding with research involving human and non-human subjects, so also astrobiologists should have ethical standards for which they are held accountable. Developing ethical standards is simply part of becoming a mature science; without established ethical norms and guidelines astrobiology will never take its rightful place among the other disciplines as a serious and respectable science.

In this essay, I will focus on three ethical challenges that can be framed as moral questions of duty or obligation:

1. *What ethical obligations do we have for the protection of Earth, our home planet?* All space missions – both robotic and human – contain some risk that extraterrestrial life forms will inadvertently be released into the Earth's ecosystems where they will substantively harm or destroy native species and ecological communities. This phenomenon is called "back contamination." An important ethical issue concerns the question of an ethically acceptable level of risk that back contamination will occur.
2. *What ethical obligations would human space explorers owe to extraterrestrial life, if we discovered it somewhere in our solar system?* Would there be important ethical obligations that human explorers would owe to extraterrestrial life? Would these ethical obligations include responsibility to protect extraterrestrial ecosystems from harm or damage caused by human explorers? Finally, would ethical responsibilities change dramatically if the extraterrestrial life form proved to be a "second genesis"; in other words, a form of life completely different and independent from life as we know it on Earth?
3. *Do we have an ethical duty to promote life as much as we can?* If human explorers discover extraterrestrial life and through examination determine that it is struggling to survive, do we have an obligation to assist that ecological community become stronger? If after a thorough investigation we determine that no life exists and that a planet is nothing more than a lifeless body of rocks and dust, do we have an obligation to attempt the creation – or re-creation – of life through a process called planetary ecosynthesis? Or, do we have the opposite obligation to respect the rocks and dust and refrain from any attempts to engineer life on a lifeless planet?

At first, these ethical challenges will be the province of the experts, astrobiologists, NASA engineers, ethicists, philosophers, theologians, etc. Indeed, this is who the various contributors to this volume are. Inevitably, the ethical reflection on these three ethical challenges will also be informed by a more general public discourse. After all, space exploration and astrobiology are ultimately social endeavors. As a public agency, NASA is funded by government funds. Further, our self-understanding of ourselves, as well as our place and role in the universe, will be profoundly shaped through space exploration and the discoveries of astrobiology. We are all part of astrobiology and NASA's space odyssey, even if we are never able to experience the thrill of space travel ourselves. As participants in this ambitious social endeavor, we should be able to participate in a public discourse about the ethical principles that will guide how the endeavor is pursued on our behalf.

For Christians, ethical reflection about astrobiology and space exploration will be grounded in their religious faith. Since a statistically significant number of Americans are Christian, it is important for even non-Christians to understand and appreciate this religious perspective. This does not mean that other religious perspectives are unimportant. Quite the contrary. One of the strengths of a pluralistic society is its very diversity of ethical perspectives. Through the mutual give-and-take of public discourse we are able to learn from others with different ethical viewpoints and to gain deeper insights into our own ethical positions, perhaps even modifying or actually changing our original perspective. Everyone should be allowed – even encouraged – to reflect on the question of space exploration from their particular ethical perspective because we can each learn from one another and we each have contributions that we can make to others' thinking about these issues.

In what follows, I intend to demonstrate how Christians would think about these ethical issues. Of course, not all Christians will agree with my conclusions and I can only offer one Christian viewpoint. Nonetheless, in the resources that I use and in my methods of reasoning, I can illustrate how Christians would approach and think about these ethical questions. This should be a valuable exercise for both Christians and non-Christians alike.

In the next section, I will make a brief digression to discuss what distinguishes Christian ethics from other ethical systems – both religious and philosophical. Then, I will suggest that there are two overarching themes within Christian thought that would guide analysis of the three ethical challenges posed above. After developing these two overarching themes, I will examine each of the three ethical challenges from a Christian perspective.

15.2 What makes Christian ethics distinctive?

In many ways, Christian ethics is quite similar to other ethical systems – religious or philosophical. Christian ethicists utilize the same methods of ethical reasoning as do philosophical ethicists. Likewise, Christian virtue theory has much in common with the understanding of virtue in other religions, such as Buddhism. Many of the resources that inform Christian ethical analysis are the same resources that inform other ethical systems. For instance, the natural sciences may be used as resources to inform and influence the analysis of Christian ethicists, just as they may be resources for philosophical or Jewish ethical analysis. Finally, reason provides a critical means for discernment and judgment in Christian ethics, as well as other ethical systems.

Despite these important similarities, the core assumptions and commitments that grow out of the Christian faith make its ethics distinctive, just as other philosophical and religious ethics are distinctive in their own way. For our purposes, Christian scripture is the key distinguishing feature for its ethics. Here, it is important to recall that Christianity grew out of Judaism. Thus, the Hebrew scriptures form an important part of Christian scriptures. Along with the Hebrew scriptures, Christians also include the New Testament texts about Jesus and the development of the early Church. Scripture provides critical resources for ethical reflection, but more significantly it provides a distinctive foundation and parameters for Christian ethics. That is to say, in order to be "Christian," ethical claims must be consistent with Christian scripture and tradition.

Most Christians recognize that their scripture is bound to a particular time and place. Given this reality, scripture cannot always provide direct ethical mandates for many of the ethical challenges confronting contemporary Christians. Certainly, this is the case for the ethical questions under consideration here. Christian scripture does not directly address the possibility of life beyond our home planet or the ethical obligations that we might owe to extraterrestrial life. Nonetheless, there are certain recurring and overarching themes within scripture that should shape and guide Christian thought on contemporary ethical challenges [1]. In the ethics of space exploration, two overarching themes are especially critical for a Christian perspective.

15.3 God's preferential option for life

The three ethical challenges outlined above share important similarities with Earth-bound environmental ethics. Both focus on ethical questions related to the protection of vulnerable species and ecosystems. While Christian scripture does not directly address astrobiology and space exploration, it does substantively address ecological issues on planet Earth. Thus, I will begin by identifying two core, paradigmatic themes within scripture that must inform any Christian environmental ethic. After developing these two themes, I will use them to address our ethical questions concerning astrobiology. Both themes concern relationships with Creation and nature.[1] The first theme focuses on God's relationship with Creation, while the second focuses on humanity's relationship with nature.

---

[1] "Creation" is the theological term for all that exists; that is all of "being." It includes all living organisms but also includes all non-living objects, such as the rocks and dust on Mars or the Grand Canyon on Earth. By

Both themes are grounded in the two creation stories, found in Genesis 1 and 2, the first book of the Bible. These two stories are cosmologies. Rather than attempting to provide scientific explanations of how the world began, both creation stories explore the relationship between God, humans, and the rest of nature.[2] As biblical scholar Theodore Hiebert explains:

Stories of origins such as these are not just descriptions of events in the distant past but etiologies, that is, explanations of why things are as they are in the present. They are first and foremost founding stories. In the act of coming into being, the character of the world, its inhabitants, and their relation to one another is revealed and established for all time. The realities of the author's world are in such accounts of origins both explained and validated as part of a design built into the world from its very beginnings [2] (p. 30).[3]

Although the two creation stories have important differences, they should be seen as complementing one another, rather than competing with each other, in their discussion of the relationships between God, humans, and creation. Therefore, I will use both stories in a complementary way to describe and develop the two overarching themes that are important for our focus.

The first creation story is the well-known story in which God creates the world in six days and then rests on the seventh. The story concludes with the creation of humans on the sixth day:

Then God said, "Let us make humankind in our image, according to our likeness; and let them have dominion over the fish of the sea, and over the birds of the air, and over the cattle, and over all the wild animals of the earth, and over every creeping thing that creeps upon the earth."

So God created humankind in his image, in the image of God he created them; male and female he created them.

God blessed them, and God said to them, "Be fruitful and multiply, and fill the earth and subdue it; and have dominion over the fish of the sea and over the birds of the air and over every living thing that moves upon the earth."

… God saw everything that he had made, and indeed, it was very good. And there was evening and there was morning the sixth day.

(Genesis 1: 26–28, 31)

---

contrast, "nature" refers to all living organisms, from the smallest microbe to the largest mammal. Thus, as I use the two terms, "nature" is a subset of "Creation."

[2] I am aware, of course, that creationist Christians view the first creation story as a scientific account of how the universe was actually created. However, this interpretation is rarely supported by serious Bible scholars. In this essay, I intend to work within the mainstream of the scholarly community.

[3] Another biblical scholar, Gene McAfee, adds: "The Genesis creation account is placed first in the biblical canon not because of its chronological priority, but because of its theological formulation of the relation of the divine to the mundane: every aspect of existence is the result of divine will and activity. … reality is a seamless whole created by God." (G. McAfee. Ecology and biblical studies. In *Theology for Earth Community, A Field Guide*, ed. D. T. Hessel (Maryknoll, NY: Orbis Books, 1996), p. 37.)

While the first creation story is the better known, the second creation story is actually chronologically older, appearing nearly 500 years before the six-day creation story.[4] The second story appears in Genesis 2: 4–24, and reads, in part:

… then the Lord God formed man from the dust of the ground and breathed into his nostrils the breath of life; and the man became a living being. And the Lord God planted a garden in Eden, in the east; and there he put the man whom he had formed. Out of the ground the Lord God made to grow every tree that is pleasant to the sight and good for food … The Lord God took the man and put him in the garden of Eden to till it and keep it. … [and] out of the ground the Lord God formed every animal of the field and every bird of the air, and brought them to the man to see what he would call them; and what ever the man called every living creature, that was its name. … the Lord God caused a deep sleep to fall upon the man, and he slept, then [God] took one of his ribs and … made [from the rib] a woman.

(Genesis 2: selections)

In both creation stories God is presented as the Creator of everything – both animate life and inanimate structures and objects, such as rivers, canyons, and rock formations. Further, a prominent characteristic of both stories is the fundamental goodness of all creation. In the six-day story, God repeatedly looks back on what has been created and judges it good (Genesis 1: 4, 10, 12, 18, 21, and 25). At the end of creation, God looks back upon all that has been created and declares it "very good" (Genesis 1: 31). As John Barr observes, the six-day account is insistent that all of creation is good [3]. This affirmation includes stars, planets, moons, meteorites, and other extraterrestrial formations. The inclusion of extraterrestrial bodies and formations is explicitly included in the six-day story: "God made the two great lights – the greater light to rule the day and the lesser light to rule the night – and the stars. God set them in the dome of the sky to give light upon the earth … And God saw that it was good" (Genesis 1: 16–18, selections). Although extraterrestrial life is not explicitly considered in the creation stories, the logical extension of the stories' affirmations is that extraterrestrial life must also be valued as part of God's good creation.

While God affirms the goodness of all Creation, including non-living objects, I will argue that God "makes a preferential option for life."[5] By "preferential option

---

[4] Scholars of the Hebrew scriptures in the Old Testament believe that the text that we have now is actually a composite of four different sources, known as the Yahwist, Priestly, Elohist, and Deuteronomic sources. The book of Genesis contains creation stories from both the Yahwist and Priestly sources.

[5] I have adopted the language, "preferential option for life," in a deliberate effort to mirror the original claim of Latin American liberation theologians that God makes a "preferential option for poor humans." See G. Gutierrez. *A Theology of Liberation: History, Politics, and Salvation*, trans. Sister C. Inda and J. Eagleson (Maryknoll, NY: Orbis Books, 1973, 1988). In mirroring the phrase from Latin American liberation theologians, I do not intend to minimize in any way their crucial call for justice for the poor and marginalized in Latin America and throughout the world. Rather, I want to suggest that God makes a preferential option for all of life, especially when it is vulnerable to powers of exploitation and oppression. Poor and marginalized persons in Latin America and in other parts of the world struggle constantly against this vulnerability and

for life," I mean that God prefers, values, and privileges life over non-life. Further, God intends for all of life – both human and non-human – to thrive and flourish as both individuals and as ecosystems. This claim is well grounded in the Garden creation story.

Whereas the six-day creation story in Genesis 1 depicts God as the cosmic creator, the Garden creation story in Genesis 2 depicts God as "the divine farmer" [2] (pp. 66–67). After first creating the human, "Adam," God next plants a garden and causes life to spring forth abundantly. Then, God creates the animals, brings them to Adam, and asks Adam to name them. Finally, God creates the woman, "Eve," as a soulmate for Adam. Again and again and again in this creation story, God creates and nourishes life. This does not mean that God devalues inanimate aspects of Creation, such as canyons, rock formations, or beautiful sand beaches. It is not that inanimate objects are not good or important. They are. But rather, God still makes this preferential option for life and its flourishing.

From a Christian point of view, God's preferential option for life means that all of life has intrinsic value. By this I mean that all living organisms, as well as their ecosystems, are entitled to a basic, underlying level of respect – and, even reverence – by humans. Every living organism is good in and of itself, regardless of the instrumental value it may have for humans. The claim that all living organisms have intrinsic value is not absolute, trumping all other claims. That is, the claim of intrinsic value is not intended to protect non-human organisms from certain legitimate uses by humans, such as consuming plants and animals for food. It is not as though there is no hierarchy of value and we have no important decisions to make, concerning how we value and make use of different species and even individuals within species. It is at this point that we move from the first overarching theme, concerning God's relationship with the natural world, to our second overarching theme, concerning humans' relationship with the natural world.

Humans' relationship with the natural world is shaped by predation. For our purposes, predation may be defined as *the forced forfeiture of the life or well-being of one organism so that another organism may live and flourish.* An obvious example of predation occurs when humans consume plants and animals for food, but predation may also occur for other reasons. For example, we may cut trees in order to have lumber for building houses as shelter against the elements. Alternatively, we may fumigate mosquitoes because they carry malaria and threaten human health.

the prospect of being extinguished. But, as a result of human activity, much non-human life is also vulnerable to the threat of being extinguished. Thus, it is fully consistent with Christian teachings to claim that God "makes a preferential option for poor humans" and also a "preferential option for all vulnerable life." Among Latin American liberation theologians, Leonardo Boff has argued very persuasively that concern for poor persons and vulnerable non-human life are complementary, rather than competing, Christian concerns. See, for example, Boff's *Cry of the Earth, Cry of the Poor*, trans. P. Berryman (Maryknoll, NY: Orbis Books, 1997).

Scientific research involving living organisms is also a form of predation, when it involves the forced forfeiture of an organism's well-being so that we can learn through scientific study. For instance, medical researchers may use animals to test new drugs. In this case, the researchers may take the animal's well-being or life in order to improve the health and flourishing of humans through a new drug. Even pure scientific research, pursued with no predetermined benefit for humans, is a form of human flourishing for the scientists who intellectually flourish through their pure research. To reiterate, predation is an integral component of humans' relationship with nature. While predation is inevitable, it may be either ethical or unethical, depending upon how it is pursued. The two creation stories establish criteria for ethical predation in their explication of humans' relationship with the rest of nature.

The six-day creation story acknowledges the reality of predation when it notes that humans were given "dominion" over the rest of nature. Human "dominion" grows out of the story's claim that humans are created with the *imago dei*, the image of God. That is, God sets humans apart as special and different from the rest of nature, and God gives them special responsibilities for the stewardship of nature. For most biblical scholars, the interpretative key to this passage is the "royal metaphor" that was prevalent in the dominant Egyptian and Mesopotamian cultures that surrounded the Hebrew people at the time. In these cultures, the king was described as possessing the "image of that culture's god." As biblical scholar Bernhard Anderson observes, the *imago dei* described "the relationship of the king to the deity and specifically his function as the god's representative in his royal office" [4] (p. 41).[6] Representing the god's authority on Earth, the Egyptian or Mesopotamian king was responsible for insuring the safety and security of the people and their land. Also included within this kingly authority was responsibility for insuring justice and protecting those who were politically and economically weak and vulnerable.

According to biblical scholars, Genesis 1 takes this common "royal metaphor" and "democratizes" it. Rather than ascribing the *imago dei* to one particular individual who was the king, the first Genesis story ascribes the *imago dei* to all of humanity. Anderson writes, "Using mythical language, the priestly writer affirms that Man stands in close relationship to God (the Heavenly Council) and is the agent of the divine rule on earth" [4] (p. 42).[7] In this capacity as God's agent on

---

[6] For a discussion of the *imago dei* image as it appeared in Egyptian theology, see J. Atwell. An Egyptian source for Genesis 1. *The Journal of Theological Studies*, **51**(2) (2000), 463–467.

[7] Writing 35 years ago, Anderson uses the term "man" and masculine pronouns to refer to human persons or the human species in general. Although this was the conventional practice when Anderson wrote, since then many writers have consciously adopted less gender-specific terms, such as "humans," instead of more sexist terms, such as "man." In quoting Anderson, I have kept his original language. However, from the rest of my text, it should be clear that as a writer I strongly support inclusive language.

Earth, humans are to serve as stewards, or caretakers, responsible for the care of non-human nature and the ecological communities in which we live together. Anderson continues: "Thus Man's special status, as the image of God, is a call to responsibility, not only in relation to fellow men but in relation to nature. Human dominion is not to be exercised wantonly, but wisely and benevolently so that it may be, in some degree, the sign of God's rule over his creation" [4] (pp. 44–45). Viewed in this way "dominion" places a special responsibility upon humans to be good stewards of the rest of nature. Finally, this role of caring for nature must be complemented by an equally important responsibility to respect the dignity and insure justice for all human persons, just as the Mesopotamian and Egyptian kings – as representatives of their nations' gods – were responsible for protecting and insuring justice for the most vulnerable persons in their societies. All humans bear the *imago dei* and thus all humans are entitled to respect and justice.

The six-day perspective on humans' relationship with nature must be read in conjunction with the view in the Garden creation story, which deepens and elaborates the six-day story's claims. As we've already seen, the Genesis 2 story depicts God as a farmer, who plants a beautiful Garden for humans to live in and cultivate. When Adam is placed in this paradisaical Garden, God instructs him to "till it and keep it." The Hebrew verb used at this point for cultivating (or tilling) the Garden is *ā bad*, a verb which customarily means to "serve," as in the servitude given by a servant to his or her master [2] (pp. 65–66). As biblical scholar Theodore Hiebert explains, "humans are regarded as subservient to the soil upon which their survival depends" [2] (pp. 65–66). That is, from the perspective of the Garden account, *humans were created by God to serve nature, rather than nature being created to serve humans.*

While the Garden creation story does make some distinctions between humans and other animals, it is rather the similarity between human and non-human life that is the most striking. Humans, animals, and plants all share a common origin: the arable soil. This is important because it puts all of life on the same plane, without a hierarchy of importance. Further, it suggests that all of life – human, animal, and plant – belongs to a single community of life. As Hiebert observes: "The human being is viewed more as an ordinary member of the community of life than as a privileged being set apart from it" [2] (pp. 65–66). For the writer of the Garden account, humans may be viewed as older siblings of the rest of nature.

Taken together, these two creation stories paint a very distinctive portrait of humans as "servant-stewards" for the rest of nature. Created in the *imago dei*, humans are set aside as unique from the rest of creation. However, the two stories are careful to underscore that this special status does not place humans on a pedestal, with the rest of nature serving them. Instead, it is the exact opposite. God singles out humans in the same way that a parent singles out older brothers and sisters to watch over their younger siblings. Similarly humans are to watch over and take care of their non-human siblings. Indeed, humans are to serve their non-*human* siblings. Equally important, humans are to serve and protect their *human* siblings, especially those persons who are the most vulnerable and who suffer injustice.

These perspectives on the relationship between God, humans, and nature provide a clear framework and criteria for Christian thinking about ethical predation. God's preferential option for life and the intrinsic worth of all living organisms establishes the context in which ethical decision making must occur. Within this context, a profound respect and reverence for the living organisms that must forfeit their well-being form the foundation for any justification of predation. Associated with this respect and reverence would be an authentic desire to promote the flourishing and self-determination of all organisms as much as possible.

For predation to be judged ethical, three additional criteria must be met: first, there must be well-grounded reasons for predation. That is, the forced forfeiture of living organisms' well-being should not be done lightly or without good cause. Second, when predation cannot be avoided, it should be minimized as much as possible. Questions of proportionality are crucial, here; a substantial forced forfeiture for a superficial gain would not be ethical. Third, some form of reparations should be considered, even when predation is ethically justified. For instance, the destruction of a certain, vulnerable species' habitat in order to build a needed hospital to care for humans may require the setting aside and development of other areas so that this particular species will be able to continue to flourish.

From a Christian perspective, the theme of God's relationship with nature, especially the preferential option for life, and the theme of humans' relationship with nature, including criteria for appraising the appropriateness of predation, provide the parameters and the analytical tools for thinking about the three ethical challenges raised by astrobiology and human space exploration. Let us turn now to those issues.

15.4 Ethical obligations for the protection of Earth and its ecological communities

In terms of space exploration, *back contamination*[8] offers the major threat to Earth, its inhabitants and their ecological communities. Back contamination refers to the accidental introduction of extraterrestrial life into some or all of the Earth's ecosystems. The concern is that released extraterrestrial life forms will have pathogenic or large-scale ecological effects, killing Earth-based organisms and destroying

---

[8] For an informative discussion of back contamination, please refer to Margaret Race's chapter in this volume.

ecological communities. Potentially, such extraterrestrial pathogens could even kill humans. We already have models for this risk in the invasion of non-native Earth species into new ecological communities. Examples of non-native species that currently threaten US ecological communities include the Northern snakehead fish, zebra mussels, Africanized (killer) honey bees, and the fast-growing vine Kudzu.[9]

The protection of the Earth and its ecological communities has always been an important concern for space exploration. The Outer Space Treaty of 1967 requires that space exploration be done "so as to avoid … harmful contamination and also adverse changes in the environment of the Earth resulting from the introduction of extraterrestrial matter" [5]. More recently, the NASA Astrobiology Roadmap lists planetary protection as one of four fundamental operating principles for the development of the new science: "Astrobiology encourages planetary stewardship through an emphasis on protection against forward and back biological contamination and recognition of ethical issues associated with exploration" [6].

From a Christian ethical perspective, there are several important considerations concerning back contamination. God's preferential option for life would affirm the intrinsic value of all life and not discriminate between native Earth species – including humans – and invading alien species. All of life has intrinsic value. While this consideration establishes the context in which we think about back contamination, it is not the over-riding moral claim. From a Christian perspective, humanity's role as "servant-steward" would be the principal moral claim. Humans have been entrusted with the responsibility of caring for and promoting the flourishing of all living organisms and their ecological communities here on Earth. Therefore, humans have an obligation to protect Earth-based life and their ecosystems from threats and risks posed by the introduction of extraterrestrial organisms. This obligation is especially critical when the extraterrestrial risk arises as a result of human space exploration.

The prevailing scientific assessment is that the risk of back contamination on Earth is extremely low [7] (p. 4). NASA and the other space agencies have been very vigilant, going to great lengths to insure that extraterrestrial microbes will never be released into the Earth's ecological communities. For instance, in its plans for new missions to Mars, NASA will require that the return of Martian samples be in " … a fail-safe, durable container that can be remotely sealed, cleanly separated

---

[9] One model of this risk is the invasion of the air-breathing freshwater snakehead fish into US waterways. These fish are originally from Asia, where certain species are highly valued for food. They were imported into the USA for food and as aquarium fish. Released into US aquatic ecosystems, the snakeheads proved to be "voracious predators, feeding upon other fishes, crustaceans, frogs, small reptiles and sometimes birds and small mammals." Federal officials believe that if the snakehead fish become established in North American ecological communities, "their predatory behavior could drastically modify the array of native species. As a result, they could disrupt the ecological balance and forever change native aquatic systems." See the National Invasive Species Council, Northern snakehead (*Channa argus*). Available online at: http://www. invasivespecies.gov/ismonth/archives/snakehead/snakehead.html, accessed 12 August 2005.

from Mars, monitored en route, and opened in an appropriate sample receiving facility on Earth" [7] (p. 4).

While NASA and the other space agencies have certainly maintained due diligence in protecting against back contamination, there remains a significant moral issue that I have not seen addressed in any of the literature. The risk of back contamination is not zero. There is always some risk. In this case, the problem of risk – even extremely low risk – is exacerbated because the consequences of back contamination could be quite severe. Without being overly dramatic, the consequences might well include the extinction of species and the destruction of whole ecosystems. Humans could also be threatened with death or a significant decrease in life prospects.

In this situation, what is an ethically acceptable level of risk, even if it is quite low? This is not a technical question for scientists and engineers. Rather, it is a moral question concerning accepting risk. Currently, the vast majority of the people exposed to this risk do not have a voice or vote in the decision to accept it. Most of the literature on back contamination is framed as a discourse among experts in planetary protection. Yet, as I've already argued, space exploration is inescapably a social endeavor done on behalf of the human race. Astronauts and all the supporting engineers and scientists work as representatives of all human persons. From the Christian perspective, the claim of the *imago dei* requires that all human persons be treated with respect and guaranteed justice. In this situation, to treat persons with dignity and justice means that everyone must have an opportunity to voice their opinion concerning whether humans should accept the risk. Thus, we might suggest four criteria that would be essential for any judgment that a certain level of risk was ethical. These four criteria are:

1. The best practices of planetary protection must be followed. As I suggested above, NASA appears to be exercising due diligence in its plans for preventing back contamination when samples from Mars are eventually returned to Earth. Yet, pursuing best practices by itself does not necessarily guarantee an ethically acceptable level of risk.
2. There should be opportunities for open comment by those individuals or groups that have concerns about the risks of back contamination. These comments should be taken seriously and NASA should publicly respond to these concerns.
3. A committee of neutral or disinterested persons should review the planetary protection measures for the return of spacecraft and samples. This committee should include persons with a diversity of expertise, including ecology, biology, chemistry, specialists in risk analysis, and ethicists. The ethicists should represent a diversity of philosophical and religious perspectives.
4. The entire process of soliciting comment, analyzing the risk factors, and deciding on whether the risk levels are ethically acceptable should be transparent to the interested public.

15.5 Do we have ethical obligations to extraterrestrial life?

While encounters with intelligent extraterrestrial life have become prominent in popular culture through movies, such as *Contact* and *Independence Day*, we are probably more likely to first discover unintelligent extraterrestrial life somewhere in our solar system. This life form would most likely be microbial, although conceivably it could be a visible but still fairly small life form, such as lichen on a rock or perhaps some small but visible organism swimming in extraterrestrial water.

Based upon our understanding of life on Earth, we would expect to find extraterrestrial life near a source of liquid water. As Chris McKay observes, "When we scan the other worlds of our solar system, the missing ecological ingredient for life is liquid water. It makes sense, then, that the search for liquid water is currently the first step in the search for life on other worlds" [8] (p. 1260). There are several places in the solar system that have water and are, therefore, the most likely candidates for harboring extraterrestrial life. The planet Mars offers good possibilities for the eventual discovery of life [9] but there are other locations, as well.

If extraterrestrial life is discovered, it could be very similar to life as we know it here on Earth [8] (p. 1262). Alternatively, we could discover extraterrestrial life that is radically different from life on planet Earth. As McKay observes, "Other worlds may have a different chemical baseline for life. The usual speculation in this area is that the presence of ammonia and silicon, rather than water and carbon, might be preconditions for life on other planets" [8] (p. 1261). Radically different extraterrestrial life would represent a second form of life; that is, a second "genesis." There are several possible discovery scenarios, depending upon the life form discovered. If the extraterrestrial life form is microbial, then it would most likely be found in an Earth laboratory during examination of extraterrestrial samples returned from a space mission. Alternatively, if the life form were larger and more visible, such as the lichen or water-borne organism suggested above, then it is possible that the discovery would be made *in situ*, during a robotic or human mission.[10]

What obligations do we owe to extraterrestrial life, if it is discovered? Within this broad issue of duty there are several important questions. From a Christian perspective, these questions must be addressed within a context of respect and reverence for all life. God's preferential option for life grounds the claim that all of life has intrinsic worth and that God intended for extraterrestrial life to flourish and be self-determinant.

One question concerns our moral obligations to extraterrestrial ecosystems. Mirroring concerns with *back contamination* on Earth, there is a complementary

---

[10] For a more detailed examination of various discovery scenarios, see M. Race. Communicating about the discovery of extraterrestrial life: different searches, different issues. SETI Paper #: IAC-04-IAA.A.A.2.02 presented at the International Astronautical Conference, Vancouver, Canada, October 2004.

concern with possible *forward contamination* of extraterrestrial destinations by spacecraft originating from Earth. The most likely scenarios of forward contamination involve "hitch hiker" microbes from Earth traveling to an extraterrestrial destination, such as Mars, and subsequently escaping onto the Martian surface or into the atmosphere. The prevailing scientific assessment is that the probabilities of forward contamination are extremely low [7], and NASA appears to be very vigilant in protecting against this concern: "Depending on the mission, a variety of elaborate procedures are used such as cleanroom assembly of the spacecraft, sterilization of landing spacecraft, special planning for orbital lifetimes to avoid premature impact on the planet, and heat treatment or special packaging of scientific instruments to further reduce the bioload or number of microbes on the outbound spacecraft" [7] (p. 3).

In its efforts to prevent forward contamination of Mars and other extraterrestrial destinations, NASA appears primarily motivated by its concern to protect the integrity of human science rather than protecting the integrity of the extraterrestrial destination. As Margaret Race and I observed, "... all the key NRC reports on Mars exploration and sample return are written from the perspective of ensuring the integrity of scientific experiments to detect and characterize possible extraterrestrial life, rather than any concern *for* that life. Part of the reason for this position is pragmatic: it is important to preserve life and the ecosystem in a manner that insures they will be available for research by future scientists of all nations" [10] (p. 1583). In other words, NASA's chief concern appears to be guarding against "false positives" in the search for extraterrestrial life.

A Christian perspective would share the scientific concern for protecting the integrity of extraterrestrial ecosystems – but for different reasons. Whereas the scientific perspective informing NASA concerns the importance of preserving integrity for future scientific research, the Christian perspective would center on God's preferential option for life and the importance of promoting the flourishing and self-determination of all living organisms. Further, humans have special obligations as "servant-stewards" of creation. As humans are transformed into a space-faring species, the scope of our stewardship responsibilities would increase with our reach into space. Thus, human explorers would be responsible for nurturing and protecting extraterrestrial ecosystems, especially when it is our own space exploration that poses the threat to their ecological integrity.

This leads to a related question concerning the return of living samples to Earth under those circumstances, where the specimens are discernable *in situ* to robotic or human explorers. As noted above, a possible example of this scenario would be the discovery of a colony of lichen living among the rocks on Mars or an organism swimming in Martian water. From a Christian perspective, the ethical response would be governed by an initial precautionary principle that precluded simply picking up a specimen *in situ* and returning it to Earth for further research. As we have seen, human servant-stewards must treat all other living organisms with reverence and respect in a way that promotes their flourishing and self-determination. Upon first discovery, humans would not have sufficient understanding of the extraterrestrial ecosystem in order to understand what actions might harm or destroy it, let alone understanding how to promote flourishing and self-determination of extraterrestrial ecosystems. Similarly, human servant-stewards would not have adequate knowledge at the level of individuals, either. Thus, upon first discovery, humans would need to gather information about the extraterrestrial species and its habitat through observation and study before engaging in direct contact and manipulation. This information would include some understanding of how the extraterrestrial flourishes and what threatens the extraterrestrial's well-being. We would need to gain an understanding of the range and characteristics of the extraterrestrial's habitat. Further, human explorers would need to develop some understanding of what actions might damage or destroy that habitat. To the degree possible, we would need a rough census of the extraterrestrial's population. Once sufficient study had been performed, the transport of living samples to Earth would be ethical provided that the three criteria for predation were met, since it is unlikely that the specimens would ever be returned to their Martian habitat unharmed.

Of course, the more likely discovery scenario is that we will find extraterrestrial microbes on Martian samples returned to Earth. From a Christian perspective, God's preferential option for life would extend to microbial life as well. Thus, the handling and treatment of the extraterrestrial microbes should be performed with the same attitude of respect and reverence. Once discovered in an Earth laboratory, study of the new life form should initially focus on understanding how the microbial organism flourishes and what harms it so that human scientists and space explorers can understand how to care for the microbial extraterrestrial in the laboratory and in its indigenous extraterrestrial habitat. God's preferential option for all of life would mean that even microbial organisms should be treated with respect and reverence as living organisms with intrinsic value.

Inevitably, the discovery of non-sentient extraterrestrial life will raise questions concerning whether predation is justified for scientific – or perhaps, eventually, commercial – purposes. From a Christian perspective, such predation may be ethically justified provided the three criteria for ethical predation are met and provided it occurs with an attitude of respect and reverence. However, a final question remains concerning the moral status of extraterrestrial life, if it represents a "second genesis" of life. In an early article, Margaret Race, Chris McKay, and I argued: "From an ethical point of view, the need to preserve a life-form, however lowly, must be more compelling if that life-form represents a unique life-form with an evolutionary history and origin distinct from all other manifestations of life" [7] (p. 3).

From a Christian perspective, the designation of "second genesis" is purely a matter of Earth-bound humans' perspective. We would expect God's preferential option for life to extend across the universe, encompassing all life forms, from the most common Earthly microbial organism to the most "exotic" organism living on the far side of the universe. At the foundational level, all living organisms have intrinsic value and no special status would accrue to a "second genesis" organism. While all living organisms should receive a foundational reverence and respect, the three criteria for ethical predation would appear to offer "second genesis" organisms special protection from predation, even if they were just an extraterrestrial microbial organism. In such a scenario, it would be difficult to meet the first (well-grounded reasons) and second (minimal harm) criteria in the case of "second genesis" organisms. Meanwhile, the third criterion (some form of reparation) would presumably be impossible to meet.

15.6 Do humans have an ethical duty to promote life?

Our third cluster of ethical issues concerns human duties to promote life. Specifically, we will examine large-scale planetary engineering undertaken, either, to assist struggling extraterrestrial life or to establish extraterrestrial life where it currently does not exist. These possibilities have been discussed within NASA for well over 25 years and various terms have been used to identify the possible procedures. For our purposes, I will use the term, "planetary ecosynthesis."[11] Planetary ecosynthesis refers to the process of improving or creating the requisite conditions for sustaining life on another planet through intentional human engineering, so that a permanent, life-giving environment is created. Chris McKay argues that the most likely planet in the solar system for planetary ecosynthesis is Mars [11]. Consequently, we will use Mars as a specific example in the following discussion. (The technical details of planetary ecosynthesis are discussed by McKay in his chapter in this volume.)

If planetary ecosynthesis proved feasible, what are the ethical considerations concerning the creation of conditions for life on a lifeless planet? Chris McKay and others have observed that the ethical considerations of planetary ecosynthesis hinge on how we assign value. On the one hand, if value is construed to include only biological life, then planetary ecosynthesis is ethically justified. Indeed, planetary ecosynthesis may even be obligatory because value accrues only to life. On the other hand, if value is construed more broadly so that it includes not only biological

---

[11] The term "planetary ecosynthesis" is suggested by Christopher P. McKay and Margarita M. Marinova in their paper, The physics, biology, and environmental ethics of making Mars habitable. *Astrobiology*, **1**(1) (2001), 90. They argue that this term is preferable to "terraforming" because it "avoids the misleading comparison with Earth."

life but also inanimate landscapes, such as dirt, rocks, and geologic formations, then planetary ecosynthesis is unethical. Indeed, we are obligated to respect the dust and rocks of a lifeless Mars and leave Mars alone. With this as a general framework for thinking about planetary ecosynthesis, let us examine the two scenarios outlined above.

In the first scenario, human explorers discover life on Mars or some other extraterrestrial body. Unfortunately, the extraterrestrial ecosystem is very fragile and vulnerable. The living organisms within the ecosystem are desperately struggling for survival. It appears likely that the ecosystem will eventually collapse and this form of extraterrestrial life will become extinct. Suppose that through observation and study we develop some insights into why Martian ecological communities and inhabitants are fragile. Further, we develop a plan for planetary ecosynthesis that should rejuvenate the Martian ecological communities and promote a renewed flourishing of indigenous life. As a space-faring species, do we then have an ethical obligation to help or enhance extraterrestrial life through planetary ecosynthesis?

From a Christian ethical perspective, a strong case for planetary ecosynthesis could be made under these circumstances. God's preferential option for life indicates that God prefers, values, and privileges life over non-life. God intends for the Martian ecological community and all of its life forms to grow and flourish. Further, God has set humans aside and given us special responsibilities as "servant-stewards." Humans are to serve the rest of nature by protecting and nurturing it. As a newly established space-faring species, our servant-steward purview now extends beyond our home planet into space, including Mars. If it were technically possible to rescue a struggling Martian ecosystem, then there would be strong ethical support for planetary ecosynthesis from a Christian perspective.[12] Yet at the same time, Christian ethical support for promoting the flourishing of life on Mars could well be checked by an equally important ethical concern for the flourishing of life on Earth.

Although the United States is the wealthiest and most affluent society in the history of planet Earth, a significant proportion of its population exists in poverty, without adequate nourishment, clothing, shelter, or healthcare. For example, 16.7% of the total population of children (12.1 million) in the United States lived below the poverty line in 2002, according to the US Census [12]. In 2003, 41.6 million Americans under the age of 65 (16.5% of the population) reported having no

---

[12] See C.P. McKay. Ethics and planetary engineering, does Mars have rights? In *Moral Expertise, Studies in Practical and Professional Ethics*, ed. D. MacNiven (London and New York: Routledge, 1990), pp. 186–192; R.O. Randolph, M.S. Race, and C.P. McKay. Reconsidering the theological and ethical implications of extraterrestrial life. *The Center for Theology and the Natural Sciences Bulletin*, **17**(3) (1997), 5–6; and C.P. McKay and M.M. Marinova. The physics, biology, and environmental ethics of making Mars habitable. *Astrobiology*, **1**(1) (2001), 105–106.

healthcare coverage [13]. This number of poor is literally obscene, given the wealth and affluence of the American society. Recall that the Christian conception of the "servant-steward," especially as developed in the Genesis six-day creation story, incorporates both care for nature and care for other persons. From the perspective of the six-day creation story, all human persons bear the image of God. All persons must be respected and receive justice. In the US context, genuine justice would mean that all persons should have access to sufficient nourishment, clothing, shelter, and healthcare. From a Christian perspective, promoting the health and flourishing of our human neighbors here on Earth would appear to be a greater responsibility than a planetary ecosynthesis project to promote the flourishing of non-sentient extraterrestrial life on Mars. Robert Haynes appears to disagree with this ethical judgment. He writes: "…would not such a venture [planetary ecosynthesis] inevitably erode the level of public funding that otherwise might be devoted to more immediate earthbound priorities…for example the alleviation of disease, poverty and famine?…ethical questions of this latter kind are by no means novel. Unfortunately, they have little force in affecting public policy, as so amply demonstrated in the enormous amounts of money that even poor countries lavish on their military establishments" [14] (p. 176). From a Christian ethical perspective, Haynes' argument is flawed. For the most part, space exploration is a public endeavor, supported by public monies for the entire society. A profound economic injustice occurs if planetary ecosynthesis is funded, while innocent children and adults live in poverty. After all, the entire space program, as a public trust, is pursued on behalf of the poor along with all of the other citizens of American society.

Haynes appears to think that funding planetary ecosynthesis is permissible anyway, if there is little reason to believe that public policy makers will address the problem of children in poverty. But, proceeding on with planetary ecosynthesis as though there is no problem with poverty actually serves to enable and exacerbate the economic injustice suffered by the most vulnerable members of our society. By ignoring the problem of poverty and economic hopelessness among so many in our society proponents of planetary ecosynthesis, such as Haynes, become co-opted as part of the cause of economic injustice that persons in poverty suffer. From a Christian perspective, the flourishing of extraterrestrial life on Mars must also be linked with the flourishing of human life here on our home planet.

The second scenario envisions a Mars completely devoid of life in any form. Here, the goal of planetary ecosynthesis would be to create life where no life previously existed. That is, the point is to develop an environment that sustains the conditions necessary to support living organisms. As Chris McKay and Margarita Marinova explain, planetary ecosynthesis would take at least 100 years on Mars and the newly created environment produced would sustain plant life only.[13] Once a sustainable environment has been established, new plant life from planet Earth could be successfully introduced.

As we have already seen, the two creation stories that are normative for Christians both affirm the goodness of all creation, including those inanimate objects such as the dust and rocks on a lifeless Mars. At the same time, the stories suggest that God makes a preferential option for life. This preferential option should not be interpreted as devaluing the goodness of inanimate rocks and dust, but it does suggest that humans, as servant-stewards, are to encourage and promote life whenever and wherever they can. Just as the Genesis 2 Garden Creation story depicts God as the "divine farmer" planting a garden for humans, so also planetary ecosynthesis could be viewed as humans faithfully "planting a new garden" on a lifeless planet. Of course, planetary ecosynthesis would require great effort and expense because of the distance and the numerous unknowns and risks associated with such an undertaking. Given these difficulties, planetary ecosynthesis could be permitted – but not required – from a Christian ethical perspective.

Before this second scenario could be ethically permitted, however, two key questions must be addressed. First, as with the first scenario, a Christian ethical perspective would require that we link the creation of a new garden on Mars with better attainment of flourishing by poor humans on Earth.

The second question concerns the assumption in this scenario that Mars would be completely lifeless before beginning planetary ecosynthesis. Robert Haynes frames this question nicely when he writes: "Short of totally ransacking the planet searching for indigenous biota, an undertaking as damaging as it would be costly, it is difficult to see how one could ever convince uncompromising critics that the planet is totally sterile everywhere" [14] (p. 178). While it is possible to prove that Mars does have life, strictly speaking, one can never conclusively prove that Mars is lifeless. This problem is further complicated because the possible life on Mars could be a "second genesis" life form, completely different from life as we know it and therefore potentially more difficult to recognize.

The ethical concern here is that we may unwittingly destroy or irreparably change an indigenous Martian ecological community through a process of planetary ecosynthesis. From the Christian perspective, humans' role as servant-stewards protecting and nurturing ecological communities is pivotal to thinking about our duties and obligations as a space-faring species. As we have already seen in our

---

[13] Here, a caveat is in order. I am imagining a scenario in which extraterrestrial life forms are struggling because of the imminent collapse of a well-defined, specific ecological community. I am not thinking of the situation where an extraterrestrial life form is in danger of extinction through the evolutionary process in which one species gives way to a newer species. That is to say, I am not arguing that we have an ethical obligation to interfere with the evolutionary process – either on Earth or another planet. Rather, I am arguing that we have an obligation to promote the flourishing of life by creating healthy ecological communities. This may require planetary ecosynthesis to achieve this end.

earlier discussions of both back contamination and forward contamination, this Christian vision for ecological stewardship tends to privilege indigenous living organisms and their ecosystems over alien organisms encroaching from another planet. On this basis, our first concern in planetary ecosynthesis should be preserving any indigenous Martian ecological communities and promoting the health and flourishing of the indigenous Martian species that are the citizens of those communities. Planetary ecosynthesis designed to create an environment which is sustainable for new life introduced from Earth should be pursued as the "last choice" option. Thus, insuring that Mars is completely devoid of life becomes a critical issue. In a fascinating article, Chris McKay has offered an interesting proposal for chemically determining whether Mars is a genuinely "dead" planet [8] (p. 1262). However, there is much technical and scientific work remaining on this question. Thus, before planetary ecosynthesis would be ethically *permissible* in this second scenario, we must develop a stronger scientific consensus that Mars is completely devoid of life.

God's preferential option for life over non-life grounds a general Christian predisposition in favor of planetary ecosynthesis. However, as we have seen in both of the scenarios considered here, several important issues must be addressed further before planetary ecosynthesis would be permissible and supported from a Christian ethical perspective.

15.7 Conclusion

Over the past 50 years, humans have made remarkable progress in our quest to become a space-faring species. This current moment in our history is especially exciting because we stand on the threshold of new exploration possibilities and the development of a new space science, astrobiology. As we continue to develop and mature as a space-faring species, we will begin to encounter ethical questions concerning our conduct in space. As I have argued throughout this chapter, space exploration should be conceived as a social endeavor that astronauts, astrobiologists, engineers, and others conduct on behalf of the entire human species. Accordingly, the space ethic that guides our explorations should be informed by a public discourse that includes a wide diversity of perspectives.

In this chapter, I have offered a preliminary examination of three ethical areas that we will certainly have to address as a space-faring species: What are our duties and obligations to any extraterrestrial life that we may encounter in our space travels? What obligations do we have to the extraterrestrial places that we visit? What obligations do we owe to our home planet and all of its inhabitants, both human and non-human, as we return from space? I have examined these issues from a particular perspective, that of Christianity. I have tried to demonstrate how a Christian might think about these ethical questions. Since Christians form a statistically significant percentage of the American population, it is important for both Christians and non-Christians alike to know and understand the Christian ethical perspective on these issues. But, it is also important that other perspectives – both religious and non-religious – enter into this public discourse. Thus, in addition to laying out a Christian perspective, I also hope that I have provoked a broad and diverse public discourse on these ethical issues that we will face as a space-faring species. I hope that this public discourse will occur both inside and outside astrobiology. If I am a bit successful in provoking this public discourse, then in a way I will have made my own small contribution to our development as a space-faring species [15].

References

[1] B. C. Birch and L. L. Rasmussen. *Bible and Ethics in the Christian Life*, revised and expanded edition (Minneapolis, MN: Augsburg Press, 1989), pp. 186–188.

[2] T. Hiebert. *The Yahwist's Landscape, Nature and Religion in Early Israel* (New York and Oxford: Oxford University Press, 1996).

[3] J. A. Barr. Man and nature: the ecological controversy and the Old Testament. *Bulletin of the John Rylands University Library of Manchester*, **55** (Autumn 1972), 31.

[4] B. W. Anderson. Human dominion over nature. *Biblical Studies in Contemporary Thought*, The Trinity College Biblical Institute, **10**, ed. M. Ward (Somerville, MA: Greeno, Hadden and Company, Ltd., 1975).

[5] United Nations. *Treaty on Principles Governing the Activities of States in the Exploration and Use of Outer Space, Including the Moon and Other Celestial Bodies*, available online at: www.unoosa.org/oosa/en/SpaceLaw/gares/html/gares_21_2222.html; accessed 17 December 2007.

[6] NASA. *Astrobiology Roadmap*, available at: http://astrobiology.arc.nasa.gov/roadmap, accessed 12 August 2005.

[7] R. O. Randolph, M. S. Race, and C. P. McKay. Reconsidering the theological and ethical implications of extraterrestrial life. *The Center for Theology and the Natural Sciences Bulletin*, **17**(3) (1997), 1–8.

[8] C. McKay. What is life – and how do we search for it in other worlds? *PLoS Biology*, **2**(9) (September 2004), e302. Available online at: http://biology.plosjournals.org/perlserv/?request=get-document&doi=10.1371/journal.pbio.0020302; accessed 11 August 2005.

[9] NASA. *The Vision for Space Exploration* – February 2004 (Washington, DC: NASA, 2004), p. 9. Available online at: www.nasa.gov/pdf/55583main_vision_space_Exploration2.pdf; accessed 11 August 2005.

[10] M. S. Race and R. O. Randolph. The need for operating guidelines and a decision making framework applicable to the discovery of non-intelligent extraterrestrial life. *Advances in Space Research*, **30**(6) (2002), 1583–1591. (Emphasis in the original.)

[11] C. P. McKay. Ethics and planetary engineering, does Mars have rights? In *Moral Expertise, Studies in Practical and Professional Ethics*, ed. D. MacNiven (London and New York: Routledge, 1990), p. 185.

[12] B. D. Proctor and J. Dalaker. US Census Bureau, Current Population Reports, P60–222, *Poverty in the United States: 2002* (Washington, DC: US Government Printing

Office, 2003), p. ix. Also available online at: http://www.census.gov/prod/2003pubs/p60–222.pdf, accessed 15 August 2005.

[13] National Center for Health Statistics. *Chartbook on Trends in the Health of Americans* (2005), available online at: http://www.cdc.gov/nchs/data/hus/hus05.pdf, accessed 15 February 2006.

[14] R. H. Haynes. Ecce ecopoiesis: playing God on Mars. In *Moral Expertise, Studies in Practical and Professional Ethics*, ed. D. MacNiven (London and New York: Routledge, 1990), pp. 161–183.

[15] C. P. McKay and M.M. Marinova. The physics, biology, and environmental ethics of making Mars habitable. *Astrobiology*, **1**(1) (2001), 90–104.

16

Comparing stories about the origin, extent, and future of life: an Asian religious perspective

Francisca Cho

16.1 Paradigms of cosmic meaning

The purpose of this chapter is to compare and contrast Western and Asian *ways* of telling stories about the world rather than the actual stories they tell. The hope of this exercise is that by understanding the logic that drives Asian storytelling, we can see a new way of getting past the inevitable differences between the stories that religion and science tell. This way of approaching conflicting stories, it will be argued, offers significant improvements over two popular strategies for reconciling religion and science: first, the argument that they tell completely different and non-overlapping stories; and second, the attempt to integrate religious and scientific stories into one seamless whole. In contrast to these approaches, the logic of Asian storytelling recognizes the importance of having many different and overlapping stories about the same event without feeling compelled to reconcile them to each other.

The voyages of discovery that spanned the globe and mapped new continents in the early modern era of European history were underwritten by faith in progress and the capacity of human knowledge – which were, in turn, rooted in a Judeo-Christian philosophy of history. In the present age, which has already embarked upon interplanetary exploration, we are situated in a vastly more complex ethical and religious world: the prior age of exploration and intercultural contact has revealed the diversity of religious and philosophical perspectives within Earthbound cultures.

The premise of this essay is that it is possible to take advantage of this new complex order by triangulating scientific, monotheistic, and Asian religious perspectives in order to gain new knowledge. I do not mean by this the "add on" method of determining how a Buddhist or Confucian would respond, for instance, to the

Exploring the Origin, Extent, and Future of Life: Philosophical, Ethical, and Theological Perspectives, ed. Constance M. Bertka. Published by Cambridge University Press. © Cambridge University Press 2009.

ethical questions raised by the prospect of reviving, transferring, and/or sustaining life on Mars. While these questions are not irrelevant, the point is to go beyond the additive process of compiling the responses of the world's religions. Such a survey is interesting, but its ultimate purpose needs to be clarified. Is the point, for example, to decide which religious view will be authoritative – for if there is one dominant lesson from the prior age of discovery, it is that religious formations vehemently disagree with each other. Shall we proceed on the principle of "majority rule" and confer the advantage to Christian values, based on its numerical and cultural domination of American society? Or should we advance more pragmatically, selecting the tradition in most sympathy with modernism and scientifically based values?

In this essay, the purpose of comparative study is to illuminate the parameters of our existing knowledge by demonstrating alternative ways of framing our questions. The dialogue between science and religion is organized by the specter of competition and conflict, as evidenced by much of contemporary political, ethical, and policy discourses. In this context, to interject, say, Buddhist cosmological views is likely to create an exotic interlude whose relevance to the principal conversation is unclear. The comparative method provides a targeted objective, however, and that is to accentuate the most deeply held and axiomatic values of the original culture. Comparison, in other words, makes visible the invisible yet central assumptions that create our knowledge and our disagreements. To become aware of these assumptions makes them less inevitable and expands our possibilities for knowing.

On the issue of terraforming Mars, science and religion must jointly confront relevant ethical concerns: are all forms of life equal, or can we differentiate based on sentient capacities, such as reason? What is the human relationship and responsibility to nature – is all of creation of equal worth, or do human beings possess a special latitude, by virtue of our intelligence, or by virtue of divine will? These are relatively concrete questions that bear directly on our possible interplanetary actions. Aside from secular ethicists, who hold that reason is a sufficient resource for resolving ethical dilemmas, most people look to religious precepts as a guide. So, very quickly then, biological questions about what life is, as well as what is alive, and ecological questions about the ontology of nature press religious worldviews into service.

For monotheistic traditions, belief in a creator God and the role of human beings within that creation are the most relevant considerations in making ethical choices. Their applicability to the present context is clear: God created human beings in his own image (*imago dei*), which means that we are the stewards of creation, understood as partners of God's purpose. We possess dominion over nature because we are set apart from it, but this dominion also entails an ethics of care. So far, so good. If the sole purpose of the inquiry is to answer the limited question of how to conduct our research on and interaction with Mars without transgressing the ethical sensibilities of Christians, Jews, and Muslims, the answers seem easy enough. The totality of the universe, with planets other than ours, and even the possibility of life on other planets, in other solar systems, can be subsumed in the Genesis account of creation. Interplanetary exploration that expands the magnitude of human interaction with the universe poses no rupture with the Biblical worldview. The ethics of stewardship can extend to interplanetary settings, such as Mars, particularly given the fact that, so far, there are no signs of life on that planet equivalent to or approaching the level of human complexity and intelligence.

Neither scientists nor religious people, however, are content with pragmatic problem solving. Although it appears quite possible for Christians and astrobiologists to agree on how to proceed in relation to Mars, discussions between religion and science have a way of exceeding practical questions to the thornier ones of truth and meaning. As a result, some scientists challenge the need for God and other supernatural principles in order to explain the origins of the universe and of life. Counter challenges, entailed in such ideas as "Intelligent Design" and the "anthropic principle," question the presumption that naturalistic explanations can tell us everything. Despite the appearance of divergence and opposition here, a comparative religious perspective demonstrates the more important commonalities that bind these disputants together.

The assertion of this chapter is that it is a single, shared view of narrative that creates the tension between monotheistic and scientific perspectives on cosmic meaning. The physicist Steven Weinberg's speculation that, in the light of scientific discoveries, the universe appears "pointless," and human existence the "farcical outcome" of blind and random forces [1] is no less an embrace of the Christian sense of narrative than the affirmation of divine providence. And as the Indian philosopher Anindita Balslev observes, in a thoroughgoing analytical (and comparative) context, Weinberg would need to explain and defend his narrative assumptions [2] (p. 60).

A Chinese Confucian would find Weinberg's conclusion rather astonishing at face value. This is not because she implicitly sides with religion and disagrees with the assertion of cosmic meaninglessness. Rather, it is because Weinberg's conclusion entails an either/or supposition that is not self-evident to someone from a different culture. Jeffrey Russell conveniently sums up this widely shared supposition in a recent volume of essays on *Science, Religion, and the Human Experience*:

Now, it may be that the universe is exclusively physical. ... Such a universe, the product of randomness and causation, is without inherent meaning or purpose. Or it may be that the universe includes both the physical *and* the spiritual and ideational entities that exist, relate, or occur; these are not limited by space-time, and they are not exhaustible by physical explanations. Such a universe has intrinsic order and meaning [3] (pp. 111–112).

But are we really limited to these two options? It is difficult to think so within the context of a culture that does not posit a divine creator. As Pascal Boyer notes, "Most religious systems in the world are not about an eternal Creator," and, "In fact, the creation of the universe is of limited interest to most people in the world" [4]. The ancient Greeks and Indians, as we will see below, believe the universe has no beginning. The Confucians do not even ponder the question of beginnings, focusing instead on the natural order, or "Way," that patterns the universe. Hence it is possible to discern order and meaning in the world apart from a creator God. These are startling facts to keep in mind when the debates between evolution and Creationism function as the proof text for the conflict between religion and science.

Against the canvas of comparative religions, then, the "conflict" of religion and science turns out to be a rather limited dispute between the opposite ends of a shared model of cosmic meaning. This model can be traced to monotheism itself, with its historically unique emphasis on a supreme God who creates the world *ex nihilo* as part of a grand, providential scheme. The Jewish scholar Daniel Matt deems this kind of myth making essential for a purposive human life:

A myth is a story, imagined or true, that helps us make our experience comprehensible by offering a construction of reality. It is a narrative that wrests order from chaos. We are not content to see events as unconnected, as inexplicable. We crave to understand the underlying order in the world. A myth tells us why things are the way they are and where they came from. Such an account is not only comfortable, assuring, and socially useful, it is essential. Without a myth, there is no meaning or purpose to life. There is just vast emptiness [5].

In contrast to the fundamentalists, Matt is more concerned with the functionality than the literal truth of the Biblical creation narrative. What he articulates, then, are the conditions for meaningful existence writ as the demands of the human psyche. Catholic theologian John Haught, who is explicit in his disavowal of both Creationism and Intelligent Design, also attends to this particular human requirement:

I have to ask whether I can surrender myself completely to any reality that in its deepest strata lacks the qualities of intelligence, concern or the capacity for self-sacrificing love and promise keeping that I associate with the very highest ideals of my own species ... I fear that I cannot bow religiously to a finally unpromising, impersonal universe, no matter how far its boundaries in time and space extend, without becoming less than a person myself [6].

In contrast to the beginning-less universe of Confucianism, Judeo-Christian cosmology and its attendant mode of meaning-making might be called a narrative mode, with the classic, plot-driven profile articulated by Aristotle in his *Poetics*. Basic to the Aristotelian requirement is the feature of integration, in which each element of the tale plays an important part in the articulation of the whole story. Hence there can be no episodic or random event, dangling in isolation from the totality of the narrative. Plots must feature a beginning, middle, and end that evolve intelligibly to comprise a meaningful whole.

The linear cosmology of monotheism proposes that the world happens only once, and the events that take place within it participate in and fulfill a much larger scheme of significance. It is also notable that aside from the creation of the world itself, the most important narratives of the Abrahamic faiths – God's covenant with Israel, the event of Christ, the witness of Mohammed – take place within human history. It is common speculation that this tendency to sacralize history, down to its political life, is an important impulse behind the revolutionizing social and political reforms that spawned the liberal democracies of the modern West. History is the site in which human salvation unfolds.

Narrative plots are a way of articulating the meaning of individual lives and sub-communities as well as the meaning of history at large. These personal stories unsurprisingly model themselves after canonical ones: "Martin Luther King, for example, understood himself in the light of the Exodus and the crucifixion, and these motifs of liberation and self-sacrifice come to us in turn through the story of King's life, not through theological propositions" [7]. Biblical tales model personal ones, whose heroism enlivens – and keeps alive – the Biblical paradigms. The plot-driven narrative is dominant down to contemporary literature and film. The recent completion of the *Star Wars* hexalogy attests to our desire to complete the story; our sense that meaningfulness is a function of knowing the meaning of each element in relation to the whole.

When it comes to the purpose of human history in relation to the vast cosmos that science is deciphering, narrative and plot are essential for constructing a comprehensible reality that has meaning for us. In our storytelling tradition, one either assents to the myth and affirms the inherent value of life, or one dissents and is trapped in the remainder of meaninglessness. Either the story is true, or it is not true. Not all cultures share this logic about storytelling, however. It may be worthwhile to look at how stories function in other places in order to learn something about our own practices. Indian storytelling, for example, might provide a way out of our apparent dilemma.

16.2 Indian stories about the world

Comparative ethicists have noted the central role a culture's view of creation plays in its ethical reasoning. If stories about how the world came to be influence our thinking about what is moral, it is because a cosmogony offers "a pattern for human choice and action that stands outside the flux of change and yet within the bounds of human knowing" [8]. I concur that cosmogonies offer stable patterns for shaping a world of meaning. But to plumb the wisdom of this observation means that we need to look beyond the *content* of the stories and observe the logic of the storytelling itself. In India, there is a discernible purpose in having multiple stories about the same event.

In Indian mythology, there are many different stories about the creation of the world. One common cosmogonic theme is divine incest, versions of which can be found in both the *Vedas* and the *Brahmanas*:

Prajāpati approached his daughter; some say she was the sky, others that she was the dawn. He became a stag and approached her, as she had taken the form of a doe [9] (p. 21).

In addition to sexual cosmogony, we have the reverse logic of creation from the powers of asceticism:

Prajāpati, wishing very much to have progeny, practiced asceticism. As he became heated, five were born from him: Fire, Wind, Sun, Moon, and the fifth, a female, Dawn [9] (p. 31).

Another common myth found in the *Vedas* entails the sacrifice of a cosmic Man, whose body forms the world, including the social classes:

When they divided the Man, into how many parts did they disperse him? ... His mouth was the Brahmin, his arms were made into the nobles, his two thighs were the populace, and from his feet the servants were born. ... The moon was born from his mind; the sun was born from his eye. ... From his navel the atmosphere was born; from his head the heaven appeared. From his two feet came the earth, and the regions of the sky from his ear. Thus they fashioned the worlds [10] (10.90).

The imagery of egg and seed provide yet another Vedic creation myth:

In the beginning the Golden Embryo arose. Once he was born, he was the one lord of creation. He held in place the earth and this sky. Who is the god whom we should worship with the oblation? [10] (10.121).

The god born of the embryo is unnamed. With the accretion of Indian literature, however, the names and the principles of creation multiply. Hence we also get creation from the blending of primal spirit (*puruśa*) and matter (*prakrti*), and creation as the awakening of the god Visnu, or sometimes Brahmā, who emanates from the former.

The mythology of later literature (the *Purānas*) exhibits some attempt to reconcile different views, but exhibits a "preference for the synthesis of disparate views into a larger whole rather than the rejection of apparently dissident elements in favor of a single view considered to be exclusively true" [11]. This is perhaps because the universe has many lives, evolving through endless cycles of creation and destruction. This seems unlikely, however, since the myths do not bother to tag themselves to any particular cosmic cycle – to assert, for example, "This is the creation myth of *this* particular universe, at *that* particular time." Another explanation for the multiple cosmogonic myths of India might be the cumulative nature of its sacred literature. As more stories get written, more creation myths get added. In any case, the endlessly cycling cosmology of India finds no dissonance in the episodic and even contradictory stories of cosmic origins. Given the beginning-less and endless repetitions of world cycles, whether the creator god is Visnu, Brahmā, or Prajāpati depends on the occasion of the telling. The cosmogonic stories, like the world cycles themselves, are variations within a possible range of options.

Given the cosmogonic pluralism of India, it is interesting to speculate on how well and in what way scientific cosmology – such as Big Bang theory – might be integrated. From the Indian perspective, Big Bang theory simply adds another story to the mix. But this is not to say that science is reduced to mythology, or that it simply does not matter what stories you tell. The operative logic here is to ask what a particular story buys you, and the whole point of having different stories is that one can do different things with them. What Big Bang theory buys us is a mathematical and chemical reconstruction of the beginnings of the physical universe. Traditional cosmogonic myths, on the other hand, enable Indians to do other things, such as to visualize the social order, affirm the dominant ritual practice of sacrifice, and divide the world into organizing principles – spirit and matter – in order to understand the liberative capacities of the self. These are not conflicting narratives, just different ones.

Indians are not the only people who tell different stories to accomplish different ends. We do that too, but the prevalence of Greek logic in Western culture (to be examined in the next section) tends to make us renounce the practice. We place a premium on the one, true story and consider multiplicity a form of contradiction. That is why we seriously entertain such claims as "Reduction of all ideas to neural impulses means that no moral or ethical concept is better than any other" [3] (p. 121). The presumption is that there can be only one story, and brain science is a story that obliterates other stories that tell us about who we are as moral beings.

The multiple cosmogonies of India encourage an understanding of different stories as doing different things. This shares a degree of similarity with the "non-overlapping magisteria" (NOMA) thesis, made explicit by Stephen Jay Gould, in which science and religion are distinct and hence non-conflicting domains of knowledge [12]. According to NOMA, science studies the "ages of rocks," and religion the "rock of ages," otherwise abbreviated into a string of oppositional pairs: the objective versus the subjective; the outer versus the inner; the empirical versus the symbolic. But the similarities between NOMA and Indian storytelling run aground here. A problem with the NOMA schema, relative to the Indian perspective, is that oppositional pairs such as "inner" and "outer" implicitly assume exhaustiveness, as if they have covered all there is. My point is not simply that science and religion do not account for everything. Rather, it is the presumption of wholeness that needs to be scrutinized.

NOMA presumes the world is like a pizza pie from which science and religion can be apportioned as separate and non-overlapping pieces. This is, for some, an improvement over the conflict model that deems either religion or science, exclusively, is the whole pie. But the improvement is more imagined than real because of the mutual and limiting "pizza presumption." If knowledge is nourishment for living, our kitchen is full of pies: in addition to pizza, there is quiche, chocolate cream, chicken pot pie, key lime, shepherd's pie, etc., and all the delicacies yet to come. The point is to get beyond the idea of the complete pie. No one just eats an entire pizza – this is boring and nutritionally unsound; on the other hand, no one concocts a pie from slices of all existing pies – this is just plain unappetizing. Rather, people eat slices of different pies in serial and random order, at the same sitting or between uneven and intermittent stretches of time. The way we tell and use stories is more akin to the idiosyncratic and unpatterned, or individually patterned, ways we eat different foods. The point of having different stories is not to stitch them together into one big complete story. No one can eat that way, and neither do we live that way.

The difference between storytelling in the contemporary West and traditional India is an implicit understanding of what language is really doing. The plot-driven narrative assumes that language – and the story – is exhaustive, and therefore it cannot tolerate contradiction or multiplicity. The narrative pluralism of India, on the other hand, assumes that any given story is a function of specific context and purpose. More importantly, no one story is required to serve all purposes. The myth of the exhaustive story is driven by a logic that is wedded to a particular conception of meaning-making. The mechanics of this logic originated, surely enough, in cosmological speculation, by way of the Greeks.

16.3 The logic of origins

The ancient Greeks shared the Indian belief in cosmic cycles. Parmenides (540–480 BCE) provides the basic reason – that "what-is is ungenerated and imperishable" [13] (8.3), following the dictum of *ex nihilo nihil fit* ("from nothing, nothing can come"). From this comes the Greek belief in the circularity of the universe, including its notion of the "eternal return," or the belief that all phenomena are fated to an exact and endless repetition. In late antiquity, the question of cosmological origins was taken up again by the Middle Platonists, who looked to the *Timaeus* to postulate that the universe indeed had an origin in time. It was this speculation that created much common cause between Greek philosophy and Christian thought. However, it ultimately provided an occasion for Christianity to break from Greek views and forge a distinct theology of creation.

Up until the middle of the second century, early Church thinkers conformed to Plato's notion of creation as the ordering of a chaotic but pre-existing substrate. Although the *Timaeus* broke with Greek tradition in asserting the temporal beginnings of the world, it did not violate the principle of *ex nihilo nihil fit*. The concept of *creatio ex nihilo* that appeared in 2 Maccabees 7:28 of the Hebrew Bible was also understood in this light – that "creation from nothing" was creation of an *order* that was not there before, but not creation from literal nothingness. This understanding is affirmed in the Genesis account as well, where God creates the world from primordial chaos and darkness rather than from nothing. In the second century, however, Theophilus of Antioch unequivocally stated that matter cannot be eternal and coeval with God because this compromises God's absolute sovereignty. From here on, the *ex nihilo* view, defined explicitly as the creation of Being from non-Being, is finessed by Church apologists as the orthodoxy of Christian monotheism.

The *ex nihilo* doctrine visualizes a sovereign God who is responsible for creation. Hence creation is *contingent* and absolutely *dependent* on God for its existence. On this point, Augustine takes up the question of why God was moved to create, and the problem of reconciling this movement with God's own immutability: "if we ascribe to God's works a beginning in time, we obviously suggest the idea that in some way he disapproved of his own previous eternal inactivity, and condemned it as sloth and idleness, and therefore changed his ways!" [14]. Augustine resolves the problem with the following declaration:

But in God there was no new decision which altered or cancelled a previous intention; instead, it was with one and the same eternal and unchanging design that he effected his creation.... he did not stand in need of his creation, but produced his creatures out of pure disinterested goodness [14].

The goodness of God, as well as of his creation, is the point here, in keen refutation of Gnostic dualism and its problematizing of evil. The existence of evil does not make the world evil, according to Augustine. If God allows for the existence of evil, it is because it is a part of his providential design. But the "eternal and unchanging design" that gave rise to creation means that the world has a purpose and moves towards a meaningful end.

Christian creation theology unfolded in an historical context that indelibly shaped it. The controversies with Gnostic heresies that challenged the singularity, omnipotence, and goodness of God were immediate triggers, but more broadly, it is the tradition of Greek logical discourse that made the lasting imprint. If *ex nihilo* theology decisively breaks with the Greek adherence to *ex nihilo nihil fit*, it nevertheless adheres to the rules of logic that Parmenides assumes in his cosmological reflections. These are the famous Laws of Thought, as later formulated by Aristotle: the Law of Identity (A is A, and not not-A); the Law of Contradiction (nothing can be A and not-A at the same time); and the Law of the Excluded Middle (everything is either A or not-A). With these principles, Parmenides argued a monistic view of the world as absolute Being that cannot change, which is unoriginated, unending and timeless. More importantly, he offered the first exercise in systematic metaphysics by asserting a necessary correspondence between thought and reality: "It must be that what is there for speaking, and thinking of *is*, for it is there to be" [13] (6.1). This has been the ancient Greek legacy to the world – its demonstration of "the potential power of philosophy as possessing a magic even beyond that of religion and mythology" [15].

The Greek legacy is apt to be overlooked if we focus solely on the content of Christian cosmogony: its linear creation overtly contradicts Greek cyclical thinking. The greater commonality must be understood, then, not as a matter of actual beliefs but rather the language game – the particular art of storytelling – that is being practiced. For if there is anything that the Christian response to scientific cosmology demonstrates, it is the conundrum of the logical demand that the world either *is* or *is not* created by God. To resort to a metaphorical interpretation of Genesis or to consign Christianity to the "symbolic" as opposed to "factual" domain of knowledge is simply to obfuscate the question of truth. But the either/or demand, as Parmenides formulated it, is not the self-evident way of the world that we take it to be. As Thomas McEvilley notes, "Logic is projected outward as ontology" [15]. In other words, our belief about the nature of things is dependent on our logical language.

For Parmenides, this belief might be characterized as a faith in the power of definitions. "Being," by definition, *is* and hence cannot not be [13] (2.3). Furthermore, Being is complete and whole, and therefore does not countenance division or change [13] (8.22–28). This leads Parmenides to deny the reality of our everyday experience, in which mutability is the norm. By virtue of a definition – Being as that which does not change because it would no longer be itself – the whole world of phenomenal reality is banished into non-existence. Theological language about God proceeds along the same definitional logic. In fact, "God" can easily substitute for the "Being" of Parmenides to the degree that it is a function of postulation. Anselm's ontological argument for the existence of God is a useful example. The postulate of "that which none greater can be thought" features as a part of its definition the demand that it exist: "For something can be thought of as existing, which cannot be thought of as not existing, and this is greater than that which *can* be thought of as not existing" [16]. By definition, then, God, as that which none greater can be thought, must exist. To contradict the conclusion violates the rules of logic just as surely as asserting that bachelors are married males. The whole point of a definition is to stipulate. From there, one need only to add the faith that what can be thought must *be* [13]. The greater weight of Greek philosophy, and Christian theology after it, chooses to begin with an a priori assertion of the truth of definitions over the messy and indeterminate world of sense experience.

Let us now turn to the Indian experience. The logic of Indian cosmology shares a key premise with that of the Greeks: "Of what is not there is no becoming; Of what *is* there is no ceasing to be" [17] (2.16). The *Bhagavad-gita* articulates the principle of *ex nihilo nihil fit* as an absolute consensus within Indian thought. Something cannot come from nothing, hence the universe is eternal. The cosmos, or the ordered manifestation of the universe, on the other hand, cycles through periods of creation and destruction very much like the cycle of seasons. They are imaged most often as the alternation of form and potentiality:

> For a thousand ages lasts
> One day of Brahma,
> And for a thousand ages one such night:
> This knowing, men will know what is meant by day and night,
> At the day's dawning all things manifest
> Spring forth from the Unmanifest;
> And then at nightfall they dissolve again
> In that same mystery surnamed "Unmanifest."
> Yea, this whole host of beings
> Comes ever anew to be; at fall of night
> It dissolves away all helpless;
> At dawn of day it rises up again [17] (8.27–19).

In Indian theistic traditions, God is coeval with, but never a creator of, the universe, for this would make him liable for the inequities of creation. This is the very logic that drove Gnostic thought – that is, the problem of evil. It is an interesting divergence of history that the problem of evil never drove Indians to a final monotheistic solution. The world is simply accounted as the by-product of divine play (*lila*) rather than purpose or design, and evil is the inescapable progression of a devolutionary universe. Of the four periods, or yugas, into which creation is divided, our present world is always described as the Kaliyuga – the last and most degenerate. The predictable entropy of the universe does not beg to be pondered any more than this.

The alternative virtue of Indian philosophy, to counter its lack in monotheism, is its focus on individual consciousness and experience, as Balslev notes. Despite the inexorable destiny of the universe, individual salvation is still possible by attending to one's own mind. There is in this orientation a tendency to privilege consciousness as more important than the phenomenal world itself, for our perception of this very world is but a telling symptom of the diseases of the mental life. For example, "some Indian thinkers thought that the very idea of fulfillment of a purpose [in the cosmos] presupposes a sense of want, a lack of some sort, that is felt by an agent in question" [2] (p. 67). That is to say, the perception of meaninglessness betrays a frame of mind that is more interesting, more critical, than the mind-objects themselves. This insight is the point of departure on the path to religious liberation.

In the Buddhist system, this unrelenting focus on our own infelicitous perceptions of the world gives rise to an acutely pragmatic sense of pedagogy, expressed as caution and flexibility in the use of words. The principle here is the priority of the needs of the audience over the literal meaning of words: differences in consciousness mandate different – even conflicting – teachings. Conversely, the audience is warned to not literalize the Dharma (the Buddha's teachings) in the famous simile of the raft ("It is for the purpose of crossing over, not for the purpose of grasping" [18] (22.13)), and the simile of the snake ("Those teachings, being wrongly grasped by them, conduce to their harm and suffering for a long time" [18] (22.10)).

It is critical to mention here that the Buddhist sense of words is intimately bound up with its unwavering focus on the phenomenal world of sense over and against the definitional world of essences. The impermanence and instability of this phenomenal world is the starting point of the Buddha's philosophy, much like his Greek counterpart Heraclitus, whose thinking stands in structural opposition to Parmenides'. And the following description of Heraclitus' premises is equally applicable to the Buddha:

Things that are constantly changing are not susceptible of definition, since a definition which applies one moment will not apply the next. Since things lack definition, every kind of metaphysical declaration is compromised [19].

We have, as a result, the Buddha's famous injunction against speculative views about the world – that it is or is not eternal, for example, and his constant skepticism about declarative assertions. If the Law of the Excluded Middle announces that everything is either A or not-A, the Buddhist Middle Way is classically expressed through the tetralemma, or the four-fold negation of not-A, not not-A, not both A and not-A, not neither A nor not-A [20] (1.2.27, 15.32). The point is to avoid the dichotomy of A or not-A because observation tells us that nothing *is*, in the sense of Being, but rather is constantly *becoming* in the unending liminality of both *is* and *is not*. Personal experience and observation – not logic – are the final court of appeal in this declaration. As the Buddha notes repeatedly, even well reasoned arguments can produce wrong conclusions [18] (76.27). The constancy of logic may be a comforting language game, but in an unsteady world, words are equally volatile in what they signify and how they work. In that case, the use of words must be dictated by social context.

This understanding of utterances as "speech-acts" that *do* things which are much more important than what they actually *say* has come into greater recognition in modern Western philosophy of language.[1] Its appreciation in ancient China, particularly in the Confucian lineage, also demonstrates that conscious use of speech-acts is not the monopoly of anti-essentialist philosophies. Confucius staked his teachings on the idea that the "Way" of the sages lay in a timeless set of principles that included "rectifying names," or educating people into the proper definitions of social roles – ruler, minister, father, son, elder, younger, husband, wife, and friend. There is no room for paradoxical "both/and" assertions here: the father is always the father relative to the son, and such consistency is essential to maintain the social and natural order. The stability and timelessness of social identity, however, are transmitted via an active, performance oriented pedagogy. Consider this well-noted passage from Confucius' *Analects*:

Tzu-lu asked, "Should one immediately put into practice what one has heard?" The Master said, "As your father and elder brothers are still alive, you are hardly in a position immediately to put into practice what you have heard."

Jan Yu asked, "Should one immediately put into practice what one has heard?" The Master said, "Yes, One should."

Kung-hsi Hua said, "When Yu asked whether one should immediately put into practice what one had heard, you pointed out that his father and elder brothers were alive. Yet when Ch'iu asked whether one should immediately put into practice what one had heard, you answered that one should. I am puzzled. May I be enlightened?"

The Master said, "Ch'iu holds himself back. It is for this reason that I tried to urge him on. Yu has the energy of two men. It is for this reason that I tried to hold him back" [21] (11.22).

As Confucius explains, the content of his teaching is determined by the character of the student. It is not that his principles change, but rather his audience does. The *Analects* is a collection of fragmentary dialogues, anecdotes, and proverbial utterances, but in overview, the text can be stitched together as a narrative of the Master's interactions with his disciples. Unlike the plot-driven narrative, which emphasizes the continuity and internal coherence of its events, the *Analects* coalesces around its speech-acts, and emphasizes the priority of context for exercising care in the use of words. This care attends to the pragmatic effects of language but does not entail reverence towards its internal grammar

---

[1] The phrase "speech-acts" is taken from J.L. Austin. *How to do Things with Words* (New York: Oxford University Press, 1962). This text, based on Austin's 1955 William James Lectures, is a seminal work in modern philosophy of language.

and logic – "The Master said, 'It is enough that the language one uses gets the point across' " [21] (15.41).

16.4 The comparative application

In the preceding pages, I have sketched out a cosmology with a sense of meaning-making and storytelling that presents an alternative to both the monotheistic and modern scientific sense of narrative. It is now time to make explicit what this comparative exercise might offer, relative to the ethical questions at play in astrobiology and the matter of terraforming on Mars.

The logic of storytelling in Asian culture suggests a way past the potential conflict between monotheistic religious values, on the one hand, and scientifically based views, on the other. It does this, first, by cautioning that there are limits to the applicability and usefulness of already existing stories, *both* religious and scientific. Our interplanetary ventures may lead us to unprecedented encounters and situations. Preserving the integrity and value of our existing stories may very well depend on our ability to leave them aside at the right time, so that we are able to make use of *all* our most valuable and hard won stories. Secondly, Asian story-telling logic exhibits its own sense of consistency. But this consistency does not derive from telling the same story in every situation. Rather, it entails choosing the most useful story in each context so that we may act consistently with its demands. Actions demand singularity – we must choose either to terraform or not terraform on Mars, and we must choose one way to go about it. Right actions are the result of acting consistently with the best story for the job.

At the heart of astrobiology is the question of life beyond Planet Earth, either nurtured or discovered. Central to this possibility is the question of how to define or recognize life. We begin, naturally, with the norm of terrestrial life forms and seek evidence of extraterrestrial life, for example by scrutinizing Martian soil for evidence of bacterial organisms or by measuring the Martian environment against our standard of life sustainability. While this is a logical point of departure, there may come a time when it will behoove scientists to radically redefine existing biological conceptions of life. The impetus to do this will be thoroughly practical and ethical: in the event that we do not find anything that looks like us (in either elemental or complex form), we will be faced with the choice of rethinking our concept of life or concluding that we are the only ones alive in a universe of inert and dumb matter. We may even conclude that nothing is alive, including ourselves. Our choice of story will be hugely consequential for determining our subsequent actions.

The inevitability of such choices is clear, given how science and technology is already complicating the question of life. Medical procedures sustain the "life" of persons in a persistent vegetative state who are apparently reduced to a symbiotic

machine limited to the most basic autonomic functions. Even as medical procedures sustain life, they create ethical and legal confusion regarding the nature and
boundaries of life. On the other hand, artificial intelligence, robotics, and cognitive
and neuro-psychoanalytical computing are evolving machines to the point where
scientists predict that passing the Turing test is in the not-too-distant future, and
humans will not be able to tell whether they are interacting with a machine or
another human. The Japanese penchant to humanize their robots is probably a sign
of things to come [22]. The popular Zen adage that even rocks have Buddha nature
may prove a useful story in turn. From any number of directions, we are already
confronting the need to expand our repertoire of stories about the origin, extent,
and future of life. In this context, the eliminative bent of demanding only one story
is destined to create confusion and conflict.

In spite of its own eliminative tendencies, science embraces new theories as a
matter of principle, and looks forward to radical shifts in how we model our empirical world. But what about religion, which proffers timeworn wisdom and whose
evolutions constantly negotiate the subtle balance between history and orthodoxy?
This is a particularly pressing question for evangelical Christianity, whose principle of Biblical inerrancy tends to overlook the multiple and conflicting stories
already existing within its own scriptures. It is here that the point about putting
one's story away at the appropriate times in order to preserve it is vividly apropos.
Forcing stories into places where they do not belong is a surefire way of destroying
them. This can be illustrated by another current strategy for reconciling religion
and science.

Next to the idea of the independence of religion and science, the pursuit of
integration is perhaps the second most popular tactic of reconciliation. It is particularly favored by theologians who dominate the religion and science dialogue,
and it entails accommodating scientific stories, particularly evolutionary biology,
into the religious one. Just like the NOMA hypothesis, the integration project
turns on the presumption of the single constitutive story. The ideal of integration
expresses the desire for intellectual unity, based on the faith that wholeness is the
only way to make life meaningful.

In a recent essay in the popular Catholic magazine *Commonweal*, however,
Peter James Causton observes that "The twin gods of evolution, as it is currently
understood, go by the names of chance and necessity. Considering all this, is nature
really where we want to go to find reliable evidence of the Divine?" [23]. Causton's
point is to ask how much of this godlessness the traditional story of an all powerful and loving God can accommodate. The attempt at integrating evolution pushes
the Biblical story perilously close to losing its import altogether. The purpose of
religion is not to survive at all cost to its fundamental message, or to account for
scientific knowledge. At times, religion must indeed take what science reveals into

account, but this process should not obscure religion's own purpose. The Biblical story should be left untampered so that it can do what it was meant to do – to provide a moral compass via a meaningful sense of our place in the world.

This does not mean embracing a Biblical fundamentalism that rejects science. Nor does it mean segregating religious stories from the new worlds that science helps to engender. As I speculated in the opening discussion, the Biblical narrative of creation may very well provide a useful guide for current astrobiological ventures. But the point is to exercise the kind of reflection exhibited by Causton – that is, to acknowledge when stories have fragmented beyond the point of integration. The lesson from Asian storytelling practice is that when this happens, we need not submit to the impulse to glue and patch, which only creates the most unpalatable of dishes. We need to preserve our functional stories wherever they come from. In our culture, that certainly includes the tale of the creation of life by an all powerful God, in contrast to the NOMA restriction that only science can tell the story about life. The use of the religious story, on the other hand, should not preclude Christians from telling the scientific one about the age of rocks. Certain stories persist because they have a purpose, and stories multiply because our purposes are many. The demand to logically align or reconcile our stories into one big seamless entity is a peculiar prejudice.

Finally, let us address the matter of rationality and choice. The prescription to embrace many stories without reconciling them is not a recipe for haphazardness or indifference. Much like science itself, the program hinges on sound judgment with a keen eye for practical problem solving. It is in the nature of science to contest theories and to seek the best theory to explain an event because explanations matter very much. Was the plane crash due to pilot error, or a defective part? Between these two options, the answer matters, and scientific investigation means choosing what we hope is the right answer. There is always something at stake in choosing a story, and we aim to choose well.

Choosing well, however, also means knowing when stories do not compete, thereby eliminating the need for choice. Was the crash an act of God? For the regulatory agency, the insurance company, and the scientists, the answer is irrelevant to their respective purposes. For the bereaved survivor, it may be the only question worth asking. Neither conflict nor integration is an appropriate description here. It is not a contradiction to tell different stories to attend to different purposes. The capacity to hold to more than one explanation – that it was pilot error *and* an act of God – is not integration of science and religion either. We only muddle things if we try to tell two stories at once, just as we ruin the dish if we mix pizza and cream pie on the same plate. Consistency is a matter of avoiding conflicting recommendations for action. The "act of God" theory, for example, does not prohibit nor determine how frequently airplane parts should be checked for stress fractures. Holding to multiple stories does not, in and of itself, violate the principle of consistency.

Undoubtedly, there are already voices within our society that speak the kind of storytelling logic I have extracted from Asian practices. I have posed this study as a comparative cultural analysis, however, because the weight of cultures shifts in different directions. The impact of Greek logical discourse beginning with Parmenides makes it difficult for the alternative thinkers in our culture to be understood. To bring in the edifice of another civilization helps us to open up our thinking and hear those among us who sing in the same register. In this chapter, the tune I have attempted to transmit is the idea that narrative wholeness is a way of making certain intellectual practices meaningful, but that life itself has many more requirements. Germane to life as a whole, a lot of people around the world do not believe in the possibility or wisdom of the definitive utterance. Skillful living is a matter of performative speech-acts, or knowing what to say in which context. Only the boor, or the bore, insists on repeating the same story everywhere.

References

[1] S. Weinberg. *The First Three Minutes: A Modern View of the Origin of the Universe* (New York: Basic Books, 1976).

[2] A. Balslev. Cosmos and consciousness: Indian perspectives. In *Science and Religion in Search of Cosmic Purpose*, ed. J. Haught (Washington, DC: Georgetown University Press, 2000).

[3] J. Burton Russell. Science, religion, metaphor, and history. In *Science, Religion, and the Human Experience*, ed. J. D. Proctor (New York: Oxford University Press, 2005).

[4] P. Boyer. Gods and the mental instincts that create them. In *Science, Religion, and the Human Experience*, ed. J. D. Proctor (New York: Oxford University Press, 2005), p. 238.

[5] D. Matt. Kabbalah and contemporary cosmology: discovering the resonances. In *Science, Religion, and the Human Experience*, ed. J. D. Proctor (New York: Oxford University Press, 2005), p. 133.

[6] J. Haught. *Deeper than Darwin: The Prospect for Religion in the Age of Evolution* (Boulder, CO: Westview Press, 2003), p. 38.

[7] I. Barbour. *Religion in an Age of Science* (San Francisco, CA: Harper and Row, 1990), p. 72.

[8] R. Lovin and F. Reynolds. *Cosmogony and Ethical Order: New Studies in Comparative Ethics* (Chicago, IL: University of Chicago Press, 1985), p. 5.

[9] W. Doniger O'Flaherty. *Hindu Myths* (Harmondsworth, UK: Penguin, 1975).

[10] W. Doniger O'Flaherty (trans.). *Rig Veda: An Anthology* (Harmondsworth, UK: Penguin Books, 1981).

[11] C. Dimmitt and J. A. B. van Buitenen. *Classical Hindu Mythology* (Philadelphia, PA: Temple University Press, 1978), p. 16.

[12] S. J. Gould. *Rocks of Ages: Science and Religion in the Fullness of Life* (New York: The Ballantine Publishing Group, 1999).

[13] D. Gallop (trans.). *Parmenides of Elea: Fragments* (Toronto: University of Toronto Press, 1984).

[14] H. Bettenson (trans.). *Augustine: City of God* (Harmondsworth, UK: Penguin Books, 1972), 12:18.

[15] T. McEvilley. *The Shape of Ancient Thought: Comparative Studies in Greek and Indian Philosophies* (New York: Allworth Press, 2002), p. 54.

[16] J. Hick (ed.). *The Existence of God: A Reader* (New York: Macmillan Publishing, 1964), p. 26.

[17] R. C. Zaehner (trans.). Bhagavad-gita. In *Hindu Scriptures* (London: J. M. Denton and Sons, Ltd., 1966).

[18] B. Ñānamoli and B. Bodhi (trans.). *Middle Length Discourses of the Buddha* (A translation of the *Majjhima Nikāya)* (Boston, MA: Wisdom Publications, 1995).

[19] T. McEvilley. *The Shape of Ancient Thought: Comparative Studies in Greek and Indian Philosophies* (New York: Allworth Press, 2002), p. 37.

[20] M. Walshe (trans.). *Long Discourses of the Buddha* (A translation of the *Digha Nikāya)* (Boston: Wisdom Publications, 1995).

[21] D. C. Lau (trans.). *Confucius: The Analects* (Harmondsworth: Penguin Books, 1979).

[22] T. Hornyak. *Loving the Machine: The Art and Science of Japanese Robots* (Tokyo: Kodansha International, 2006).

[23] P. Causton. Darwin's ghost: can evolution and Christianity be reconciled?" *Commonweal: A Review of Religion, Politics, and Culture,* **83**(17) (2006), 15.

Index